# RIDE TO GLORY

*The People v. Charles Robert Darwin*

## An American Novel

## Warren LeRoi Johns

Author of *Dateline Sunday, U.S.A.*
Editor of www.CreationDigest.com

1stBooks - rev. 05/03/02

*Premier Edition*
November 24, 1999

*Paper Back Edition*
2002


**General Title Inc.**
**P.O. Box 344**
**Brookeville, Maryland 20833**


E-mail: *CDigets@aol.com*
Web Site: *http://www.CreationDigest.com*


Printed in the United States of America

# ACKNOWLEDGEMENTS

A salute of honor is due 19th century Austrian monk, Gregor Johann Mendel, whose 1867 report, virtually unnoticed at the time, opened doors to the science of genetics; and to the James D. Watson and Francis Crick team, credited with unraveling DNA's double helix mystery in 1953. The discoveries of these gifted scientists shifted scholar's attention beyond the macro realm of rocks and fossils to the micro sphere of genes, setting the scene for a twentieth-century assault on mega-evolution's Achilles heels.

Informal consultations with an array of independent scholars guided the framing of threshold issues addressed in *Ride to Glory*. Robert H. Brown, Harold G. Coffin, David P. Genter, Robert V. Gentry, Jim Gibson, P. Edgar Hare, Ariel A. Roth, Warren Harvey Johns and Varner J. Leggitt inspired insight. Encouraging words from Henry M. Morris and John D. Morris, champions of the truth about origins, assured focus; scientists Jim Gibson and Duane T. Gish graciously reviewed the draft manuscript for scientific accuracy; and trial attorney John R. Marcus evaluated courtroom procedures.

The published works of Michael J. Behe, Raymond G. Bohlin, Michael Denton, Duane T. Gish, Phillip E. Johnson, Lane P. Lester, Marvin L. Lubenow, Norman Macbeth, Richard Milton, Henry M. Morris, Hugh Ross, and Ian T. Taylor opened invaluable research trails. Irrespective of whether these giants of scientific scholarship agree or disagree with the book's theme or format, the author is indebted to the profound contribution of their collective works as well the other sources referenced in the bibliography. Rregardless, these references cited should not be construed to imply endorsement of the author 's view of the issues.

The paired profile drawings of Darwin, by C.H. Jeens (1897), and the ape "Mafuka," from a Richard Lydekker publication (1901), were made available courtesy of Ian Taylor, TFE Graphics. The artistic skills of the talented Guy Henry and his artist wife, Marilyn, fulfilled the author's vision of a dust jacket; Regina R. Hayden and Tanya Holland combined word processing and computer skills to perfect the book's format; the critical copy edit process thrived under the painstaking oversight of Beverly Rumble; and caricaturist Dan Ginter gave pictorial life to the author's cartoons spoofing evolution.

Thanks to the hospitality of Fred and Winona Humphreys of Johnson City, Tennessee, it was my privilege to meet and interview Fred's beautiful, 91-year-old mother, Wilma Humphreys, living eyewitness to the 1925 *Scopes* Trial. The storyline references this charming, southern lady and excerpts one or two of her vivid memories. I'm grateful, too, for background perspectives of Dayton,

Tennessee's rich lore of the Scopes Trial legal history shared by Steve Snider and Richard M. Cornelius.

Scientists, like lawyers, share the joy of advocacy derived from a quest for truth that leads to the discovery of enlightening evidence but with interpretations vulnerable to a passionate subjectivity when fueled by inherent bias. Explanations for the origin of organic life differ radically due to conflicting threshold belief systems.

Frank Haskell's description of the Battle of Gettysburg grips the reader with graphic poignancy. Riveting emotions were captured in a diary written days after the battle by a soldier who survived as an eyewitness to history. Later, Haskell lost his life in the Battle of Cold Harbor. No other witness to the bloodbath between the states lives to contest his account.

No *Homo sapiens* living on Planet Earth witnessed the how and when of the appearance of original life. Yet, non-witness, Charles Darwin, conjectured that spontaneous generation produced the first-ever living cell from random chaos and that all plants and animals evolved by multiple, incremental modifications over millions of years from that first life. In contrast, the Christ of the Bible, an eyewitness to pre-history, confirmed the reality of Creation and the deluge that followed. To embrace Darwinian evolution as "fact" implicitly rejects the testimony of Jesus as "fiction."

Mega-evolution's dogma projects a secular religion that worships at the shrine of faith in random chaos. The *RTG* author cherishes belief that life exists as a gift of an Infinite Creator who endows each soul with free choice and eternal purpose. The God of the universe declared: *"It is I who made the earth and created mankind upon it. My own hands stretched out the heavens; I marshaled their starry hosts."*\*\*\*

**Warren L. Johns**

*To the Memory of Varner and Charlene*

**Champions of All Things Good**

The tickets read:

Mid-Atlantic University
Presents

Orchestra
SEC 2
ROW G
SEAT 12

2:00 PM    September 1, 1996
Chamberlain Playhouse

The People v. Charles Robert Darwin

The People v. Charles Robert Darwin

September 1, 1996
Chamberlain Playhouse

## *"In the beginning..."*

Life had a beginning! Debate rages as to the how and when of the event. All rival explanations for life's origin require faith! The myth of mega-evolution roots its faith in what Darwin himself described as *"...a mere rag of an hypothesis with as many flaw[s] and holes as sound parts."* *

In *Darwin Retried*, Harvard educated Attorney Norman Macbeth referenced Tennessee's 1925 Scopes show trial, noting the "burden of proof" trap set by Clarence Darrow.

*"The proponents of a theory, in science or elsewhere, are obligated to support every link in the chain of reasoning, whereas a critic or skeptic may peck at any aspect of the theory, testing it for flaws...The winner in these matters is the skeptic who has no case to prove."* Attorney Macbeth suggests that the trauma inflicted on William Jennings Bryan *"...would have been equally disastrous for Clarence Darrow if he had tried to discharge the burden of proof for the other side."* **

**Ride to Glory** uses a make-believe courtroom setting and a lawyer's "chain of reasoning" to unmask evolution's Achilles' heels masquerading as science. To make its case, **Ride to Glory** cites updated factual data in endnotes blending academic credibility with fictional drama. Evidence is presented to readers, as to a jury, and the "burden of proof" shoe is put on mega-evolution's foot.

Lawyers share the professional challenge of all scholars: Collect the best evidence; marshal the facts; and seek honest conclusions. The results can be harnessed for presentation in book form by activating the computer's electronic wizardry. Without the input of intelligent design, no computer could create a book—even in 4.55 billion years. Darwin's *Origin* didn't write itself; **Ride to Glory** didn't evolve.

The DNA code vested in Galapagos finches observed by Darwin carried the inherent information that powered the diversity seen in different-shaped bird beaks. Finches never parent eagles. This quite real genetic adaptability cannot be extrapolated to prove make-believe, molecule-to-man traveling mega-evolution's mythical mutation/natural selection route.

Four presidential profiles peer from the granite crags of South Dakota's Black Hills. An army of stone soldiers has been excavated from China's earth.

No student of history claims these classic works of art designed themselves. Yet, mega-evolution asserts the living human models for this art appeared through mutation and natural selection—without intelligent design!

The premier publication of **Ride to Glory** coincided with the 140th anniversary of the sale of the initial 1500 volumes of *Charles Darwin's Origin of Species*—two days before the November 24, 1859 publication. This paperback reprint enters the twenty-first century, accessible on the world-wide web under the auspices of www.CreationDigest.com and *1ˢᵗ Books Library.* ᵀᴹ

Logic encourages faith in the Creative power of an Intelligent Designer—the eternal God of the universe. The alternative: worship at the secular shrine of speculative myths' unknown and unknowable god.

**wlj**

# I
# Day One: Sunday

# *Fact-Free Science*

FACT FREE "SCIENCE"

**"WHAT DO YOU MEAN 'PROOF'?
I *ADMIT* I'M RIGHT!"**

*7:33 a.m., Sunday*
*September 1, 1996*

# 1

Scouting for fossils always fired the imagination of Joshua Chamberlain Ryan. Family excursions to Cumberland Cave and Calvert Cliffs, Maryland's treasure troves of pre-history, whetted his appetite and fed childhood dreams. In a favorite fantasy he saw himself as intrepid explorer, scaling rock-strewn slopes of remote landscapes and stumbling across long-missing transitional life-forms predicted by his boyhood icon, Charles Robert Darwin.

In time, the juvenile hobby matured into serious academic study. Josh's passion for learning led to a career path in geology/paleontology and earned coveted status as a Ph.D. candidate at internationally acclaimed Mid-Atlantic University. But by the time the young scholar hit MAU's prestigious hallways, too many unanswered questions left him disenchanted with Darwinian thought and his ardor for old bones cooled. And ever since Harvard law student, Traci Kilburn, entered the mix, youthful aspirations of prying secrets from the sequestered past had taken a back seat to obsessive preoccupation with the living present.

Her scintillating apparition, floating languidly across the rough-cut hillsides creasing Ryan family farmland, routinely invaded his sleep. Dream sequences featuring an ephemeral Traci pre-empted notions of adventurous treks through fossil graveyards.

This night, the phantom dipped so close, Josh imagined inhaling intoxicating perfume. Flaunting a come-hither smile, the tantalizing vision approached to the rhythm of exotic drumbeats. The pounding percussion intensified until the racket obliterated the dream. Still, the pesky beat rattled the senses without mercy!

Half-awake, he cleared the cobwebs enough to determine that the intrusive clatter came from incessant knocking at the mill-house door. The cranky scholar staggered reluctantly into Sunday's dawn, grabbed a robe, and approached the front door grumpily.

"Did I catch you dreamin' about that sweet thing again, Dar? It's time you joined the world and strutted into the spotlight waiting you this afternoon at Chamberlain Playhouse!"

Jonathan Thomas Daniels, lifetime neighbor and pal, had a way with words and an ebullient personality to match. Years before, JT had branded his sidekick with the nickname Darwin, recognizing his zest for fossils. JT brushed by with a fast-food tray of coffee decorated with a couple of copies of an embossed brochure worthy of a presidential inaugural.

A grumpy Josh rubbed his eyes, demanding, "What's goin' on, Bro? You and the Montgomery County sheriff are the only people cruisin' the scene at this outrageous hour. You oughta' be in church with the sinners...You could use some of your granddad's preachin' to rid you of those lawyer-like flaws!"

JT grinned, slipping the scholar a steaming cup of Sunday morning brew.

"You're the story, man! Look here! There you are—center spread, just across from a full-page photo of His Honor, the legendary Judge Edward Anthony Stone—squeezed in, big-as-you-please, between yours truly, and that glamorous dream girl of yours...uh, what's her name? T.r.a...sorry, Pal, you know me...no good at names."

JT jabbered on, stalling to give the scholar/dreamer time for a reality check. After a mandatory grunt and a few "harrumphs," Josh Ryan, alias Darwin, recaptured his gentle disposition. Then sipping coffee, the two scanned the hot-off-the-press program for Mid-Atlantic University's six-day reprise of the 1925 *Scopes* "trial of the century. "

Sure enough, the printed program plastered Hollywood-style photos of the three cohorts, front-and-center. The previous April, Traci Kilburn had recruited Josh to take the stand as expert witness. JT Daniels, senior law student from USC, had been snared with the same brassy shenanigans. Both would share the stage with the make-believe-Darwin for the unscripted, unrehearsed show.

Josh admired the brochure's professional touch but could manage only a left-handed compliment.

"Looks like Ol' Pace Terhune did a number on the hype. A slick brochure like this costs big bucks...but his budget can afford it. We downtrodden hired help aren't getting paid a sweet nickel!"

"I know better than that, Dar. Once Ms. Kilburn hypnotized that poor excuse of a mind, you would have given the deed to the family farm just to play a part in the show!"

"You're a certifiable loony-tune, man! She doesn't give me the time of day except for this cockeyed show-biz scheme!"

"Listen to you! You're pathetic! You've wasted twenty-three years of prime time huggin' on old bones. As to women, you don't have a clue. How you gonna' fly unless you crank up the jets?"

Josh enjoyed the exchange, welcoming the semi-scold as intended jest.

"Don't forget, Bro, thanks to her...and you, my alleged friend—willing co-conspirator I might add—my Ph.D. candidacy hangs by a thread. In the bargain, the fossil find of my lifetime has mysteriously disappeared. I have to reach up to touch bottom!"

"That's peanut stuff, Dar. Look at the big picture. Today a star is born...As we speak, the big time invites you to appear! The spotlight awaits. Shine it on, man, shine it on!"

4

Josh retaliated by swatting his pal's shoulder and stammering something undecipherable. At that, the two cracked up in boisterous laughter, in the finest tradition of all spirited young men. On the brink of unsought notoriety, the carefree, bogus, Darwin had no way of knowing that within the week, his cut-and-dried scholar's life would change course forever.

*******

## December 1995

Billed as *The People v. Charles Robert Darwin,* the make-believe campus show trial to be staged the first week of September 1996 promised to be the highlight of a series of centennial events celebrating Mid-Atlantic University's 1896 founding. The university administration locked onto the idea, satisfied that the *Monkey II* trial extravaganza could showcase campus scientific disciplines.

Touted shamelessly as fictional heir to the original Dayton, Tennessee, 1925 *Scopes* "trial of the century," the week-long event debuted as the brain-child of Pace Terhune, 41-year-old former ace Hollywood promoter, currently presiding as MAU's Vice-President for Public Relations and Development. Unencumbered by false modesty, he boasted a shirttail relationship with Albert Payson Terhune, author of *Lad, A Dog* and other yarns trumpeting the virtues of the majestic collie. No creative slouch himself, Pace excelled at crafting premium spectacles.

Ferreting out a thousand guests daily, each willing to shell out 50 bucks to witness a two-hour unrehearsed amateur drama, neither daunted nor deterred the indefatigable ringmaster. He knew how to pull out the stops, whetting consumer appetites by mixing syrupy words with eyebrow-raising action. Long before the September 1 premiere, the lead story fashioned for the Entertainment section of *The Washington News Press* appeared with this provocative headline:

### Is Evolution a Fact?

Edward Anthony Stone, retired Federal judge, university trustee, and senior partner with the law firm Cabot, Calvert, and Stone, had been cast as the event's centerpiece. Intimate friend of U.S. presidents and confidante to politicians of both parties, the barrister's career success approached the legendary.

Informed of the dubious assignment, the judge rolled his eyes and chuckled, "You've got to be kidding!" Ed Stone's words flowed like honey, projected by a mellow timbre that echoed the gentle nuances of fifth-generation Marylanders. To dilute resistance, the wily Terhune sugar-coated the pitch.

"It should be a day at the beach for you, Judge Stone. Rules of law will take a back seat to drama and educational enlightenment. Of course, you'll have total

5

freedom to call the procedural shots." Sheepishly, Terhune added, "Let's face it. Entertainment is the name of the show-biz game."

Unimpressed, the judge countered, "So what else is new?"

Terhune tried another tack.

"No cost has been spared. The stage will replicate the 1925 Dayton, Tennessee courtroom...lots of oak furniture...Bryan and Darrow could cross swords here without realizing Maryland had replaced Tennessee as backdrop."

Judge Stone didn't bat an eye, maintaining a "So what?" expression. Cool and casually confident, Terhune seized the moment to unveil his big gun, venturing a nonchalant, "When I ran the idea and your anchor role by the Duchess, she blessed the part as 'an honor the judge deserves.'"

Dropping the name "Duchess" rang bells and made dogs bark! Lady Regina Ann Morgan-Carrington commanded attention!

Granddaughter of Mid-Atlantic University's founder, Sam Houston Morgan, Lady Carrington was the seventy-eight-year-old widow of an Englishman who manipulated his title to match an American fortune. The imperious socialite ruled her domain from Morgan Manor, the family's Victorian mansion, just off the MAU campus. Undeterred by advancing years, the grand heiress, dubbed "the Duchess" by those who dared, swung a big stick and inflicted punishment without qualms. One folk tale alleged that the aggrieved Duchess had purchased an auto dealership solely to fire a discourteous service manager who publicly humiliated her chauffeur. The Duchess intimidated even those who saw her as snooty and spoiled rotten.

Unintimidated, the judge offered solace. Ed Stone protected the personal interests of Lady Carrington and oversaw the legal demands of her family's Pan Oceanic Petroleum, an international conglomerate. The two friends served as MAU trustees and enjoyed a half-century social/business relationship based on trust. If the Duchess saw merit in Terhune's Centennial scheme, so be it—the judge would play the game. Shaking his head with the hint of a grin, the legal sage pronounced Terhune victorious.

"OK, you win! But I'm going to hold you to that 'anything goes' entertainment and 'full discretion' concept."

Convinced of an assured future for *Monkey II*, Terhune sensed a winning streak and pressed his luck.

"Incidentally, as duly appointed presiding judge at the next *trial of the century*, you have the privilege of picking the brightest law student clerk from Cabot, Calvert, and Stone as the event's coordinating trial counsel."

"You're an unconscionable California rascal! If I had my way, I'd sentence you to a lifetime of clearing sagebrush trails in the Sierra for the convenience of the coyotes," Judge Stone jested.

Truthfully, he enjoyed the brassy wit of the canny entrepreneur. Edging away, he decreed, "Traci Kilburn." Noting the question mark in the gaze of his nemesis, he tossed back an over-the-shoulder explanation with judicial finality. "She's the best and the brightest to arrive in our office in years. Harvard Law, graduation scheduled 1997, articulate writer, captivating public speaker, and local name recognition—her father's a hotshot Johns Hopkins surgeon."

Judge Stone's eyes twinkled. "Mr. Promoter, I hope you won't hold it against this fine lawyer-to-be, but even in your jaded West Coast perception, I'm sure she'll come across as poised and glamorous—likely to steal your show!" As Terhune beamed, pleased with the pay dirt, the judge returned a laser look, displaying shrewd insight.

"And oh, Terhune, in case you thought I bought that snow job about the Duchess endorsing the alleged honor—forget it. If she gave a hoot, she would have told me herself—long before now!"

Unfazed, Mid-Atlantic University's consummate organizer nodded benignly, aware that with Ed Stone on board, the rest of the casting should be a cinch.

Jessica Saunders, a petite, 39-year-old CPA with a karate black belt and FBI connections, had agreed to serve as court clerk. She didn't jump for joy at the news, but Pace predicted correctly she would be a good sport and go along. The astute lady functioned as Terhune's senior associate in the university's public relations and development office.

Terhune ensured his own prominence at all six presentations by playing bailiff. What better way to capture the essence of each day's drama, spinning headline news from the juiciest tidbits of raw information.

Selection of an eight-person jury from willing locals proved a cakewalk: Marie Mackin and Ira Lonergan, MAU campus activists; a couple of senior citizens happy to escape Leisure World retirement routines; the owner of a fast-food franchise; Lyle Grant, 50-year-old Hampshire Country Club golf pro; a Sherwood High teenager; and Slade Lassiter, MAU science grad student.

Months earlier, Terhune had handpicked Dr. Karl Striker, flamboyant campus scientist, for the show's top billing. Blunt and hard-edged, Striker had a reputation that spread far beyond the university's boundaries. The forty-six-year-old Striker's rocket scientist father had fled Germany after World War II in time for the professor to be U.S. born. Swarthy-complexioned and black-haired, the distinguished scientist would make a handsome witness. But when informed the *Monkey II* stage would feature the university division chairman, Judge Stone warned that controversy loomed.

"You get what you pay for, Pace! From every rumor I've heard, you may have found yourself a zealot who worships at the shrine of Charles Darwin."

Terhune shrugged indifference, choosing to ignore the yellow flag. Still trying to be helpful, the judge continued, "Hearsay has it your expert witness sports a rocket scientist mind trapped in a *sieg heil* personality. There's reportedly a high mortality rate for his division's Ph.D. candidates who don't pledge allegiance to Charles Darwin."

Unimpressed, Terhune laughed it off.

"Thanks for your concern, Judge, but I don't have a clue or a care about Darwin or his theories. All I know is that evolution guarantees hot debate—and controversy sells!

"I've heard the same stuff," Terhune continued. "Last spring, campus newspaper editor Ira Lonergan ran a student opinion poll. Anonymous responses took poor Karl to the woodshed: *arrogant, flaunts authority, exploits limelight, cunning, opportunistic, manipulative politico*—everything I need to stir controversy and hawk tickets to this shindig."

Though revered for a juggernaut mind paired with immaculate social skills, Karl Striker's abrasive tactics earned few kudos from students. What's more, the unpopular scientist didn't give a hoot! Nor did Pace Terhune! Striker generated star-quality presence, certain to grab an audience and to guarantee great press reviews. With his nose for show biz, Terhune pursued with gusto.

Initially, the articulate Striker welcomed the prospect of national plaudits emanating from the showcase event. But some months after endorsing the trial and consenting to display his academic prowess, he had second thoughts and his enthusiasm soured.

Alarmed at hints of defection, a dismayed Terhune approached Striker with statesmanlike persuasion, to no avail. The professor reneged, abandoning Terhune and *Monkey II* fame. Striker's ego refused to submit to unscripted questioning from the probing minds of law students. The prospect of stardom tempted his vanity but could never compensate for the risk of marring a meticulously manicured public image.

Striker could say "No" in multiple dialects. Fluent in both unaccented German and English, the professor spurned Terhune's overtures, barking a "nein" while shaking his head ferociously. As the refusal reverberated, he haughtily flaunted his credentials.

"You've heard, have you not, about the rock found in the Antarctic that exhibits what some claim to be the fossil remains of primitive micro life from Mars? While I'm quite unconvinced, Houston will be my address the entire summer of 1996. Perhaps you are aware that I am one of a select few nationally prominent consulting scientists to NASA!

"No way will I submit to a harebrained scheme that trivializes the esteemed reputation that MAU's science disciplines enjoy!" The words thundered with immaculate precision, bitten off in no-nonsense syllables.

Believing Terhune duly awed, he signaled his contempt with a snort. "Scientific research outranks staged Barnum-and-Bailey stunts. Go find somebody else who wants to make a monkey of himself!"

Terhune wasn't aware of the NASA rock research, nor was he awed. But he couldn't miss the sharp edge in Striker's voice. Under the guise of helpful counsel, the professor uncorked a stream of insults at a subordinate.

"Why don't you invite Augustus Morley to be your star? He thinks he knows it all and has always craved recognition...Besides, Gus rarely seems overworked. He has loads of time to squander!"

That was that—case closed!

Terhune took the snub without flinching. But his grandiose scheme for replicating the "trial of the century" teetered perilously close to extinction.

Any fool could recognize the limits of Dr. Augustus Morley, faithful supplicant to Karl Striker. He could add little luster to a staged confron-tation dramatizing conjectural science. Bald, corpulent, with squinty blue eyes peering from behind steel-rimmed, half-lenses, Gus Morley's reedy voice lacked resonance, and his scholarship rarely drew rave reviews.

Terhune sensed that this underling harbored resentments toward the division chairman, having been passed over for promotion by the university administration in favor of the younger and brassier Striker. Still, his good-old-boy personality offered saving graces: affability, a sympathetic ear and ready accessibility to students, and a reputation for fairness. This second echelon of academic authority reacted attentively to all comers, so Terhune approached on eggshells. The ringmaster diplomatically downplayed Striker's rebuff and skirted the embarrassment of inviting the jolly Dr. Morley to volunteer. Aware that a stage appearance by the colorless figure would spell disaster for *The People v. Charles Robert Darwin*, Terhune came right to the point.

"Need your help, Doc. I'm looking for a witness for *Monkey II!* From your administrative perspective, I'm wondering who you would judge the most brilliant scientist on the MAU campus—excluding yourself and Dr. Striker, of course?"

After sober reflection, Gus Morley waxed eloquent. "What would you think of a grad student? I can vouch for his academic skills."

"Tell me more."

"Ph.D. candidate, double major in geology and paleontology. Young Joshua Chamberlain Ryan has never received less than an 'A' at Mid-Atlantic U. He thinks fast on his feet and devours public dialogue. The boy hurls sharp words with reckless abandon—without fear, favor, or guile."

"And?"

"The lad's obsessive demand for proof above tradition gives him an irritating, in-your-face style. Tends to put a burr under the saddle of the powerful—if you know what I mean."

"Is that bad?"

"Only that his wise-guy, too-big-for-his-britches attitude annoys senior faculty. Good witness material—he has the steel-trap mind of an intellectual and speaks the language of the street!"

The fate of *Monkey II* hinged on the untested gut reaction of the gregarious Gus Morley, paired with the intrepid good luck of ringmaster Pace Terhune.

The amiable teacher continued, "Joshua Chamberlain Ryan is your man!" Morley salted his response with a knowing wink.

"Sounds close to what the doctor ordered. Anything else?"

Morley piously unloaded a detailed spiel. "While it's reported that Josh Ryan's not on speaking terms with his grandmother, Lady Carrington, you might find his participation conveniently newsworthy in this centennial celebration year. Young Ryan happens to be the great-great-grandson of Sam Houston Morgan, Civil War hero and, coincidentally, Mid-Atlantic University's founder."

The next revelations captured Terhune's fancy.

Joshua's mother, Leslie Anne Carrington-Ryan, had died shortly after his 1973 birth; father Michael Joseph Ryan lived in a state of virtual excommunication from the Morgan-Carrington family. Grandson Joshua felt the sting of being shunned by a matriarch who carried chips on both shoulders. Fascinated, Terhune concluded he had underestimated the plodding professor. Envisioning the headline opportunities, the ringmaster's mind churned with schemes to snag this Ryan fellow. The informant waited for the understated magnitude of his gossip to sink in, then dropping his voice to a whisper, added a final clincher.

"One more thing you may find timely. It's been rumored that while still a youngster, this grad student answered to the name Dar, short for Darwin, thanks to a childhood hobby of scouring the hinterlands for fossils. Hearsay has it that he dug for dinosaurs as far away as Wyoming and Colorado."

The MAU promoter had heard what he needed and put all systems in go. Backing out the door of the academic's office, he rendered profuse "thank you's," sensing a casting coup in the alias *Darwin*.

Unbeknownst to Terhune, the benign Gus Morley had conveniently concealed the fact that, of all the students under the rigorous discipline of Dr. Karl Striker, none rankled the academic top dog more than Joshua *Darwin* Ryan.

Thanks to his irrepressible wit masked by a choirboy face, master prankster Dar Ryan owned the classroom. Driven by a genuinely inquiring mind, the happy-go-lucky student never hesitated to interrogate a teacher or to dispute a professorial opinion. Mere days into his grad study program, a classroom clash offered a hint of things to come. Mimicking innocent inquiry, he reported he had "tried to teach my pet iguana to fly, but the dude refused to sprout feathers and kept crash landing!" Following the first round of classmate laughter, he quipped, "Poor Iggy never was the same after that...couldn't get him to go near a tree...and he didn't want to hang around with birds!"

Josh hit a nerve when he made fun of Professor Striker's perfunctory explanation of spontaneous generation.

"When I was a kid, I strung some wires in a cardboard box, glued on a dial, and drew a television screen on the front with a crayon. Nothing happened. No picture, no sound, nothing. Why doesn't spontaneous generation electrify the contraption and kick in a live TV picture...I can't wait a billion years!" Student mirth exploded on cue.

Dr. Striker scoffed, "That's absurd!" and tried to skirt the challenge, but Josh persevered.

"Which do you believe to be more supernatural: Organic life originating from random coincidence? Or from some master-planned, intelligent design?"

The glowering professor snapped, "If you are interested in serious science, Mr. Ryan, you can see me after class. If not, I suggest you restrain yourself and conceal your ignorance."

Gratified at the knee-jerk reaction, Josh Ryan smirked. The die had been cast! Though no feckless rabble-rouser, intellectual impertinence always lurked just beneath his facade of indifference. To the surprise of most, his straight "A" average survived the withering scrutiny of a hostile major professor, thanks to superior performance and Striker's impeccable integrity, which was as rigid and unbendable as his ideas.

Initially, Pace Terhune had no idea how many hoops he would need to jump through to land the young scholar as the *Monkey II* Darwin. By the time he parted company with Dr. Morley, he decided that he needed the talents of the resourceful Jessica Saunders.

*******

"The Morley clues check out: local kid; great-grandson of MAU founder Sam Houston Morgan; partially orphaned shortly after birth; raised by his computer consultant dad, Michael Joseph Ryan; sprouted tall on his family's 25-acre Magruders Mill spread, overlooking the Hawkings River; bright-as-a-dollar!

"Except for being motherless since birth, Josh Ryan comes across as your typical 23-year-old grad student."

The casually efficient Jessica had managed to compose and deliver to her boss the requested personal profile within the week, despite Christmas holiday distractions. An impatient Terhune asked for a bottom-line oral summary.

"Sounds like a winner!"

"Could be—no skeletons in his closet...if you don't count fossil bones."

Terhune smiled at his associate's subtle wit. "Anything else?"

"Some good news, and some bad news...depending on your viewpoint." She knew how to keep Terhune's attention by drawing out the nitty-gritty.

"Please, the good news first!"

"A Genghis Khan in the classroom behind a choirboy face, all the trappings of innocence, without a trace of guile."

"An audience will eat it up. What could be the bad news?"

"This modern Darwin thinks the author of *Origin of Species* and *Descent of Man* a perpetrator of obsolete myth! If you parade this youngster on the stand as your expert witness, your buddy Striker is not likely to dance for joy!"

Pace Terhune whooped!

"Great! Can't think of a better way to whet the public's appetite than good old-fashioned clash and clatter. Why don't you try to round him up and get him over here today if you can."

Jessica Saunders didn't flinch. She had one last tidbit of bad news.

"Sorry, Pace. No can do. This guy is off on some wild-goose chase, prying some ancient bird fossil out of the rocky strata of Middle Tennessee...Word is he'll be roughing it there until spring quarter."

"Is he some kind of nut case?"

"Only crazy to pursue fossils! Seems as if he managed to convince Doc Striker that this find could be a rare *Archaeopteryx* that deserves centerpiece booking in his Ph.D. dissertation."

"Can we reach him by cell phone?"

"Probably...but I've got a better idea."

"Let's hear it!"

"This morning, Judge Stone called to set up a time when he can introduce you to Traci Kilburn. She's home from Harvard for the holidays and will be clerking in the Cabot, Calvert, and Stone downtown law offices. He all but guaranteed that she'll agree to cross-examine your *Monkey II* expert."

"So what's the better idea?"

"Every scrap of data I picked up tells me young Ryan would be even less inclined to play the *Monkey II* expert than Striker, if that's possible. Word is, young women see him as a hunk, but he's very shy."

"I'm beginning to see the light."

"If the charming Ms. Kilburn proves to be everything you've been told, she's the one to land your quarry! I've been told that my Cherokee ancestors used live bait to attract and capture even the highest-flying eagle."

"Bingo! What would I ever do without you, Ms. Saunders?"

She thought to herself, "A ringmaster like you, Mr. Terhune, could get along with or without a sidekick, if anybody could!" Inwardly pleased with her grand strategy, she couldn't quite suppress the hint of a smile.

*April 9, 1996*

# 2

The unpretentious Camp Ryan site perched precariously on the lip of an eroded gully. This remote chunk of Middle Tennessee's rocky top had once sheltered five generations of Magruders—Grandmother Ryan's family. Broken bits of stone foundation and a haphazard array of chipped bricks littered the rough-cut landscape—lonesome remnants of a long-gone agricultural heritage.

The cedar-studded terrain would have continued to nourish the industrious Magruders but for the Tennessee Valley Authority's pursuit of rural electrification. The property could still be accessed via the helter-skelter ruts of unmapped, tortuous dirt roads—a ritual trek undertaken at least once by each new branch of the Magruder family tree, intent on tapping their roots. Here, midst the rocky scrabble of a remote cove feeding Tennessee's Center Hill Lake, an adolescent Josh Ryan had discovered a marine fossil bonanza, swathed in limestone-lined gullies, routinely eroded by rivulets of rainwater that exposed stratified sediments, containing the secrets of pre-historic life.

When routine scrapings unveiled the outline of an ancient bird, the astonished student reported the find to Dr. Striker, who endorsed the field project as the basis for a doctoral dissertation. The student's frenzied passion for discovery could not be discouraged by brutal winter storms that lashed Middle Tennessee and deterred the faint-hearted. Curious locals dropped by from time to time to chat with the solitary camper, only to leave, shaking their heads, convinced that "Grandma Magruder's Yankee grandson is a sure 'nuff nut case."

Win or lose, at least the diametrically opposing views of teacher and pupil found respite from confrontation in the excavation's common ground. Dr. Striker believed that production of a rare fossil bird-like specimen could confirm his commitment to the evolutionary bird-from-dinosaur theory. Every bit as resolute, Joshua Ryan pursued the dissertation project convinced that his fossil could contribute to the school of thought that held that birds not only were unrelated to dinosaurs, but also appeared concurrently and coexisted with the extinct reptile behemoths.

Down and dirty from his lonely three-month vigil, Josh Ryan paused to savor the reward of discovery and to inhale the rich April aromas. After enduring grueling days of sweat and grime in the ice-frosted trenches of winter, he believed fervently that within hours he would triumphantly carve the bony fragments of an ancient *Archaeopteryx* from the earth.

His spirits soared.

\*\*\*\*\*\*\*

14

A slightly disheveled Josh neither noticed nor cared about the few sidewise glances from more carefully attired patrons bustling through the Nashville airport. Anticipating the arrival of JT Daniels, he hummed a country ditty, occasionally lip-synching words he had never bothered to memorize. He chuckled aloud at his arm-twisting success in persuading JT to exchange a week of basking in the sun at California's Laguna Beach for the arduous soil-sifting drill south of the Mason-Dixon Line. Unlike the inquisitive Josh, JT made no secret of his personal boredom with fossilized traces of prior life. No way would he sacrifice his spring vacation to roughing it in a remote gully of rocky-top country except for his loyalty to Dar Ryan. JT gritted his teeth, faking pleasure at the prospect of trading sun and surf for crumbling slivers of rock.

The two had shared similar childhoods, exploring the deer trails that traversed the hardwood forests marking the banks of Maryland's meandering Hawkings River and the boundaries of adjoining family farms. Born in 1973, they both had grown up as partial orphans: Josh's mom, Leslie Anne Carrington-Ryan, died days after his birth; Parker Daniels, father of JT, had died a hero's death as a helicopter pilot flying rescue missions in Vietnam jungles. Michael Ryan routinely invited JT to join the Ryans on business jaunts to Hawaii and California's Silicon Valley. To reciprocate, JT's mom regularly performed double duty, looking after Josh as she would a second son.

Each youngster boasted a grandparent with a nationally recognized name. The imperious Lady Regina Ann Morgan-Carrington ruled the roost of Montgomery County's social elite with an iron fist, but refused to give the time of day to Grandson Josh Ryan. In contrast, the two pals flourished under the benevolent oversight of JT's granddad, Bishop Brock Daniels, a gutsy civil-rights leader who had marched boldly at the side of Martin Luther King and proclaimed righteousness from national pulpits with persuasive power.

Twenty-three years of friendship built on the happenstance of geography and the commonality of tight family traditions forged bonds rivaling the affinity of blood brothers.

After an uneventful landing, JT found Josh pacing patiently at the arrival gate. In a flash they dodged through the crowd to the parking lot, climbed into Josh's jalopy, and roared off across the fragrant Middle Tennessee landscape. Cruising through the blossoming Southland spring in a rusted Mustang vying for its own fossil status, the guest from the West teased Josh about the ramshackle condition of the car, appropriately marked with a bumper sticker warning:

*"Do Not Paint—Rust Test in Progress."*

The duo had not enjoyed a face-to-face chance to jest and reminisce since the previous September. JT broke the spell.

"Walking through the airport, you seemed to be limping more than I remembered." JT had resurrected a delicate subject, off-limits to all but himself.

As teenagers, the twosome had played on Sherwood High's junior varsity football squad. Then one autumn afternoon, as they were homeward-bound on bicycles following an exhausting practice, a reckless driver sideswiped the pair. JT survived with a badly bent bike frame. No such luck for Josh. He had suffered a bone-crushing injury to his left leg that terminated forever any prospect of future football fame. The brutal blow mutilated bones, muscles, and ligaments. Medevac transported the young victim to Baltimore's Johns Hopkins Hospital where a top East Coast surgical team spent the better part of an afternoon trying to reassemble his leg.

JT refused to leave his friend's side until he was wheeled to a surgical unit. Despite excruciating, gut-wrenching agony, Josh's pale face betrayed not a tear. His vow as a five-year-old, never to cry again, bolstered his stoic reserve. But once the doors to the operating room swung shut, JT himself broke down in anguish for a pal in pain. Now JT quickly withdrew the intrusion, concealing the painful memory with down-home parlance.

"One thing about it, Mr. Ryan, that double-hitch in your git-a-long never slows you down!"

Josh responded with a lopsided grin and a mishmash of Southern-style lingo mixed with a slang phrase of camaraderie picked up in Hawaii during a travel stint with his dad.

"Don't y'all worry, Bro! No need to go carryin' on about me, ya' hear?" Daily he rejoiced in the gift of walking, thanks to a reconstructed limb. He brushed off the concern brashly with a playful swat to JT's shoulder. "The old gimp sets in when I squat cramped on a rock and forget to stretch now and then. A little limp never hurt anybody. Maybe with the luck of the Irish, some glamour gal may mistake me for a war hero!"

JT looked dubious, so Josh turned to a more cheerful topic.

"Anyway, all leg bones turn to dust sooner or later...unless they fossilize to confuse paleontologists." Josh guffawed at the blatant absurdity. JT smiled, choosing to drop the unpleasant memory.

"What ya' doin' anyway, Dar, wastin' spring sunshine in a gravel pit?"

"*Archaeopteryx*, Bro, *Archaeopteryx!* Just think...in the next few hours, you may become famous for just bein' on hand to scrape up the pieces!"

JT chortled, "More like bein' seen as a nut case for passing up Laguna Beach for a week, diggin' up stuff no one gives a tinker's toot about. You've always been a head case, Dar—it's time you abandoned that Latin gibberish and got a life."

16

"Look, Bro, if Archy proves to be the real thing, maybe I can unscramble a raging dispute." Josh ignored the friendly jab and moved to serious mode.

"The Doc Strikers of the world believe birds came from dinosaurs, betting that *Archaeopteryx* is an intermediate life form. Others see it as nothing more than an extinct critter with feathers, unrelated to modern birds. Another view brands Archy as a hoax—merely a cat-sized dinosaur with fossil feathers counterfeited to enhance marketability."

"In other words, Sir Darwin, this expedition is for the birds!"

They both laughed easily then reverted to discussing old times for the remainder of the short drive. Within the hour, the Magruder stomping ground overlooking Center Hill Lake loomed ahead. Come morning, they would tackle Josh's pet project—the scrupulous excavation of strata in order to introduce ancient fossilized remnants to the probing scrutiny of April sunlight.

While this was no Laguna Beach, the comforts of Camp Ryan proved more promising than JT's pessimistic expectations. A beat-up camper with narrow bunks provided shelter from the vicissitudes of Tennessee spring rains. The tacky camper's cooking facilities risk the reps of even the world's most skilled gourmet chefs. And the cramped semblance of a privy barely qualified for in-house status.

The tight quarters had been already pre-empted by stacks of files, cardboard containers, notebooks, and word-processing equipment, including an experimental portable micro-mini TV recorder, all scattered helter-skelter throughout the camper. Although exasperated at the sight, JT groaned in resignation.

An experienced light packer, it took him but a few moments to toss a sleeping bag on the vacant bunk and to unpack essential gear. The newly recruited employee/guest quickly brandished his freshly honed legal skills by initiating mock capital/labor negotiations in the name of the downtrodden.

"Seems as though this job is akin to that of a fireman's—round-the-clock, 24-hour duty. At $11.25 an hour, that comes to something like $270 a day for five days or $1,350 for the week."

A measly $11.25 an hour for the alleged privilege of scraping rock and sandstone in an unpredictable spring climate in rugged Tennessee back country held zero attraction for the negotiator. Money never drove JT's ambitions. Blandishments of a multimillion-dollar career in the NFL had so far failed to divert him from his ambition to become a trial lawyer. He expected to graduate from Southern Cal's Law Center in the spring of 1997.

Josh rolled his eyes, faking despair. "Well, would you listen to the wind blow!"

JT ignored the impertinence and pressed his case for marketplace equity. "Then when you remember that two-thirds of that is overtime and worth time-and-a-half, the tab tops $2,000. Add another couple of grand for first-class, round-trip fare to California befitting the lofty status of any aspiring lawyer, plus another $1,000 or so for the extra-hazardous duty of this rocky slope, and I figure a flat $5,000 will just about do it."

The uncrowned king of the makeshift fossil capital pretended to struggle with the computations, relying on his brain as personal computer. Apart from acquired skills in the rock-related sciences, the young scholar had built a reputation as a mathematics whiz, known for frequently testing that *whizdom* at the twenty-one tables in Vegas and Atlantic City.

"Your figures make sense to me!"

An impassive JT waited for the second shoe to drop as pompous, make-believe management resorted to company-store exploitation. An attentive JT recognized the dropping shoe before it landed with a thump on the conversational floor.

"Of course, we need to deduct a few bucks for on-the-job training for a scientific novice...not to mention the extravagant fare of country-resort-style room and board delivered courtesy of Camp Ryan.

"The way I see it, that would justify a net balance due labor in the range of $123.32 for the week—unless you persist in driving me nuts with these petty attempts at extortion."

At this juncture, the sparring negotiators broke into peals of laughter, slapping knees and shoulders in outrageous jubilation reminiscent of childhood rituals. Suddenly JT straightened, brushed himself off, and with all the seriousness he could muster, demanded coffee-break time, declaring this exercise as pivotal to his pending contract.

"While it's too much to expect of a tyrannical sweatshop, I hope you have a stock of semi-decent coffee grounds and a beat-up coffee pot that survived California's Gold Rush."

Beckoning JT to follow him back to the camper, Josh bowed low in a sweeping gesture, inviting, "Be my guest. I knew you were coming, Mr. Football All-American."

Miraculously, tucked in the corner of the cupboard, JT spied a brand-new coffee brewer and a couple of pounds of freshly ground Sulawesi blend. Before he had time to ask where he could tap a dose of muddy $H_2O$, the host pointed to a cache of several gallons of sparkling bottled spring water.

Soon the two sat on jutting boulders, inhaling the rich blend of fresh brew while sipping in silence. It had been a glorious afternoon. All the loose ends of family happenings had been spun together in a tidy fabric. Eventually, lengthening shadows concluded the day, postponing serious excavations until dawn.

Easing into the evening, accompanied by the monotonous hissing of a Coleman lantern, Josh launched into an easy banter, briefing his guest on project details. He confided that the more he studied Darwinian thought, the less corroborating evidence he found. JT sat passively, half-listening to his host while trying to choke down an unappetizing, prepackaged meal.

"I'm not prepared to call Darwin a fraud, but from everything I've been able to read and see for myself, his theory keeps unraveling."

JT paid little heed to the spiel. The shadows had hardly disappeared into darkness before he stood, stretched, yawned, and departed in quest of a bunk's refuge.

"Us hired hands need plenty of rest to rally our bodies to earn the big bucks!" he jested. "I can't wait to start scratching stone at the crack of dawn. Maybe we should loiter another hour or two, searching in the dark, considering how many billions of years Mr. Fossil Bird has been waiting to be discovered."

With that, he disappeared in the darkness.

Left alone with the rhythmic harmony of a cricket chorus, Josh dug a hand deep into the soil's damp grit, scooping out a fragment of ancient history. Sifting the grains through his fingers, he imagined the intoxicating secrets these bits and pieces of inorganic matter could tell.

The Magruder clan had once managed to carve its niche while eking out a living on the still fertile hillsides sheltering the raw ravines feeding Center Hill Lake. This brush with hidden history sent a shiver through his body. Too excited to sleep, Josh ambled in the direction of the derelict camper, fully expecting to be out scratching stone at sunrise, with or without the California labor negotiator.

## *April 10, 1996*

# 3

Shaking jet lag, neophyte paleontologist Jonathan Thomas Daniels arrived a little late at the digs the first morning on the job. But within moments, he found himself scratching gravel like a pro, succumbing to the infectious enthusiasm of his mentor. Josh's imagination leaped as they mined the Pleistocene strata.

True to form, the restless Josh had risen at daybreak, donned a hard hat painted the color of burnished sun-rays, and begun the delicate excavation, temporarily postponing a first cup of coffee. Mounted unobtrusively beneath his helmet's visor, a revolutionary micro-minicam recorded every sight and sound of the expedition. Cutting-edge technology, the gadget had been installed as a test project underwritten by a corporate client of Michael Ryan's. Each day's recordings were assembled in pre-addressed packs, driven to nearby Smithville, and shipped overnight to the deep-pocket sponsor.

By mid-week, the student team had proved successful beyond expectations. The mysterious feathered creature from ancient times awaited its debut, Josh hoped, to the plaudits of a headline-hungry Karl Striker.

Just a few scrapes away from pay dirt, the industrious stone cutters glanced up, startled by a throaty feminine voice wafting down from the crest of a sheltering ridge. The unscheduled intrusion introduced a stunning apparition, back lighted by the mid-morning sun.

"Hey, Darwin, find any stupendous fossils lately?"

Since before he could remember, no one except JT had called him Darwin. Josh didn't recognize the nuance in the musically sexy tone. The puzzlement on JT's face mirrored his own.

Without waiting for an invitation, the young woman skipped nimbly down the slope like a deer, chestnut hair flying. The descending vision of loveliness drew the curious audience of two to their feet. Spellbound, they watched the apparition navigate the slope. Attired in boots, tight-fitting blue jeans, and matching denim jacket, up close she turned out to be an exceptionally attractive woman approximately the age of the excavators!

Whoever she was, she moved with purpose and arrived with impact.

Picking her way methodically through sharp outcroppings of granite and sandstone, punctuated by patches of rugged native brush, she tripped on a protruding root and finished her descent by stumbling not ten feet from the base of the incline. As she tried to catch herself and preserve a shred of dignity, her momentum propelled her forward instead. Arms flailing for balance, she was headed for a face-down arrival but for the comforting salvation of Josh's willing arms. Despite his spontaneous gallantry, the collision's force sent them both sprawling in the dust in an awkward embrace.

He managed only the tired cliché, "We'll have to stop meeting like this. Do you want to try that again?"

At that instant, Josh Ryan totally forgot his preoccupation with fossils!

Recovering her poise, the alluring lady scrambled to her feet and tried brushing away the thick Tennessee dust. Still entranced, the young men watched as she rearranged errant strands of a chestnut-colored mane that framed sparkling hazel eyes set in a glowing olive complexion.

"Manna from heaven," thought Josh. "Dial 1-800-AWESOME!"

"I'm just collecting autographs. America's Darwin and his football jock sidekick, I presume?" Cool and collected, she extended her hand, revealing a crisp, business demeanor.

"Hi! I'm Traci Kilburn, Harvard law...hacking my way through this jungle on behalf of that citadel of learning, Mid-Atlantic University!"

Still mystified, the two targets cautiously acknowledged their identities, waiting for more. She enjoyed their bewilderment, toying with their psyches and rationing clues to her mission.

"By the way, I've seen you both before...years ago. My dad is Dr. Percy Kilburn, the physician who welcomed you to a Johns Hopkins surgery unit when you survived a bicycle accident."

Double-takes revealed her listeners' continuing confusion.

Still in command, she laughed, explaining, "No, I was not a part of the hospital team, only waiting in the lobby to hitch a ride home to Columbia with my dad. He was just going off duty when you guys showed up."

After accepting her host's offer of a cup of steaming coffee and joining the earth diggers for a get-acquainted confab on some fallen logs drawn into a circle, Traci finally got to the point. Concluding the small talk, she beamed triumphantly, "Looks like I've come to the right place." She proceeded to bring her quarry up to speed with a short summary shared with matter-of-fact candor.

Harvard law; Cabot, Calvert, and Stone clerk; *The People v. Charles Robert Darwin* assigned as her personal project; her previous day's trip to Dayton, Tennessee, to research the 1925 *Scopes* courtroom. She vibrantly described plans for *Monkey II* and its scheduled September staging in MAU's Chamberlain Playhouse.

Then, matter-of-factly, she dropped her bombshell. "Of course, Darwin, you will be the star witness!"

Before the embarrassed Josh could react with a, "No way, Jose," JT broke into a raucous guffaw. Catching his breath, he warned, "If you're seriously looking for another Darwin as an expert for *Monkey II,* are you sure you want a creationist?"

Although tongue-in-cheek, the appraisal drew daggers from Josh and a question from the guest.

"Are you really a creationist?"

A subdued Josh sidestepped deftly, returning candor for candor.

"I'm a wanna-be scientist in the making. I've never joined, or attended a church. I've never opened a Bible. I doubt I can even define *creationist*."

"But do you believe in God?" she persisted.

Josh stiffened, then replied. Reading his body language, JT realized that the moment was inappropriate for levity.

"I can't swallow religion packaged like a pain pill and stamped with an exclusive brand name reserved for a dues-paying elite. And if I don't shape up and pay, I get to burn in hell fire? Doesn't sound much like a God of love!"

Before she could comment, he added a disclaimer. "But hey, I'm a theological illiterate. What do I know? Perhaps JT's clergyman granddad can clue me in someday!"

With that, JT wisecracked, "For sure, this guy can use some tutoring on something beyond geologic columns!"

"Then you are an evolutionist?" Traci asked.

Josh's mood changed abruptly. Uncharacteristically open under the spell of this inquisitive lawyer-to-be, unguarded thoughts spilled out, like floodwaters bursting the seams of a decrepit dam.

"Sure I believe in relentless changes in life...Label that evolution if you like. But that's not what the Darwin crowd hypes!"

"So you believe in evolution, but you don't really believe in evolution?"

He grimaced at the subtle twist. "You lawyer types are all alike."

Unruffled, she shot back, "Why don't you say what you mean?"

"Humans change constantly: bodies age, deteriorate, then wither away; during the same time span, character trends move graphically—some upward-bound, for better, while others slide downhill to nowhere."

"Anything else?"

"The gene pool within each prototype life form drives kaleidoscopic change and spectacular biodiversity!"

"And that's evolution, right?"

"Not by a long shot! Every prototype life comes endowed with its unique genetic code. The genetic pattern guarantees a wide range of variety within the prototype...but not new and different prototypes!"

"What about Darwin's Galapagos finches?" Her eyes sparkled as she sparred with her quarry. Josh met the challenge with a cocky élan.

"A pair of Galapagos finches possess the genetic information enabling them to parent offspring with different-shaped beaks. But descendants of Galapagos finches remain finches forever—never eagles, much less turtles."

"Dare I ask why?" This time the tone carried a touch of tease, but he detected her sincerity.

"The genetic data in a prototype create diversity but never generate brand-new genetic information that links one distinct prototype to an entirely different one—never! Genetic info may be damaged, scrambled, or lost forever—but entirely new information is never added to the gene pool!"

Then came the crooked grin of a mathematician, wise to the world of gamblers' odds. Intrigued, she let him ramble.

"You'll never see a professional roulette-wheel gambler put serious money on the color orange or the number 999. The bouncing ball could never land on a spot that doesn't exist—even with a billion wheel spins a day for 4.55 billion years!

"A bouncing ball to nowhere guarantees lose-lose!"

The silence that greeted his spontaneous logic left Josh hoping that he had resolved the matter. He longed to move to another topic: the life story of the enchanting Harvard law student. But the Camp Ryan guest wasn't ready to abandon the probe of her witness candidate. Before she could continue, JT joined the act.

"You'll have to excuse my pal's tendency to lapse into meandering rhetoric at the least hint of encouragement. He's afflicted with this photographic memory that latches on to every phrase he's ever read—like sand sopping up water. Sometimes his memory spouts high-falutin' rhetoric he's read in some book or science magazine. The guy can't help blurting out chapter and verse of all he knows, on demand—Dar is genius!"

Josh ducked his head in embarrassment at the absurd hyperbole while accepting the good-natured humor.

JT challenged, "Before you sandbag Darwin loyalists, why don't you condense the old guy's core ideas in language we lawyer types can understand?"

JT had opened the door to a threshold issue. Josh obliged, wrinkling his nose as though gulping a dose of sour medicine.

"Conjecture City, here we come...Primitive, simple, single-cell life supposedly appeared via spontaneous generation. Minute modifications, strengthened or diminished by use or disuse, were allegedly preserved, compounded, and then transmitted genetically to future generations, thanks to mega-doses of time.

"Voila—entirely new, complex life forms emerged gradually, via random-chance genetics; natural selection theoretically preserved beneficial alterations; survival of the fittest promised progress toward perfection.

"Given eons of time and innumerable transitionals, those spontaneously generated single-celled primitives theoretically grand parented all plant and animal life—including our own family tree!

"Now about that Brooklyn Bridge I'd like to sell you..."

"That's it?"

"You said condense! If you can stand more, try scanning the 1,477 pages of *Origin* and *Descent*...it's a load! The guy showed skill at painting over missing evidence with the broad brush of prolific verbiage! He ignored the unanswerable with silent eloquence!"

"Get real, Dar! Dump that fancy science-speak," said JT.

"Try this. Look around today. Do you see pansy seeds growing daisies? Or a confused flower, in the process of transition, halfway between a pansy and a daisy?"

"What about old-bone evidence?" challenged JT.

"Take a look at those billions of fossils dug up since Darwin's day. Where are those *innumerable* transitionals he insisted must have existed?"

"Go on!"

"If Darwinism is more than wishful thinking, we should find fossils of intermediate species by the bazillion! And just because two creatures share an efficient leg bone design doesn't prove genetic relationship!"

"So if evolution isn't operational today, and can't be proved from pre-historic bones...?"

"That the point! Evidence of evolution doesn't surface in the living world or in written history. The fossil record shows abrupt appearances and sudden extinctions. Those so-called simple cells, allegedly ancestral to all life, still thrive, without being wiped out by superior descendants—and the molecular world of genetics points to a predictable consistency for all life forms rather than random fluctuations!

"Where are the quantum changes demanded by Darwinism? Plain and simple, mega-evolution doesn't fly, walk, swim, or even move!

"If you buy into these quaint daydreams, you must believe with a straight face that the human brain, heart, and lungs originated by sheer coincidence from some single cell generated spontaneously in primordial muck, thanks to billions of years and millions of transitional steps—without design or designer!

"Darwin managed to conjure up a human ancestry that theoretically includes both a pointed-ear quadruped with a tail[1] and some single-sex hermaphrodite of the aquatic persuasion!"[2]

With a sly wink aimed toward Traci, Josh editorialized, "Maybe Charles Robert could fall in love with himself but that single-gender business doesn't cut it with any Magruders Mill boy I know—present company in particular...Perhaps, during those five lonely years on the *Beagle,* Charles inhaled strange smoke that showed him the light?"

This seamless nonsense drew a hoot from JT. Traci rolled her eyes.

Having made his point, Josh scrambled to his feet, beckoning the two law students to follow. He lead them to the excavated handiwork. Kneeling, he pointed proudly to the trophy fossil. In honor of its hoped-for future identification as a rare specimen of the long-extinct *Archaeopteryx,* he had dubbed the critter Archy.

"An impassive Sphinx stands silent guard, mute sentinel to what used to be. The likes of Charles Darwin and Doc Striker do not pretend that this edifice shaped itself from random cuts of wind and millennia of eroding rains."

The listeners waited for Josh to translate the relevance of what he had said to this Tennessee moment. Searching their faces, he obliged.

"The same goes for the Mount Rushmore presidential profiles! The identity of the creative artisan is public knowledge. Yet Striker leads a pack of bright observers convinced that the genealogy of guys like Archy sprung from a single-celled animal that appeared spontaneously in ancient muck without intelligent design—thanks to the whims of random chance. Hello??"

Standing again to favor his aching knee, Josh saved his last line for Traci.

"See why I would make a lousy star witness for your show? I don't run with the pack trailing Drs. Striker and Morley. Besides, I can't act worth squat."

The enterprising Traci sidestepped rejection. She countered that Josh Ryan's unimpeachable credentials deserved a ticket to stardom, riding the innovative turns of *Monkey II*'s yellow-brick road.

"All the more reason for you to test your wings. You've mastered the subject, and your controversial views deserve public attention."

"And you can burn some bridges of academia ta' boot," JT mumbled in an unintelligible whisper.

She pushed the sale with a provocative twinkle. "Besides, I can't think of anyone I'd rather have available this summer to tutor me in the mysteries of ancient origins."

The subtle innuendo persuaded JT. Convinced that Josh would find himself in good hands, he jumped on board! Now it was two lawyers against one reticent scientist.

"Listen, Dar, you owe it to this lady after making her gulp your rotten coffee for the last couple hours. When you're famous, I promise to frame your autograph. And I'm beginning to think if you don't say 'Yes' soon, she's liable to hang around here all summer, until you surrender—just like General Grant."

He drew the line, inches short of mutiny. "I'm not about to give up my pathetic sleeping quarters to another guest just because of your infernal stubbornness."

"Don't stress that lawyer brain of yours, Mr. Daniels! I'm not above sacking out on the camper floor." The lady understood one-upmanship. Sensing an opening, Traci embraced the moment to corral her target.

25

"Your USC buddy is loaded with wisdom, Mr. Ryan."

Flaunting mock seriousness she assured, "Never fear, Darwin, I'll not embarrass you...On the other hand, this is only the start of spring vacation, and I'm not planning to leave without a contractually enforceable 'Yes.'"

The clincher went straight to the Achilles' heel of Josh's commitment to truth...and nothing but. "Just think, Josh, you claim not to be a supplicant at the feet of Darwinian tradition. How many people of independent thought with the courage to express it are ever given a gratuitous stage to challenge unscientific ideas like the flat Earth or the sun and the universe circling Planet Earth?"

By the time she began to suggest that the name Joshua Chamberlain Ryan might rank with the likes of Copernicus and Isaac Newton, Josh had had enough. But before he could wave the white flag, the mischievous guest added a zinger. "Oh yeah, lest I forget—Tennessee is the volunteer state. You both could enjoy a place in history if JT volunteers to play lawyer and joins me as co-counsel to examine the *Monkey II* witnesses."

The invite drew vigorous protest from JT. "Hey, wait just a doggoned minute. I didn't swap a week at Laguna Beach for this!"

The protest echoed feebly. Unbeknownst to Josh, JT's trip to Tennessee had detoured through Washington, D.C., where he had sealed an offer to affiliate with Cabot, Calvert, and Stone. Trapped, JT groaned, reluctantly *volunteering* to strut his stuff on the stage.

At last, capitulating to feminine wiles, Josh grinned. "OK. You win! But in the words of you lawyers, my surrender comes with conditions: Attorney-to-be Thomas must appear as co-counsel; no strings or limitations on anything I might say; and MAU's geology/paleontology faculty must consent to the insanity."

Already JT had signaled his assent to the first condition and Traci stipulated to the second while nonchalantly sloughing off the third as "no problem." A fledgling star, known as "lamb-to-the-slaughter Darwin Ryan," had been born. Exultantly, Traci shifted to all-business mode. Whipping out a cellular phone, she confirmed to someone on the other end of the line, "It's a done deal!"

Then, adapting a line from astronaut language, she added, "The *Eagle* has been landed."

"Crash landed," lamented JT.

The next instant she scampered up the steep incline from which she had made her debut. She didn't look back until she mounted a ledge atop the ridge. Sweeping her arms in a grand gesture matching the drama of her arrival, she called, "Nice meeting you boys. See you next summer."

With that, she disappeared.

The boys, wrung out from the encounter, busied themselves with the final details of completing the excavation. Within minutes, a jubilant Josh had packed the rare specimen in a cushioned travel case and stowed it in the semi-safe haven of the rickety camper, ready for the triumphant journey north to Maryland.

Amidst small talk, Josh slipped Jonathan a $500 check written and signed the week before to complement the previously delivered round-trip airfare. JT slipped the check into his shirt pocket, jesting, "Maybe on the way home I can stop in Vegas and make a fortune at the roulette table betting on an orange number 999."

Never missing a chance to tease, the flamboyant JT tested his friend's reaction to Traci. "I swear you couldn't take your eyes off that woman, Dar. From now on I'm going to call you *putty man*."

Josh surprised himself by turning crimson. He offered no defense, only a non-committal quote from his Grandma Ryan's lexicon of the old South: "I wish I may never!"

He hastily changed the subject, announcing a celebration the following day, compliments of Karl Striker's Science Research Foundation. Dinner at Nashville's Stockyard restaurant; two tickets front and center at the Opry; and two rooms at the Airport Marriott, a welcome perk for JT, since it promised a restful night prior to his Friday morning return flight to California and a chance for Josh to relieve the ache of a crippled left leg prior to the ten-hour drive home to Magruders Mill, Maryland.

This left all day Thursday as bonus time. Josh suggested an adventure.

"Since we have a day to waste, why not further your legal history education with a jaunt to Dayton's Rhea County Courthouse, site of the *Scopes* trial. Think about it, Jonathan Touchdown, a rare chance for you to learn something about the clash of the titans, Darrow and Bryan."

With a bored yawn, JT asked, "Do I have a choice?"

Seeing the enthusiastic expression on Dar's face, JT recognized the *no* answer. Knowing his host was a stickler for hard data, JT accepted reality, knowing Josh would feel compelled to inhale the authentic flavor of *Scopes*, trapped as he was in Pace Terhune's plan to stage a reprise of the original trial of the century.

Abandoning Camp Ryan early Thursday morning, they headed southeast, bound for legal and scientific history. By mid-morning they arrived at Dayton's town square, dominated by the red-brick courthouse with its cupola-topped clock tower. Parking proved no problem.

Inside the century-old Victorian building, the locals greeted the Yankee tourists with cordial nonchalance. Daytonians were used to a parade of curious tourists traipsing through the town's showpiece.

The courthouse basement housed a "Scopes Trial Museum." A broad stairway, just inside the main entry, led directly to the second floor and the oversize courtroom where the verbal jousting took place. Jurors in recess from a trial in progress filtered through the expansive second-floor hallways. Characteristic of rural nineteenth-century American courtrooms prior to movies, radio, and TV, the Dayton courthouse doubled as community meeting center—capable of seating an indoor crowd of 300.

An accommodating presiding judge explained, "The furniture and layout are much as they were in 1925. And we still have to call time-out when the rumble of passing trains drowns out the proceedings." With a sly grin, the judge added, "Of course, the air conditioner comes as a welcome modern replacement for those funeral-home fans!"

On the trek to Nashville for an evening out, even JT admitted that the side-trip to the Dayton setting was worth the detour for its exposure to "authentic Americana."

Friday morning, with JT dropped off at the airport, a lighthearted Josh Ryan headed back to Center Hill Lake. The farewell to Camp Ryan would require nothing more than a quick hook-up to the waiting camper. Within hours he could delivery the fossil treasure to its new home—where it would grace the science archives of Mid-Atlantic University.

It was not to be!

First puzzlement, then panic swept over Josh when he arrived at what had been, only hours before, Camp Ryan. At first, he rationalized that in his exuberance he had blundered into the wrong cove. But there was no mistake. The empty trench lay open and exposed. No other trace of Camp Ryan remained.

The camper and its irreplaceable cargo had vanished!

The thieves were no ordinary pranksters. Josh had been victimized by cunning professionals. Tracks on the dusty road had been swept clean. There was no clue that any vehicle or human footprint had intruded.

At a loss to know what to do, Josh drove to Smithville to report the theft. Josh surmised that Deputy Sheriff Magruder, who took the report, might be kin, but not wanting to arouse suspicion, chose not to explore possible family connections.

Sheriff Magruder seemed unimpressed with the mysterious disappearance of an obsolete camper and its odd cargo of limestone and old bones. Nevertheless, he dutifully completed the report, assuring Josh he would be notified immediately, "If anything turns up." Disgusted, Josh walked away thinking, "Fat chance!"

With heavy heart, he went to the local Parcel Plus to fax the distressing news to Striker and Morley. Then, remembering the week-long compilation of data stored in the Mustang's glove compartment, he retrieved the last chapters of the debacle and overnighted the pack to silicone valley.

Miserable and alone, he pointed his rattletrap Mustang north to the refuge of his Maryland home base and an uncertain fate. For the moment, he could only try to reconcile what had been a brutal 24-hour transformation from exhilaration to deep despair. The rhapsody of the previous day lingered only in memory.

Seeds of destiny had been sown in the wind. Josh Ryan would reap the whirlwind with September's hurricane season.

*April 15, 1996*

# 4

Josh Ryan dreaded his empty-handed return to the Mid-Atlantic University campus. He breathed a partial sigh of relief to discover that the terrible-tempered Karl Striker was nowhere to be seen. However, hope of respite proved a mirage. When he reported the theft, even the amiable Gus Morley shed his usual kindly demeanor.

Trying to mend fences, a bewildered Josh recited repeated apologies. Morley shuffled papers absently, as he reported the collective mortification of the Science Division leadership. A stern lecture informed the luckless student that through his carelessness, "thousands of dollars" had escaped down the drain, with nothing to show for it.

But then, returning to the good-old-Gus personality familiar to students, Dr. Morley sympathized that things could have been worse. Luckily, no one had been injured or killed. Morley promised to intercede with the tempestuous Dr. Striker to assure official forgiveness for the egregious negligence, while hopefully paving the way for a replacement dissertation topic that would ensure Josh's Ph.D. candidacy.

Encouraged, Josh expressed heartfelt appreciation to the affable associate professor, who had once again interceded on a student's behalf.

Slipping Josh a copy of a one-page typewritten document, Morley casually suggested that if he would sign this "routine statement of responsibility," the reconciliation process could be set in motion, and perhaps all could be quickly forgiven and forgotten. Tossing out a feeble attempt at humor, Josh folded the page recklessly and placed it in his coat pocket, quipping, "Of course, I never sign anything without checking with my attorney."

Dr. Morley grunted without looking up, implying that the session was over. Stopping at the door, Josh remembered the third pre-condition to his participation in *Monkey II.*

"By the way, Dr. Morley, I've been asked to serve as a witness in a make-believe courtroom battle scheduled for September 1 to 6 in the Chamberlain Playhouse. I said 'Yes' subject to Science Division approval."

"I'm sure the division leadership will be comfortable with any presentation that casts favorable light on the academic credentials of the university and its scientific disciplines."

Josh interpreted this vague equivocation as assent. But the nagging memory of the grim-faced, albeit sympathetic Dr. Morley, unnerved him. He didn't need to be a 4.0-GPA graduate student to grasp the gravity of his predicament. Unless he could make amends with the hard-edged Karl Striker, plans for a Ph.D. in

1997 or anytime soon could evaporate. Grad student careers fractured by similar scenarios cluttered MAU's academic landscape, mute testimony to the high cost of displeasing the king-pin prof. The sooner Josh could arrange a meeting with Dr. Striker, the better chance to repair the tattered fabric of understanding and salvage his doctoral candidacy.

But pursuit proved futile! By June's summer break, Dr. Striker had departed for Houston, Texas, to join a consulting team of scientists looking for signs of microscopic Martian life on a meteorite chunk picked up in Antarctica. Striker had assumed incommunicado status until September.

*******

## Summer 1996

Unlike blasé routines of summers past, rapture lurked for Josh Ryan this year. The hint of prospective liaisons with Traci Kilburn stimulated his body chemistry and provided fodder for daydreams.

Presiding in his private Fort Ryan castle one early June morning, Josh polished off the last drop of breakfast coffee just before a gentle tap at the front door. True to the Tennessee precedent, Traci appeared unannounced, framed by rays of sunshine that danced on the lush foliage of the silver maples dotting the mill-house landscape.

"It's time to go to work, Dr. Ryan!"

The matter-of-fact greeting sufficed to bridge the time-gap since their April introduction. The bleary-eyed Josh could not conceal his pleasure at her presence. Straightening his lanky frame, he tried to appear suave and debonair.

"Most assuredly, history waits impatiently to decipher the results of our collaborating contribution to science and the entertainment arts!"

His feeble attempt at sophistication was delivered in a fractured German accent served up in pretended haughty demeanor. Traci brushed the pronouncement aside, deflating his act.

"Whatever you say, *Sir Darwin!*"

From the rebuilt ruins of the historic Magruders Mill, modestly restored to serve as gatehouse to the Fort Ryan estate, Josh seized the opportunity as lord of the manor to take his guest on an impromptu tour. Clad informally in blue jeans, the trapped Traci accompanied him on his rounds with bemused interest. After twenty minutes, with no end in sight, the outspoken tourist had had her fill.

"OK, I'm impressed. Now let's get to work!"

The words launched a series of summer-long tête-à-têtes. Under the silver maple guarding the shores of the Lake Never fish pond next to the mill house, the scientist-in-training and the fledgling lawyer exchanged crash courses on life's origins and courtroom tactics. Occasionally, a honking formation of Canada geese skimmed into the sylvan setting, rippling the cool face of the watery refuge. Heavy discussions focused on the Achilles' heels of Charles Robert Darwin's theory most likely to ignite the *Monkey II* stage with fiery sparks of high entertainment.

JT joined them for a mid-July reunion. The trio took a quick jaunt to Dayton's annual *Scopes Trial Play and Festival*. On the return lap they dropped by Johnson City, Tennessee, to meet eighty-eight-year-old Wilma Humphreys, an eyewitness to the 1925 trial and student in a John Scopes biology class. The charmer regaled the visitors with memories of a popular teacher who never breathed a word about "evolution" in the classroom but was picked by promoters as a prime candidate to appear as defendant in the Scopes show trial.

Shortly after they returned to Maryland, Traci prodded Josh to profile Darwin the man.

"Wasn't Darwin a virtuoso of the scientific?"

Josh snapped at the bait, missing Traci's wink and sideways glance at JT. "Darwin never trained to be a scientist! He spent three years at Christ's College, Cambridge, studying theology, preparing to serve as a rural English pastor. Instead of taking a country parish, he opted for a five-year cruise on the *HMS Beagle*. And instead of preaching religious dogma, Darwin's conjectures shaped a secular theology with articles of faith rooted in unverified random coincidence. Many of evolution's most vocal advocates dismiss religion as unscientific, then turn around and embrace faith in *speculations* devoid of corroborating evidence.

"A serious student of nature, sure...but classic scientist?"

"What then?"

"A philosopher/naturalist speculating as to novel notions!"

Josh spoke his mind with reckless abandon. He cited Darwin's cozy lifestyle buttressed by the cocoon of inherited privilege.

"The guy never had to earn a pound or a pence in a real job. Born to wealth, the Charles Darwins of 1861 thrived on income from writings and a revenue stream from family endowments upwards of $40,000 per month.[3]

"Although himself an employer of a household staff of eight, he spent a lifetime avoiding subservience to an employer. Ironically, this champion of the trappings of science managed to find ample budget to hire a cook, maid, butler, and gardener but never bothered to retain a laboratory assistant."

"Takes one to know one!" challenged JT. "Sounds to me something like the Morgan-Carrington fortune, Dar. How come a rich kid like you bothers going to school with us proletariat types?"

JT never hesitated to push his pal's buttons. True to form, Josh responded.

"Yah, yah, Bro. Imagine the headlines...*Duchess sees the light, feels abject pity for grandson!* Meanwhile, disinherited moneybags plods along on the two hundred bucks a month his dad doles out as reward for mowing Fort Ryan's green acres."

Traci savored the offbeat banter but soon returned to the bold opinions of the witness.

"Forget the man's fortune! Don't you believe Darwin's ideas are corroborated by scientific evidence?"

The student witness groaned. "Only if wishful thinking equates with *empirical evidence.*

"Look! I don't want to beat up on the memory of a bright guy who's been resting in Westminster Abbey for more than a century, but he himself admitted, *'I am quite conscious that my speculations run beyond the bounds of true science.'*"[4]

Repeatedly Josh mounted his verbal soap box, indoctrinating his friends at length by reciting Darwin's own words. Traci and JT egged him on.

"What do you consider *outside the bounds of true science?*"

"Wishful thinking, loaded with scientific lingo, qualifies as far outside. Same goes for speculations! True *science* demands testing of falsifiable evidence to separate fact from fiction. But don't hold your breath listening for believers to shout to the world that the emperor has no clothes! If Charles Darwin posed as a charismatic Vegas pit boss, my money says he could sweet-talk a crowd of the gullible to bet their wads—even if the house never lost!

"Darwin comes across as an innovative naturalist who wrote prolifically, communicating views reminiscent of his grandfather, Erasmus Darwin: Shrewd and intelligent, but advocating seriously flawed dogma. The trail he blazed was not the first to dead-end in fiction—nor the last.

"To his credit, he admitted to being a *'master wriggler,'*"[5] and honestly admitted wide-ranging uncertainties.

"So, you accept his self-deprecation, but reject his theory?"

"The *wriggler* knew how to *wriggle* with the best of them: He could spin a tale with artful equivocations. He sidestepped, ignoring the *unanswerable.*

"Take a look at several Darwin-speak *wriggles:*

*"'Best of my judgement...I think...lead us to believe...I do not doubt...it need not be doubted...as far as I can make out...it would appear...if we may trust these inferences...as far as we know...it is probable...it does not seem probable...it does not seem possible...it may be inferred...conceivable...if this has occurred...might have existed...might have been modified...no difficulty in supposing...if this had been effected...perhaps...!'"*

The audience of two listened attentively as Josh gleaned phrases from several pages of *Origin*. Traci broke in.

"What did Darwin ignore as *unanswerable?*"

"The origin of plant life, for one thing. Do plants and animals share a common ancestor? How do you get life to evolve in two completely different directions from that alleged first-ever *simple* cell? Doesn't the oldest fossil plant display meticulous complexity rather than single-celled simplicity? Why would natural selection surrender the mobility of animal life for the stationary status of a plant? And what would drive the survival instinct of a plant to produce a strawberry, only to be eaten?

"Three years before he died, Darwin confessed the futility of comprehension. *'The rapid development, as far as we can judge, of all the higher plants within recent geological times is an abominable mystery.'*[6]

"Astute observers agree!

"'*Can you imagine how an orchid, duckweed, and a palm have come from the same ancestry, and have we any evidence for this assumption? The evolutionist must be prepared with an answer, but I think that most would break down before an inquisition.'*"[7]

"Charles' mama didn't raise no fool. By focusing primarily on animal life origins, he dodged the bullet of one *abominable mystery* while capitalizing on public interest in the ancestry of mankind. Give the guy credit for being shrewd! He sidestepped *abominable mysteries.* Still, despite fleet intellectual feet, Darwin's strained interpretations have been branded *'fact-free science.'*"[8]

"Come now, Mr. Ryan, bright-minded scientists would not be hoodwinked by con artists preaching unprovable philosophy posing as science!"

"Makes sense...to a point. But even the best and brightest can be fallible and project learned bias! Like language—it's imprinted on minds by repetition! The dilemma increases when multiple generations of students confront blatant fraud!"

"Can you recite chapter and verse?"

"Easy! As we speak, some high school biology students are learning from textbooks that carry misleading embryo illustrations concocted by Ernst Haeckel, a devotee of evolutionary theory. *'Haeckel's drawings of 1874 are substantially fabricated...his oldest "fish" image is made up of bits and pieces from different animals—some of them mythical...it is the discredited 1874 drawings that are used in so many British and American biology textbooks today.'*[9]

"Exposed as doctored junk, the calculated ruse still glares from the pages of some texts more than a century after someone spotted Haeckel's hand in the cookie jar. After misleading several generations, I'm glad to say that this ruse is finally being hustled out of print."

"Do you claim only evolutionists are the bad guys?"

He threw up his hands at her exaggeration.

"Hardly! Both creationists and evolutionists sometimes grasp at straws to support pet ideas. Add big egos to brainwashed bias and you get a formula for shaded meanings and distortions."

"Are you accusing Striker of fraud?"

"He's smart! Hard-headed. Locked into pre-conceived bias, but honest!"

"Then how come you vehemently disagree with his ideas?"

"Good facts...bad interpretations!"

After several of these exploratory seminars in the cool shade of Fort Ryan's countrified setting, Traci popped a hypothetical question.

"I've been wondering what Darwin might write today if he had had access to lasers, computers, the Hubble telescope, electron microscopes, and molecular biology.

"Can't you cut some slack for a physically fragile Victorian who lacked the benefit of electric lights to brighten his four-hour work day?"

The hypothetical stumped the academic, so he pondered aloud. "Great hypothetical question!...Assuming Charles D started with a clean slate, without brainwashed bias?...Who knows?"

After a moment of hesitation, he wound up on a high note. "For sure he was one bright man. I give him that! I like to think that standing at the edge of the millennium, he would join the new generation of scientists who look beyond horse-and-buggy myths."

That was it. By August's end, the academic liaisons had covered evolution's shifting postulates. Not-so-coincidentally, Josh had maneuvered the serious academic seminar schedule to coincide with a pleasant social environment.

Traci joined him in jaunts to Baltimore's Camden Yards ballpark and the Atlantic seashore. By midsummer, a now thoroughly smitten Josh raised the stakes by hosting a day-long air excursion to Atlanta's Olympics.

One balmy August evening, after traipsing the sand and surf of Maryland's Eastern Shore, the comfort of the drive home mesmerized the young scientist. Bathed in warm darkness, he spilled his guts.

Listening attentively, Traci neither pried nor probed as he dipped deep into the rich resources of his memory bank. Leaving most details in misty limbo, he described the highlights of his parents' star-crossed romance and the Ryans' estrangement from Lady Carrington.

"My dad and mom met at an Annapolis tea dance. Best I can tell, he must have swept her off her feet, much to the consternation of her mother.

"Lady Carrington played hardball. She forbade Mom to marry a no-account sailor, and offered a folksy warning, *If you're not careful, you'll walk through the forest and end up with a stick.* But not even the Duchess' threats could slide a greased butter knife between them."

Josh wandered through a mellow haze of memories, reaching an emotional catharsis. Traci listened silently.

"My folks eloped despite the storm of opposition. The Duchess reacted bitterly. She cut off all communications and barred entrance to Morgan Manor..."

35

Not sure how to continue, Josh finally blurted out, "When my mother died hours after I was born, the Duchess blamed Dad. Her fury knew no bounds. One of Dad's friends told me she shunned the Ryans at the funeral and rode to the cemetery in her own chauffeured limousine. She let the world know she never again wanted to see or speak to any Ryan—including her grandson. Occasionally, I've spotted her picture in the paper or on TV, but we never visited."

Alarmed that his candor might have put a damper on any romantic potential with Traci, he decided to clam up. Then he felt the touch of her hand on his with the whispered words, "It's OK, Josh." Fortified, he went on.

"My mom's death tore my dad apart. Lady Carrington's vitriol poured salt in the wound. He suppressed grief by mobilizing his mathematical skills as a consultant to the computer industry...all the time doing double duty as a parent. He set up a home office at Fort Ryan so he could look after me, and took me along as he traveled the world of business technology."

"Seems like your dad did OK."

She squeezed Josh's hand, sensing he had more to share. To his own surprise, he obliged, relating a childhood incident.

"One day I came home to Fort Ryan from kindergarten with a tear-smudged face that I tried to hide from Dad. He asked, 'Did you get into a fight or something?' He wouldn't take my 'Nah' for an answer. Dropping to his knees, he hugged me and asked, 'What is it, then? What's bothering you, Son?'"

When Josh paused, stumbling for words, she stroked his arm reassuringly, "You don't need to share anything painful. I understand." But he couldn't stop the torrent of emotion. With a deep breath, he opened his heart.

"Before I knew it I was sobbing out, 'Why don't I have a mother like other kids?' He told me the tragic story as gently as he could, but when I asked, 'You mean my mother died because of me?' the dam broke. I had never seen my dad cry before! One look at the anguish triggered by my own tears did it...I vowed to never again ask about my mother and never again to shed a tear!"

Traci reached across the console to brush his cheek with her lips, saying nothing. Now she understood the scene she had witnessed in the Johns Hopkins E.R. years before—Josh's stoicism concealing the anguish of a smashed leg. The final bits and pieces of the reverie filled in background gaps in the life of the handpicked witness for *Monkey II*.

The love of an unselfish father figure; the consuming passion to explore fossil remnants of ancient life; the magnificent Fort Ryan twenty-five-acre spread where he could ride his obstreperous pet Arabian horse Turbo, short for Turbocharged, famous for unpredictable bursts of stop-and-go—because of these, Joshua Chamberlain Ryan had weathered adolescence with ease, entering young manhood remarkably unscathed.

Drained from reminiscing, Josh closed down again.

"There you have it! The story of the highs and lows of the life of Josh Ryan! You probably think I'm a nut case...spilling that stuff when I still don't know much of anything about you."

Not wanting to break the spell, Traci responded, "You may be a lot of things, Joshua Chamberlain Ryan...but for sure, you're no nut case!"

Her increasing awareness of what made Josh Ryan tick validated her confidence in their joint ability to perform on stage. Once joined by JT, the energetic team could touch the match to fireworks that should set off an academic conflagration.

But one nagging concern haunted the aspiring scientist: Why was it that despite repeated attempts, Professor Striker continued to evade and implicitly rebuff all overtures to schedule a meeting to reconcile grievances that had festered since April?

## Sunday Afternoon, September 1, 1996

**5**

Monkey II opened to an almost full house, thanks to the promotional expertise of Pace Terhune and the generosity of loyal patrons and alumni. Front and center orchestra rows cost $25 per seat, per performance, discounted to a flat $100 for all six events, beginning with Sunday's September 1 matinee. Despite Terhune's boast of a "virtual sellout," public response reflected lukewarm wait-and-see.

With trademark flourish, Pace picked Mid-Atlantic University's 1,000 seat Chamberlain Playhouse as center-stage for *Monkey II*. The campus crown jewel perched imposingly at the crest of a hill, just inside the wrought iron gates to the campus. The classic federal architecture of the playhouse set the tone for the 640-acre campus and landscape, master planned by the nineteenth-century luminary, Frederick Law Olmstead.

The playhouse dazzled with festive glory. An imposing life-size bronze casting of Civil War hero General Joshua Lawrence Chamberlain riding into battle, Toledo sword blade swinging heavenward, guarded the polished marble courtyard. An expansive double-doored entry framed access to the softly lit lobby. Directly beneath a circle of custom-cut windows, an exhibit commemorating the life of General Chamberlain commanded attention. History unfolded in a creative montage of bronze plaques and photo engravings, galvanizing attention. The tribute owed its existence to the philanthropy of Mid-Atlantic University's founder, Sam Houston Morgan.

Texas teenager and Union loyalist, the young Sam had ridden north to Pennsylvania in 1863 in quest of a blue uniform. Before he could don military regalia, he stumbled into Gettysburg's unfolding storm. None other than Joshua Lawrence Chamberlain himself welcomed him to the thin lines of a short-handed, Twentieth Maine regiment anchoring Little Round Top's lonesome flank. Awed by the valor of the Maine university professor, Sam himself underwrote the full cost of the playhouse as a permanent memorial to the old warrior whose body carried Confederate bullets to his 1914 grave.

To capture the theatrical mood sought for the fictitious trial of *The People v. Charles Robert Darwin*, the irreverent Terhune commissioned life-size cartoon caricatures of Darwin with comic relief captions to be plastered on lobby walls. The extravagant hype worked, much to the delight of some and the distress and disgust of others. Terhune considered himself neutral about the debate over the origin of life. Regardless, he knew that his professional skills would create major-league drama, fund-raising bonanzas, and front-page publicity.

Thanks to Traci Kilburn's meticulous on-the-scene reconnoitering, the stage setting scored passing marks as a replica of the original *Monkey Trial* Rhea County courtroom.

The set featured weathered oak furniture, perched on a raised platform enclosed on three sides by a wood railing. Mounted center-rear, a home-crafted bar of justice faced the audience at a height readily visible to all seats in the playhouse. A framed set of softly lighted faux Victorian windows back-lighted the judge's bench.

Center-front, a single row of eight jury seats faced the judge, backs to the audience. A spotlighted witness chair below and to one side of the judicial chair rose above the heads of the jury, assuring unimpeded scrutiny by playhouse patrons. An unobtrusive small desk for the clerk perched below and to the side of the judge, opposite the witness. A pair of counsel tables, arranged like bookends at either side of the platform, completed the picture.

Not precisely authentic, but close enough—William Jennings Bryan and Clarence Darrow could have felt comfortably at home!

Minutes before curtain time, a weather-beaten Wurlitzer theater organ in the orchestra pit blared a rafter-shaking rendition of "The Battle Hymn of the Republic." Precisely at 2 p.m., the music faded as an unseen bass voice boomed out the issue to be explored by the jury in tones befitting a daytime soap opera:

### *Is evolution a fact?*

The curtain opened to a simulated court session, Judge Edward Anthony Stone presiding. The homegrown jury waited, all eyes and ears. The cast stood in place, primed and ready.

All opening statements had been skipped; star witness Joshua Chamberlain Ryan occupied the witness box; Traci Kilburn sat at one counsel table, while Jonathan Thomas Daniels presided at the other. Jessica Saunders captured the classic pose of an attentive court clerk, and, of course, Bailiff Pace Terhune stood impassively in the uniform of a Montgomery County sheriff, alert to any newsworthy event that would add luster to the glory of the university's centennial celebration.

The cunning Terhune envisioned exploiting the staged extravaganza beyond the capital region. Three studio-quality video cameras were strategically positioned to tape every second of the six acts. All volunteer participants had signed consent forms enabling Terhune to produce a condensed, professionally edited documentary he could license to television and cable markets.

Terhune also trumpeted movie rights possibilities to Hollywood producer friend, Redondo Calizar, shamelessly promoting the "stunning caliber of the cast," "plot loaded with intrigue," and "assured public interest in the millennium's threshold issue." According to Terhune, *Monkey II* offered all the dramatic potential of *Inherit the Wind*, Hollywood's slanted screen version of the original 1925 *Scopes* trial.

Long respectful of the promotional skills of his impresario friend turned university fund-raiser, Calizar took the bait and the complimentary tickets, promising to attend with a talent scout associate. True to his word, Calizar not only showed for all six performances, but also doubled as theater organist.

By week's end, the unheralded appearance of the West coast prospector for entertainment gold would lead to a result astounding even to the flamboyant Pace Terhune. But for the moment, given the scattering of vacant seats, Terhune's hype failed to flummox the canny Calizar. The producer suspected that blocks of complimentary tickets had been given away to ensure a full-house image.

Terhune had warned the Hollywood mogul that to guarantee spontaneity, the volunteers would perform without a script. Calizar recognized that the move was intended to save bucks but also saw opportunity for realism. No one, least of all the players, knew where the evidence trail would lead.

Bathed in the intimate glow of dramatic lighting, Judge Stone made the first move to set the stage and to spotlight performers.

"Ladies and gentlemen of the jury, one or two of you may carry childhood recollections of the 1925 landmark event testing whether or not a biology Professor Scopes had the right under Tennessee law to teach evolution to high school students." Pausing to clear his throat, then casting a glance in the direction of the bailiff, he added, "Authentically convened in a county courthouse in Dayton, Tennessee, that ostensible *Monkey Trial* was nevertheless a staged showpiece, orchestrated by partisans cherishing widely divergent viewpoints."

Warming to his role, the real-life judge waxed eloquent, bowing figuratively to the legal titans of the time whose words still garner headlines.

"The presence of Clarence Darrow and William Jennings Bryan, champions of two hostile views, assured national attention. Bryan's credentials tilted to the art of oratory; 'tiger trial lawyer' characterized Darrow. Bryan and the prosecution enjoyed the triumph of a jury verdict. But Clarence Darrow's crafty performance rallied public fancy."

"*Monkey II* qualifies as a show trial of sorts—without the colorful antics of a Bryan or a Darrow. No duel of nationally known barristers is contemplated here. Although Ms. Kilburn and Mr. Daniels aspire to careers in the legal profession, their credentials are those of bright law students still learning at the feet of experienced professors.

"Since neither claims attorney-at-law credentials, maybe attorneys-ad-lib will do for this week?"

The subtle attempt at levity drew scattered titters.

"Both future lawyers hail from Maryland: Jonathan Thomas Daniels is remembered fondly for his blazing speed as a Sherwood high football phenom...and his dashing social exploits. And, oh yes, hearsay reveals that Mr. Daniels' superior scholarship as a Southern Cal law student earned him *Law Review* status!"

JT shook his head at the eulogy. To suppress his snickers, he studiously avoided looking at Josh, pretending to gaze nowhere in particular until he heard his cue to kick off witness examination.

"As for Ms. Traci Kilburn, rumor has it that she ranks in the top five of her senior class at Harvard Law...and come next summer, she and Mr. Daniels will join Cabot, Calvert, and Stone as associates. Both Mr. Daniels and Mr. Ryan were recruited for this week's trial thanks to the persuasive skills of Ms. Kilburn."

Nodding acknowledgment to Judge Stone, Traci sat with regal aplomb, waiting for the legal games to begin. Satisfied with the flowery aside to his protégés, the judge charged on to the core issue.

"Scientific evidence not accessible to the 1925 protagonists will be included. After listening to the week's testimony, you will be asked to render a simple Yes or No verdict to this key issue: *'Is evolution a fact?'*

"Your responsibility will be to judge Charles Robert Darwin guilty or not guilty of propagating *fact-free science.*" The judge scanned the audience over his half-rims, then turned to the credentials of the student witness.

"The witness representing the Mid-Atlantic University science community is a grad student majoring in geology and paleontology, boasting a perfect 4.0 GPA. The lure of fossils has intrigued Joshua Chamberlain Ryan since childhood. He strikes you as a pleasant, savvy scholar, but he has not yet earned a Ph.D., nor has his research been published in national journals."

Smiling broadly, the judge confided obscure trivia as if delivering an ultra-secret to the collective trust of 1,000 strangers. "Hearsay has it that close friends accuse him of masquerading as a make-believe Darwin."

Lodged in the so-called seat of honor, the self-conscious witness squirmed, thinking, "Enough already." The bemused audience reacted with raised eyebrows and approving nods.

"Be forewarned, the trial's primary witness thumbs his nose at Darwin's dream. What you hear may sound one-sided, but that is because a contrary voice from MAU's Science Division officialdom turned thumbs down on the invitation to appear."

The candor hit a nerve, drawing a wince from one audience observer—Dr. Karl Striker, who sat unobtrusively near the rear of the playhouse, recently returned from his Texas tour.

41

Having finished his extemporaneous hyperbole and dispensed with the opening statements, Judge Stone glanced toward JT, ordering, "Please feel free to proceed, Counsel."

Elegantly attired in fall fashion befitting a Wall Street lawyer and with a command of the English language reminiscent of his clergyman grandfather, Bishop Brock Daniels, JT opened with a rapid-fire sequence of questions.

"Have you read Darwin's *Origin of Species? Descent of Man?* The transcripts of the 1925 *Scopes* trial?"

Josh nodded and answered, "Yes," to each question.

JT opted to jump ahead to the bottom line in order to deter Sunday afternoon audience napping.

"Do you quarrel with Charles Darwin's ideas?"

"More like reject!"

Josh's blunt retort provoked no surprise from JT or Traci.

"The whole nine yards?"

"Darwin offered no explanation of where something from nothing came from; skimmed over alleged spontaneous generation of first-ever single-celled life from green slime; and extrapolated mega-evolution's unproven postulate, relying on the variety potential of everyday micro change!"

Obviously Josh Ryan planned to pull no punches. JT led him on, shamelessly demanding conclusions and opinions. Bemused, Judge Stone made no move to intervene. For now, the judge settled back, content to see if JT planned to go straight for the jugular of evolution's traditions, and wondering just how soon Traci might interrupt with an objection. At least for a few minutes in the sun, the star witness could revel in uninterrupted kid-glove treatment.

Unrestrained, JT moved to the attack, elated at the prospect of shooting philosophical fish in a rain-barrel. Summoning his cultivated skill for sounding profound, he addressed the witness point-blank.

"Mr. Ryan, this jury and this court assume you to be a serious student of geology and paleontology. Is that not a fair assumption?"

The witnessed shrugged with a confirming grin.

"You have acknowledged personal familiarity with Charles Darwin's *On the Origin of Species* and *The Descent of Man and Selection in Relation to Sex.* Given your admitted exposure to a variety of viewpoints relating to the origin of organic life, are you prepared to declare your opinion on whether or not evolution is a scientific fact?"

"Yes and no—depending on your definition."

"What's with this *yes* and *no* business, Mr. Ryan? You sound like some wishy-washy politician! You're supposed to be the expert!"

"Touché, Counsel! The capacity for diversity has been available, from the beginning, within the gene pool of every prototype. That's the *yes* answer—microevolution, change within genetic limits, is scientific fact!"

"And the *no*?"

"Mega-evolution—where given enough time, lucky mutations, and natural selection, a fish can be ancestral to a man? Big-time *No*, not a chance—never, even in 4.55 billion years! Darwin's conjecture is rife with unproven speculation: He offers glib-sounding explanations without evidence, glossed over with articulate hemming and hawing, supported by abstract ambiguity."

The witness' opening salvo continued.

"Rarely have so many been so confused by poppycock so anemic! Darwinism faces a task as hopeless as that confronting cartographers using state-of-the-art computer technology to draw maps while stuck on the faulty premise that the world is flat!"

Pretending to coax a reluctant witness to speak his mind, JT continued: "C'mon, Mr. Ryan, don't be bashful! Why don't you tell the jury what you really think? Is mega-evolution a fact?"

"Anything to please, Counsel."

Josh paused, then added: "How's this? Mega-evolution is every bit as real as the tooth fairy!"

"Anything else?"

*"Darwin's threadbare theory is to science what the Titanic was to unsinkable ocean travel."*

The brash declaration struck Karl Striker with the force of a lightning bolt—a figurative finger jabbed in the eye! It was one thing for a smart-aleck student to argue his theories in class. It was something else to flaunt impertinence in a public arena, blatantly smearing a teacher's carefully cultivated reputation.

Frustration boiled over in an avalanche of belligerence. Beside himself, Striker leaped to his feet, shook his fist at Pace Terhune, Josh Ryan, Judge Stone, and the jury, and shouted with fervent conviction: "That's a damnable lie!"

Striker's astonishing outburst turned heads! After a fit of fist-shaking hostility, the irate chairman of the university's Science Division stormed to the nearest theater exit. Faces struggled for a glimpse of the still-steaming Striker, whose stormy retreat matched the staccato cadence of Judge Stone's frenzied gavel.

Intrigued by the unexpected, spectators unacquainted with the antics of the tempestuous teacher concluded that the colorful interlude must be part of the act. Heard midst the whispered murmurs, one wag wisecracked: "The old boy must be auditioning for an Academy Award!"

The strategically placed television cameras dutifully taped the episode for TV audiences titillated by confrontation. Pace Terhune reacted with glee at the sideshow. His creative imagination conjured up a provocative headline for Monday's *Washington News Press*. Initially he leaned toward, *Darwin Branded Liar; Playhouse Erupts!*

A poised Judge Stone rapping his gavel, called for "order in the court" and declared an unscheduled recess.

The instant the curtain dropped, the Wurlitzer began majestically blasting out Broadway show tunes with Redondo Calizar at the console. Moments into the impromptu concert, MAU music major Marie Mackin temporarily forsook the jury, stepping into the spotlight to belt out favorite hits in a sultry soprano. Riveted by the musical bonus, an entranced audience remained bolted to their seats for the impromptu concert. The perplexed witness sank in his box, wondering, "For this I gave up today's Redskins tickets?"

Though visibly unperturbed, Josh didn't need magna cum laude credentials to recognize that his two-year investment of time toward a Ph.D. and slim vestige of hope for a career in geology/paleontology lay in disarray. Absentmindedly, his gaze wandered toward one of the counsels' tables, locking with Traci's for a split second.

The telepathic exchange seemed to say, "I'm sorry I got you into this," and a return, "Don't be...It's not your fault."

Both turned away awkwardly.

Pretending preoccupation with the stage curtain's unfolding descent, Josh's thoughts leaped back to the April moment when he came under the spell of this incredibly charming woman. Better than anyone else, he perceived that but for the unseen magnet that propelled him toward the fascinating personality lurking behind burning hazel eyes, the odds of his being stuck on this stage were thinner than the likelihood that the first spark of organic life appeared by random chance in some remote bed of inorganic slime.

Marooned on his private island of reflection, isolated in a sea of faces, reality sustained the young scientist: Traci Kilburn's opinion of him was infinitely more consequential than the career plan unraveling before his eyes.

A kingdom sacrificed for a dream? History reminded him that obsession with another Maryland woman had dethroned the Duke of Windsor. But before Josh could identify the Duke as a soul mate, grim realization cut the fog and snatched away this wisp of trivia.

"Wallis Warfield Simpson at least married the Duke! By the time this wretched trial is history, what's the chance Traci will still be in my life?" From here he retreated into gallows humor, asking himself, "Maybe I should at least claim some sort of royal title...the late Lord Carrington being my grandfather and all...Wonder how this peerage stuff works? Maybe I could talk my way into the line of British succession?...I could sure use the job!"

# 6

As the stage reopened, announcing the end of the expedited recess, the musical extravaganza faded as Judge Stone rapped his gavel declaring the court "in session."

During the break, Josh had reassured the legal team that the traumatic departure of the aggravated Karl Striker had not intimidated him. He still retained the courage to look for and to tell the truth. Seeking relief, his mind churned out romantic fantasies. Memory of April's Center Hill Lake rendezvous provided fuel for his imagination.

Seeing JT rise to proceed with the examination, Josh felt a surge of pride for the chum from Magruders Mill, impressed that his friend was certain to emerge as an influential figure in the legal profession. Anticipating JT's line of questioning, he stood ready to respond with his perception of truth.

The audience remained attentive, perhaps anticipating another firestorm rivaling the departure of the terrible-tempered Teutonic. Dr. Augustus Morley and other science professors remained impassively in place. Traci gave JT the signal to throw the witness another soft pitch, enabling him to unleash a further attack on presumed scientific orthodoxy.

Direct examination proceeded briskly.

"You've charged that Darwinian thought is short on evidence. Where's your evidence?"

With Judge Stone nodding assent, Josh dropped all pretense and took off the velvet gloves. Sensing he had little to lose, he launched more intellectual broadsides, hardly pausing for breath. The attack on tradition's shibboleths struck with ferocious challenge.

"What are the mathematical odds of hundreds of environmental conditions essential for organic life combining coincidentally, by random chance, in cosmic convergence—at the same instant, in the same place, in a universe of infinite space?"

JT interrupted, demanding hard data, not more mind-boggling questions. "Pray tell, what is *cosmic convergence*?"

"Organic life cannot exist without thousands of environmental conditions converging simultaneously in time and space to produce a life-friendly ecological niche."

"Can you name one or two of the *thousands*?"

"Water...atmosphere...fertile land...!"

"Darwinists would doubtless agree! What else?"

"Planet Earth moves through space in a predictable orbit with its companions the sun and the moon. Much closer to the sun, too hot; farther away, too cold. It spins without visible means of support in a pattern predictable hundreds of years into the past or future, demonstrating cosmic convergence."

The witness lapsed into hypothetical rhetoric to stimulate jury thought.

"Is the cosmic convergence that provides a life-friendly ecological niche the result of good luck? What are the chances of a blindfolded kid tossing a handful of jacks indiscriminately, and all of them landing precisely in a straight row? Or in a concentric circle? Spaced exactly fifty millimeters apart? Even after several billion tries?"

The witness studied audience reactions, then went on.

"How about a bucketful of jacks dumped from a ten-story building converging perfectly and accidentally in flawless design? Or a cargo plane packed with 100,000 jacks—from an altitude of 35,000 feet? Any remote chance they would line up like toy soldiers in precise formation?

"A world-class juggler couldn't keep miniatures of these solar system symbols spinning in perpetual orbit, floating precisely in space without supporting cables. The flicker of the simplest living cell is more than an *event*— its a tiny particle thriving within an infinite system of cosmic convergence...a single note in life's symphony of rhythm."

Judging by the shuffling in the jury box, the audacious witness assumed that his rough-spun logic had paid off. But before he could continue, bursts of coughing echoed from the back of the playhouse, interrupting the dialogue. Josh naively waited for the intrusion to pass. Rather than stepping outside, the culprit remained.

JT swiftly diagnosed the timing as more than coincidental. It took a prolonged series of gavel raps from the bench to still the distraction. Undeterred, Josh pressed on, raising his voice to be heard above the commotion.

"Compliments of cosmic convergence, ecological balance provides a Planet Earth home place for life, infinitely more complex than that simple single cell."

"Could you favor us with an example, Mr. Ryan?"

A patient JT Daniels unabashedly led the witness, anticipating correctly that sooner or later Josh would unintentionally flaunt his photographic memory, nonchalantly rattling off science literature quotes without the aid of electronic prompters.

"And what about the *Homo sapiens* brain! Thanks to the DNA inherited from two parents, all creative thought, sensory reaction, speech, and memory are driven by a personal computer—*'a three-pound brain...composed of twelve billion neurons...with...120 trillion connections'*[10] encased in a trauma-resistant, compact cranial container.

"That memory bank *'consists of about ten thousand million nerve cells. Each nerve cell puts out somewhere in the region of between ten thousand and one hundred thousand connecting fibres by which it makes contact with other nerve*

46

*cells in the brain. Altogether, the total number of connections in the human brain approaches 1015 or a thousand million million. Numbers in the order of 1015 are...completely beyond comprehension. Imagine an area about half the size of the USA (one million square miles) covered in a forest of trees containing ten thousand trees per square mile. If each tree contained one hundred thousand leaves the total number of leaves in the forest would be 1015, equivalent to the number of connections in the human brain!"* [11]

"Now consider the challenge confronting the world's most gifted electrician attempting to mastermind the wiring of that jungle of nerve cells with *a thousand million million* connections without mis-wiring, short-circuiting, or blowing a fuse!"

"If that doesn't deflate the blind ignorance of egocentric arrogance, try confining that jumble to the micro dimensions of a three-pound brain fitted snugly within the custom-designed bony cranium that offers something less than 100 cubic inches of spatial capacity!"

Imagining that this collective revelation just might have hit an academic home run, Josh continued with conviction, "So this feat of electrical/biological engineering supposedly designed its own masterpiece of a wiring diagram without so much as a master plan? And humans endowed with this living computer are supposed to believe it evolved over several billion years, beginning with some ancient microbe, thanks to untold millions of multiple lucky random mutations?

"Can independently reasoning human brains buy that Brooklyn Bridge of a mind-blower?"

He let the rhetorical question simmer. Some observers saw the emotional display as part of the act. JT and Traci knew better. Eloquent and exercised, Josh's steamy attack on mega-evolution erupted in searing bursts, pouring from his inner self in a river of intellectual magma. JT continued to mine the passionate academic's psyche, like a Forty-niner panning gold.

"Mr. Ryan, you've made clear your abhorrence for an unprovable philosophy, which alleges that unidentified single-celled organisms evolved spontaneously from prebiotic soup and then underwent millions of random mutations, thereby becoming man's ultimate ancestor. Is that not correct?"

"Yes, I've stated my abhorrence of an unproven postulate posing as scientific fact!"

"Well then, given your notoriety as a gambler endowed with exceptional mathematical skills, perhaps you could treat the jury to some appropriate illustration of the odds for or against the molecule-to-man hypothesis?"

Josh brushed off the left-handed compliment. "Thanks, I think. But I don't make my living placing two-dollar bets. Odds-making belongs to deep-pocket professionals."

Having given this disclaimer, Josh obliged with a pot shot at random chance. "Science demands reliable predictability! Where is the evidence proving that some mysterious one-celled bacterium is a quite accidental ultimate ancestor to every person sitting in this courtroom?"

"The computer my dad gave me offers a slick example of absolute precision. It features buttons and bells so complex I'm not likely to ever master all its secrets. You computer buffs know that if you're smart enough to know what keys to press and where to point the magic mouse, it can paint pictures and write books. It even converts back and forth between daylight savings and standard time—let's hope it does as well with theY2K enigma!

"No Darwinist suggests that inorganic *prebiotic soup* molecules will ever arrange themselves by chance to produce a state-of-the-art computer—and then reproduce itself ad infinitum—even in 4.55 billion years!

"Absurd! Now for a quantum leap beyond impossible!"

The animated witness was on a roll. No one bothered to intrude or object. "Marcel Schutzenberger sees odds of 10-1000 '*against improving meaningful information by random changes...The astronomers Fred Hoyle and Chandra Wickramasinghe placed the probability that life would originate from non-life as 10-40,000 and the probability of added complexity arising by mutations and natural selection very near this figure.*'"[12]

"Can you translate that equation for us ordinary people?"

Josh rose, rolled a chalkboard to center stage, and with a piece of white chalk drew a zero followed by a decimal point and the number one.

"Everybody knows that 0.1 represents one-tenth of the number 1.0."

Next, he erased the '1' leaving the zero followed by the decimal. Then he began drawing more zeros, one after the other, all the way to the extreme right edge of the blackboard so as to read.00000000000000000000!

"That's 20 zeros after the decimal point. Now add 39,980 more zeros before adding the number '1' at the very end and you have an idea of the remoteness of the odds 10-40,000."

Just in case the jury had missed the infinitesimally unreachable chance in the formula, he offered a symbol readily understood by nearly everyone—the Rubik cube. Josh fished eight of the puzzles from his briefcase, and presented one to each member of the jury, inviting their help.

"When I say 'Go,' I want you to match the colors as fast as you can. You'll have exactly sixty seconds."

Raising his arm to signal, he ordered, "Get ready." As his arm dropped, he barked "Go." A minute later he raised his right palm, calling "Stop!"

"Any winners?"

To the surprise of all, including Josh, a bright Sherwood High coed waved a perfectly color-matched cube, jumping to her feet and announcing in shrill tones, "I did it! I did it!"

Josh offered a smile and astonished congratulations before issuing the next challenge.

"Now try it with your eyes closed! No peeking—and that applies particularly to our sixty-second champion." He winked at the coed, who grinned back.

Fumbling oldies shook their heads in frustration, but all, including the Sherwood coed, obediently closed their eyes and launched gamely ahead at the "Go" command. Sixty seconds later they surrendered at the command to "Stop."

This time, unmatched color patchworks reigned. Not even the high school whiz came close to solving the puzzle without visual access. Josh thanked the jury, inviting them to keep the cubes and to continue practicing—blindfolded. Then he used the experiment to define impossible odds.

"See what random chaos achieves without plan or design? Zip!

"What are the chances that one of you could solve the cube puzzle by sheer luck—in a lifetime of tries—without looking? What about all eight of you getting lucky and stumbling onto the solution simultaneously while blindfolded?

"The odds against a living, breathing human evolving from a microbe in the past several billion years is less likely than the chance of this planet's six billion citizens, if blindfolded and given a Rubik cube, suddenly solving the intricacies of the puzzle in one simultaneous split second!

"Impossible!

"Now, how about the odds of repeating this one-time *impossible* performance again and again—multimillions of times?"[13]

An audible hush confirmed the audience's comprehension of the odds.

JT recognized that this was an opportune moment to hang it up for the day. Concerned that Josh had placed himself and possibly his career on the public chopping block, the examining law student addressed the bench.

"That's all the evidence we are prepared to place before the jury today, Your Honor."

Remembering Pace Terhune's marching orders stressing "E" for Entertainment, Traci nodded concurrence, waiving cross-examination. Taking his cue from both law-student performers, Judge Stone responded by declaring the case in recess until 8 p.m., Monday, September 2. At the tap of the judge's gavel, the curtain descended, cloaking the stage.

Given the diverse community views, Pace Terhune rejoiced at the more-than-perfunctory audience applause. Before the polite ripples of approval faded, he had begun concocting Monday's headline and action picture for release to the *Washington News Press*. Hollywood impresario Redondo Calizar offered a thumbs-up, predicting jovially, "Five days before the fat lady sings!"

In a matter of minutes, the performers had drifted away, several intent on catching at least one of the second set of televised NFL season openers. With routine good-byes to other players, Josh departed by the stage door exit, en route to the sanctuary of his mill house hideaway.

A wave of emotional exhaustion lapped at his heels, eroding the fleeting exhilaration of public debate. Entering the front door of his tiny Fort Ryan cottage, Josh prepared to unwind, embracing the cozy refuge of his secure home base.

The chatter of a noisy fax greeted his arrival. Cold chills traveled the length of his spine when he spotted the official Mid-Atlantic University seal emblazoning the letterhead. The cryptic words spit out the dreaded news:

> Sunday PM
> September 1, 1996
>
> To: Joshua C. Ryan
> From: Dr. K. Striker, Chm., MAU Science Division
>
> It is mandatory that you appear in my office promptly
> at 8 AM tomorrow, Monday, September 2, 1996.

# 7

Monday's edition of the *Washington News Press* headlined the onslaught of wildfires ravaging 750,000 acres in twenty-seven Western states; rough Atlantic surf dampening the holiday weekend of Ocean City's sun-worshiping throngs; and the disappointing NFL Redskins' season kickoff game in which the home team let a win slip away at the hands of the rival Philadelphia Eagles.

Relieved that the tantrum of the previous day's *Monkey II* hadn't made the front page, the reveries of its star witness drifted to memories as he relaxed in the soft embrace of his Fort Ryan home.

Cut into a twenty-five-acre sliver of land overlooking the Hawkings River, the site had once housed the Magruder family's pre-Civil War mill, powered by the unpredictable flow of the Hawkings. On the riverbank, a landmark chunk of granite stood silent guard beside an overgrown marker for the Underground Railway route to freedom, staked out by gutsy nineteenth-century Quaker farmers.

By the time Josh's Grandmother Ryan, born a Magruder, conveyed the battered piece of history to her son Michael and his bride, Leslie Anne Carrington, as a wedding gift, little remained to remind visitors that this centerpiece of a rural hideaway village had once provided refuge for Dolley Madison and White House treasures she had rescued before the arrival of rampaging British firebrands. Time and the voracious onslaught of vines had all but swallowed the tottering chimney and stark stone walls of the decrepit mill.

Michael Ryan restored the ruin to use as a honeymoon cottage, while he and Leslie Anne created a manor house atop the rise, overlooking the river. It was Josh's childhood imagination that provided the unlikely moniker for the spread: *Fort Ryan.*

Now the sprawling Magruders Mill terrain spilled lazily across immaculately groomed furrows of wheat, soybeans, and corn. Josh more than earned his caretaker's pay of $200 a month. Apart from crop-raising skills, he had mastered the art of cranking up and maintaining the decrepit tractor, mowing the slopes of a multi-acre lawn; tending and mending fences, repairing ruts in the earthen dam that held back the waters of a fish pond facetiously dubbed Lake Never, and leading the fruitless struggle to eradicate the omnipresent groundhogs.

Dull routines brightened when lost city types wandering Highway 97 stopped at the roadside fence to ask directions. Dressed in earth-stained jeans and mid-calf rubber boots, Josh relished hamming it up as local yokel, imparting ambiguous wisdom in a muddled Maryland accent. Lack of knowledge never hampered his courteous, albeit hopelessly inane, responses. Rustic hat tilted askew, he would gaze soberly heavenward, as though invoking help from beyond, and effect a pose of perplexity. Eventually, after this charade, he would deliver meaningless gobbledygook that defied translation.

Instructions such as, "Maybe over...yonder...let's see, now..." or "According to Great-grandpappy Ryan...hereabouts is where the bluecoats tramped north to Antietam in '62..." usually sent exasperated tourists scurrying, shaking their heads in bewilderment.

Rent-free occupancy of the rebuilt mill house, gatehouse to the Georgia Avenue access to the estate, provided bonus perk. Josh reveled in the trappings of independence implied by the title of overseer and the inherited privilege as Michael Joseph Ryan's son.

As another incentive, Josh's father underwrote the not-so-incidental cost of feed and veterinary care for the aging but still-frisky Arabian mount, Turbo. Rarely ridden, the spoiled steed roamed in meadows he considered his exclusive domain. A carefree arrogance attested to the vim and vinegar characteristic of the breed.

Leafing absently through the paper and sipping coffee, Josh noted that the weatherman promised warm sunshine trailed by mild breezes, with temperatures reaching 88 degrees. For an instant, Josh forgot his pending command appearance before the mighty Dr. Striker. Then his eye spied the headline staring from the front page of the newspaper's Montgomery County section: *Student Witness Attacks Evolution!* The subhead blared: *Darwinian Theory Branded Fact-Free Science.* A three-column-wide action shot of Karl Striker fleeing the playhouse in a state of apoplectic rage was splashed across the middle of the page.

Ace publicist Terhune understood the visceral impact of an action shot! But he had also ignored the collateral fallout—a higher bounty on Josh Ryan's head.

Until this week, Josh had managed an uneasy truce with his tempestuous mentor despite classroom antics that usually left the professor spluttering innovative profanities. The unexplained loss of the fossil Archy and the Science Research Foundation's squandered investment counted as automatic strike two. The day's headline offered up a likely strike three in the eyes of the outraged academic umpire.

Josh didn't wallow in trepidation. But it didn't take the pain in the pit of his stomach to remind him that his career at Mid-Atlantic University teetered on the brink of being called "out"—extinct as the fossil Archy.

Promptly at 8 a.m., Josh rapped at the door of the executioner's inner sanctum. No cheery voice responded. He overheard footsteps approaching. A grim-faced Augustus Morley opened the door and wordlessly motioned the student to come in.

Paneled in sculpted black walnut, the chairman's office shimmered with understated elegance. A giant, partially lighted chandelier cast shadowed ambiance. At the far end of the rectangular chamber, Dr. Striker's bulky, oversized desk anchored a raised platform. Without looking up, the division chairman peered intently at an open file. Every bit the monarch, he dispatched his royal decrees from the comfort of a padded leather throne. A back-lighted wall, draped in lavish folds of blue velvet, contributed to the intimidating atmosphere.

For some reason, the tale of the long-dead dictator, Benito Mussolini, and his marble-studded corridor of an office flashed through Josh's mind.

Rather than shift to the massive conference table, the academic chieftain remained seated at his elevated station. An administrative secretary occupied a small table to his right. Two undersized hard-backed chairs perched in lonely isolation directly beneath the professor's scrutiny. Following Morley's example, Josh stood patiently beside one of the chairs, awaiting orders.

Duly satisfied with the obeisance, Striker barked, "Sit."

Sinking awkwardly in the direction of the floor, the two could only peer up at the academic chairman over the edge of the gigantic desk. Josh instantly grasped the psychological power of the setting. Rumor had it that Striker enjoyed this ritual, having dashed the hopes of a legion of would-be scholars.

Flourishing a golden pen, Striker ignored Josh, intent on scribbling notes in the margins of official-looking documents. Papers rustled; files shuffled. Finally, he flipped one file closed with deliberate flourish. A fresh folder, bearing the name RYAN, took its place, and was ceremoniously spread open. Removing half-lens horn-rims, the academic cleared his throat. Blazing blue eyes locked with laser-beam intensity onto the unflinching student while a stiletto-tipped voice rolled through the cavern of officialdom.

"Mr. Ryan, you have been a serious disappointment to me and an embarrassment to the Mid-Atlantic University academic community!"

The big guy didn't mince words. The declaration was delivered with a muted snarl.

"So what else is new," thought Josh.

"Blessed as you are with exceptional intelligence, I've always had high hopes for your career potential—despite your occasional breaches of classroom decorum."

Unblinking, Josh betrayed neither assent nor disagreement. He amazed himself at the calmness embracing his psyche.

Striker went straight to the jugular: "But your negligent incompetence in mismanaging a Science Research Foundation project resulted in staggering financial loss." Failing to get a response, he leveled an icy stare, and added with a hint of menace, "Some wonder whether the alleged fossil find amounted to anything more than a poorly planned hoax."

Josh wanted to lash back defensively with a, "Now hold your horses! Except funding for a beat-up camper and a pittance passing for a stipend, I've never been reimbursed one lousy cent for weathering Tennessee's rocky-top winter!" But he swallowed his protest, aware that Striker would never tolerate interruption.

The onslaught proceeded unimpeded.

"Perhaps you've heard, Mr. Ryan, that doctoral candidates at this university customarily submit dissertations based upon original research. Your portfolio is missing. Despite your distinguished Morgan ancestry and political connections, I can't in good conscience make an exception and reward carelessness."

Sweating silently, lapdog Morley lacked the courage to intervene. Josh squelched a desire to retort, "Yeah, at least I don't salute a slimy prebiotic soup microbe as ancestral—like some in this room!"

Striker wasn't through. Scornfully, he unfolded then displayed the day's *News Press* headline atop the unflattering picture of himself.

"Gross negligence is one thing, but consistent disregard for professional ethics is something else. You delight thumbing your nose at the academic community. Single-handedly, you have selfishly garnered publicity for yourself while wreaking havoc on traditions dear to this university's Science Division."

Josh sat up straight as if to challenge the bitter attack on his integrity. But Karl Striker would tolerate no distraction. His verbal punch-line exploded: "You see, Mr. Ryan, it is my unpleasant duty to advise you that your academic future at MAU looks more like past history. Unless you somehow make amends, your chances of being awarded a doctorate from this university are at least as remote as the short shrift you so ungraciously give to the theories of Charles Darwin."

He motioned to his secretary, who brought him an original and several copies of a letter addressed to Joshua Chamberlain Ryan. With a single flourish of the gold pen, the angry professor terminated the academic career of its addressee. Still standing at Dr. Striker's elbow, the expressionless secretary passed the original to Josh and a copy to Dr. Morley.

Speed-reading and a photographic memory didn't soften the blow. After a couple of paragraphs of banal formalities the bottom line could be paraphrased succinctly: "Your chance for further academic study, leading to a graduate degree from this university, is as dead as the fossil bird you allegedly discovered in Tennessee."

Josh knew that strike three had been called. It was a bean ball, nowhere near the plate.

"Understand, Mr. Ryan, that I take no pleasure in this action. It comes with the mandate of the faculty committee vested with full authority to supervise your academic program."

"Sure," thought a skeptical Josh, "a committee of a dictator and two or three lackeys like poor Morley."

Mission accomplished, the secretary busied herself with outlining the day's agenda. Gus Morley arose obediently, pointing toward the door Josh had entered minutes earlier. As Josh moved to follow Morley's lead, Striker strode away with military precision, making a grand exit in the direction of a waiting classroom. Abruptly the offended academic stopped, snapped a crisp about-face, and aimed a malicious scowl in Josh's direction A final taunt ended the grim encounter:

"Oh, by the way, Mr. Ryan...," Striker's feigned casualness bristled with calculation. "Just because you and I don't agree on a dissertation should not intrude on your freedom to pursue your fetish for facts in other classrooms of this university. Far be it from me to interfere with your scholarly curiosity. You remain free to attend as many classes as you choose...as long as your energy, time, and money hold out!"

Josh grimaced from the aching stiffness of a left leg cramped in an unnatural sitting position. But except for this, Josh felt little else. Taking a couple of deep breaths, he headed for the outer anteroom and then turned into the hallway, a still-mute Gus Morley tagging along.

Striker's guillotine had maimed his psyche but could not destroy body or soul. He turned his head from side to side, reassured that it had not been severed by the blow. But his career potential looked more like the latest road-kill on Highway 97.

Josh wondered aloud, "Now where is that Laguna Beach place in southern California so highly touted by JT? Maybe I could tutor the locals in the fine art of two-person beach volleyball!" He knew it was a thought born of desperation. A wall clock showed 8:15 a.m., too early in the day to suffer the agony of defeat. Turning to go, still wondering whether he should have challenged the strident rebuke, he felt a hand on his shoulder. Dr. Morley offered solace.

"I'm sorry, Mr. Ryan." Shaking his head, he sympathized in restrained tones. "I did all I could. But you know the reputation of Dr. Striker. Several times during the summer I spoke to him in Houston by phone and was encouraged to believe that by now he would accept the loss of the Tennessee project and assign you a new dissertation subject."

His voice quavering, he reiterated the obvious. "You must realize, young man, that when you publicly challenged the traditions cherished by Dr. Striker, you chose to sit upon and then fall off the wall...like Humpty Dumpty in the nursery rhyme...not to be put back together again."

In a conspiratorial whisper, Morley confided, "Assuming you transfer to another university, regardless of any personal risk to my own career, if you ask for my help, I will do everything in my power to arrange an appropriate recommendation."

Morley extended his hand in benign benediction. Josh shook it with an appreciative, "Thank you, Dr. Morley, but it sounds like you've already stuck your neck out too far."

Walking away, he thought, "Poor old gutless Gus. He wants everybody to like him, and as a result, his clout in this cruel world is a capital 'Z' for zilch!"

It seemed a mad exercise in futility to attempt to resurrect any prospect of a doctorate at Mid-Atlantic University. The fact that Great-great grandfather Sam Houston Morgan's vision and funding founded the place wouldn't buy a cup of coffee. But it was too early to jump ship to he knew not where. MAU's school year did not start for another week, and he did not intend to jeopardize his rapport with Traci. So he remained committed to five more *Monkey II* appearances. At least until Friday, September 6, he vowed that nothing would fill his thoughts except *Monkey II*—and the intoxicating presence of Traci Kilburn.

Josh acknowledged inwardly that he had anticipated today's comeuppance as a result of his testimony. In a burst of objectivity, the chastened scholar begrudgingly recognized his own overblown arrogance. Ruefully, he acknowledged that insubordination rankles sensibilities.

"So what if Striker acts like a tyrant. He's the teacher! He has a right to demand high standards—and a modicum of respect—even from smart-alecks."

As his sagging spirits revived, he remembered a previously planned escape to another world where the Darwin name rarely surfaced. He and JT had reserved a tee time at the Hampshire Country Club golf links later that morning, for a survival-of-the-fittest competition of a different sort.

# 8

With a comfortable two hours before scheduled tee-off with JT, Josh returned to the ambience of Fort Ryan, where he could lick his wounds in solitude.

Although embarrassed to be the bearer of bad news, Josh felt obliged to track down Michael Ryan and share what had happened. He found his dad registered at Singapore's Shangri-La Hotel. The previous week, Michael had alerted Josh to his plan to arrive at Baltimore's BWI Airport early the morning of Thursday, September 5, in time for the last couple days of the make-believe trial. Josh left a brief, recorded message with a stumbling account of *Monkey II* and the disastrous morning's encounter.

Initially, Josh felt compelled to share the bleak news with Traci, then backed off. Driven by machismo, Josh strove to save face. He attributed his reluctance to altruism—his desire to avoid blaming Traci for his sorry predicament. She would find out soon enough. He had chosen to take the witness stand. For now it seemed adequate to share the grim details with his dad and JT.

With loads of time before tee-off, Josh welcomed the chance to spend an interlude with his horse, Turbo. As he ruminated, the sleek Arabian mount could be seen from the cottage window, grazing in the pasture beyond the pond. A decade earlier it was a rare day when Josh didn't ride the faithful stallion. While scrupulously tending the everyday needs of the prized but spoiled brat-of-a-pet, Josh's expanding interests had eroded this ritual to one ride a month, at most. Still, the bond between the now-aging steed and young scholar endured.

If anything had transfixed Josh's young life more than exploring for fossils, it had to be horses. This obsession flowed through the bloodline of every generation of the Morgan clan as far back as recorded family records existed.

Josh had discovered bits and pieces of the Morgan family's saga stashed away in the nooks and crannies of Mid-Atlantic University's library. A proud Lady Regina Ann Morgan-Carrington had planted the trail of information commemorating the Morgans' unbroken succession as university trustees. Her grandson's patient research had paid off in the discovery of bracing accounts of epic adventures unknown even to his father.

Great-great grandfather Sam Houston Morgan had traversed the Texas frontier, born to the saddle. When the country ripped asunder in a brutal Civil War, Texas governor Sam Houston spurned the Confederacy and suffered political ostracism. Galvanized by the example of the salty Union loyalist, the teenage Morgan, determined to join ranks with the bluecoats, risked exposure to Southern hills and backwoods trails bristling with hostile citizenry bent on secession. Riding recklessly, his audacious plunge brought him to Gettysburg,

Pennsylvania, in July 1863, astride the family's finest Arabian. Packing nothing but a side-arm, Sam Morgan felt invincible.

Family archives reverently saluted the daring dash as a *ride to glory!*

His horsemanship skills landed him a battlefield assignment as a messenger for General Meade's staff. But before he could don a snappy blue uniform, he received orders to deliver urgent information to the commander of a thin line of bluecoats deployed in a fluid, fishhook formation arcing the left flank of the Federal Army's front. Pennsylvania's Colonel Strong Vincent commanded this perilous anchor to a vulnerable rampart.

A Confederate sharpshooter missed the young messenger but struck the heart of the Arabian. The targeted victim hurtled to the ground as his mount crumpled in a broken heap. The powerful Arabian lay helpless, mortally wounded. But Sam, the lucky Texan, scrambled to his feet without a scratch.

In a fit of rage, young Morgan ran headlong into the deadly firestorm blazing on the southern down slope of Little Round Top. Driven by a combination of anger and fear, he scampered instinctively, weaving and dodging through hanging branches, scraggly bushes, and rocky projectiles.

By the time he arrived at the fragile outpost, acrid smoke, smothering the stench of death, had consumed the site. He delivered the message from HQ to the first senior officer he could find: Joshua Lawrence Chamberlain, commander of the Twentieth Maine, a pitifully thin line of blue, exposed to an onrushing wave of gray.

"Mind if I join you boys?"

Up to his ears in military crises, the cool-headed Bowdoin College professor warned the untrained volunteer to, "Keep down, Son," before wading into the tumult, waving his sword, shouting, "Let's hold 'em, boys. You know what to do!"

Armed only with his cowboy pistol, the teenaged Morgan crouched behind a jutting outcrop of Pennsylvania granite, yanked the gun from its leather holster, and took level aim at the river of gray rushing uphill at point-blank range.

Before he could pull the trigger, a sledge hammer blow spun his body in a dizzy circle, casting him violently to the crimson-stained earth. The sight of the mangled red pulp that moments before served as a sinewy left arm, sickened him. Numbly, he understood that his combat days were history. Drifting into unconscious oblivion, the young scion of the Morgan clan escaped temporarily the wrenching agony and brutal shock of manmade hell.

By the time Sam Morgan came to, all that remained of the battle at Little Round Top was the remnants of a shouting torrent of blue, bayonets fixed, cascading down the slope to envelop the last fragments of melting gray. If there was a watershed moment in the history of the Civil War, it occurred late that afternoon of July 2, 1863, on the hardscrabble slopes of Gettysburg's Little Round Top.

Joshua Lawrence Chamberlain and the decimated ranks of the Twentieth Maine regiment had changed American history.

Someone used a torn, faded-blue shirt sleeve to apply a tourniquet to staunch the flow of blood to the tattered shreds of Sam's arm and save his life. When awake enough to refocus, he felt a firm hand on his right shoulder and saw the face of the intrepid teacher/warrior, Joshua Lawrence Chamberlain.

"Looks like you took the big one, young man, but once the surgeons go to work, you should be on your feet in no time. I've seen them perform miracles. You're a brave lad. On behalf of the Twentieth Maine, thanks for coming to the party."

So said, the charismatic officer ordered a couple of still-standing bluecoats to string jackets between rifles and carry the woozy Sam to a surgeon's tent and temporary respite. Sam's only remaining battle would be fought first for his life and then to salvage his shattered left arm. Call it destiny or a tribute to primitive medical care or the Texas luck of a vigorous youthful body—Sam Morgan survived unscathed, except for an upper extremity left hanging uselessly at his side for the rest of his life.

While recovering in makeshift army hospitals, the one-armed Texan mastered the art of singlehandedly shuffling cards. As reckless in poker as in battle, Sam racked up win after win. Apart from his regular take in U.S. currency at the expense of other recuperating casualties, the swashbuckling Southern-style Yankee won and promptly recorded warranty deeds to several sections of prime wooded farmland in Montgomery County, Maryland.

By 1896, the canny entrepreneur had picked the choice 640-acre section on which to build Morgan Manor, the family's Victorian home, and broken ground for Mid-Atlantic University. He subdivided and sold the remaining acreage in 40-acre plots, reaping 100 percent profit. The transplanted Texan dug his roots deep into the rich Maryland soil, and recruited a stable of the fastest thoroughbreds money could buy. His superior business instincts were augmented by a healthy dose of luck.

Although warned that critics scoffed at the horseless carriage, Sam envisioned opportunity and made countless forays into his family's Texas and Oklahoma stamping grounds, purchasing petroleum rights before most wildcatters jumped aboard the bandwagon.

Like his father Sam, Joshua Chamberlain Morgan was born to the saddle, crisscrossing the lush, velvet green hunt country surrounding Morgan Manor soon after learning to walk. His only child, Regina Ann Morgan-Carrington, debuted to the horsey set and the glories of the Hampshire Hunts. Social extravaganzas like the annual Hunt Ball consumed the lives of the Morgans and neighboring families of privilege in rural early-twentieth-century Maryland.

Regina's daughter, Leslie Anne, surpassed the riding skills of her mother, the Duchess, and even her grandfather, Joshua Chamberlain Morgan. She ignored the prestigious Morgan Stable thoroughbreds and instead fell in love with a magnificent Arabian stallion boasting a charming personality, rugged endurance, and hair-raising rides at blazing speeds.

Leslie Anne brought to her marriage the Morgan clan's love affair with horses. At the insistence of his bride, Michael Ryan adjusted to the saddle's fit tolerably well. Although deprived of access to the Morgan Manor Stables, son Joshua Chamberlain Ryan loved the sight of horses. By Josh's tenth birthday, he had ridden regularly with Montgomery County's best in the Hampshire Hunt, racing across Maryland meadows astride Turbo, his very own mount.

Josh's mother won several awards riding Turbo's sire, but Lady Carrington sold the horse after Leslie Anne's death. Turbo's price had appeared to be a bargain, until Michael discovered that the three-year-old Arabian faced the world with the disposition of a lamb and the personality of a mule. Whether due to inadequate training or ornery genes, Turbo viewed the world upside down. He adored human companions but perceived himself as king and all *Homo sapiens* as supplicants. There were moments, sometimes days, while locked in some horsey snit, he chose not to run or even to mosey around. He wasn't lazy—just unpredictably ornery. He had better things to do than to be burdened by human baggage. To saddle the sleek stallion in the midst of one of these moods proved futile. He didn't snort or buck but stood still as a statute, waiting stubbornly for the erstwhile rider to comprehend that the train was stuck in the station, with no departure scheduled.

Dig a frustrated heel into the horse's flank, and Turbo would remember. It was unlikely that such a rider-to-be could ever hitch any kind of a ride on Turbo at any time in the future. Nothing remained but to dismount and let the fickle horse proceed with his vague plan to graze and gallop. Ultimately, Josh alone tried to ride the unpredictable mount. Even then, the irascible Turbo chose stubbornly to sit out the proposed dance more often than not.

Except for Michael and JT, no one was as close to Josh's heart as the wacky Arabian. Horse and human forged an alliance. The tyranny of equine royalty flourished under Josh's loyal oversight. Turbo wallowed in the good fortune of his kingdom and unabashedly exploited his reign. And despite the burden of an obstreperous personality, Turbo proved as gentle as any spirited stallion could.

Horse and rider carried vivid memories of a severely injured fifteen-year-old who relearned the fine art of walking after his bicycle injury. Later, Josh modified his riding skills to favor the aching left leg by mounting the horse from the right, contrary to traditional etiquette. Throughout the delicate process, Turbo never twitched a muscle in protest, instinctively sensing the injury and the need for the teenager to recover his strength. His notorious mood swings never surfaced during the ordeal.

This Monday, Josh stepped from the mill house, sauntering casually in the direction of the stable. Hearing the cottage door close, Turbo looked up, let loose a whinny of recognition, and made a mad dash to the fence to greet his private groom and personal protector. With the horse nuzzling Josh's chin and Josh's arm wrapped around the royal neck, the reunion banished nagging thoughts of academic troubles.

As Josh climbed the fence and slipped into the paddock, he toyed with the notion of riding His Majesty. Turbo vetoed the idea. Throwing up his neck in haughty response, the fickle Arabian telegraphed, "No, not today!"

Turbo wallowed in his erratic whimsies more often with age. Josh understood. Monday, September 2, 1996, had been earmarked a day for equine royalty to roam the pasture sans rider. The human companion laughed out loud at the arrogance. Resigned to the caprice of the steed, Josh took off on a brisk walk, circling the perimeter of the grassland, knowing Turbo was sure to frolic at his heels.

The sweet fragrance of a September morning invigorated. The chorus of rustling leaves hinted at the impending riot of color. For nearly an hour, the unlikely pair—cavorting riderless horse trailing carefree man on foot—redefined comic spectacle!

The only audience consisted of chirping cardinals, scolding crows, and bluebirds, cutting the air in deep, swooping swaths. Curious groundhogs peered suspiciously from afar until sent scurrying for the safety of a handy tunnel to escape death from the earth-jarring hooves of the wildly galloping Turbo.

The joyous equine repeatedly pivoted crazily, bowing in the direction of his keeper, then racing back for the mandatory muzzle pat of approval. The Arabian carved graceful arcs across the pasture's circumference that grew tighter and smaller until the wild creature surrendered, traipsing meekly at the heels of the human pathfinder. Though the ravages of time had diminished Turbo's legendary stamina, the heart of the champion roamed forever free.

Exhilarated by the sights and sounds of the home place, Josh Ryan's mental gymnastics flipped as fast and free as Turbo's mad dashes to nowhere. This day his thoughts raced to the looming issue to be examined that evening in the *Monkey II* courtroom. Josh guffawed aloud at the thought of an unknown recipe for prebiotic soup spontaneously spawning the first ever organic life form.

"Don't worry, Turbo, you *Equus* types didn't descend from some worm...or that runt *Eohippus*!" Turbo wiggled his ears intuitively, as if his royal pedigree grasped the absurdity. "That ginned-up evolutionary sequence of horses in school books is fodder for the gullible. '*[F]or most of the animals in the sequence: transitional species are not merely unusual they are missing entirely.*'"[14]

Try as he would to muster profound thoughts, a pleasant fantasy kept intruding. Fossil priorities that once seemed so certain had dissipated. By the time Josh and his horse companion completed circling the enclosure, thoughts of Traci consumed his senses.

Returning to the fence, Josh gave a farewell pat to Turbo's head before clearing the fence in a single, scissoring motion. With ample downtime before tee-off at the Hampshire, Josh retreated to the mill house, finally reconciled about sharing with Traci the bitter news that the academic credentials of her expert witness lay in Humpty-Dumpty disarray.

After four rings, the answering machine responded. The one-way conversation offered the would-be suitor an opportunity to deliver a concise account that was diplomatic but casually upbeat.

"Early this morning I met with Dr. Striker at his request. It seems my future as a Ph.D. candidate is less likely than spontaneous generation. Just wanted you to know that your tarnished witness is hangin' in."

Recognition of reality proved cathartic!

Moments before heading to the club, a message spilling from the fax machine added emotional fuel, fanning the glow sparked by the romp in the pasture with Turbo.

*Flying into BWI Airport late Wednesday evening, September 4. Don't bother to meet me since a corporate pickup is arranged. See you at the house on the hill for breakfast 9 a.m. Thursday. Look for a happy surprise. Love, Dad.*

# 9

Cruising into the club parking lot, Josh spied JT engrossed on the driving range, swinging arcs as smooth as flowing molasses. Like Josh, he had polished his natural skills at the feet of Lyle Grant, club pro and relentless taskmaster. Competitors since they were knee-high, the boys from Magruders Mill excelled on the fairways—living trophies to the patient guidance of their mentor.

Lyle spotted Josh checking into the pro shop, calling a behind-the-counter welcome. "See you on the tee at ten sharp." With a grin creasing the weathered face, he joked, "JT insisted I tag along today just to keep you honest."

"What JT really needs is a mathematics teacher to teach him to count," Josh shot back in the good-humored lingo of the golfing fraternity.

They roared in unison at the absurd accusation. Scrupulously honest, both young golfers enjoyed stalwart reputations. The pair flourished under the discipline imposed by Lyle while soaking up the finer points of the sport. Neither had ever been guilty of surreptitiously moving a ball or miscounting a stroke.

Josh Ryan hurried to the men's dressing room and opened the burnished oak locker he shared with his dad. The swinging door exposed a menagerie of golfing paraphernalia buried in a clutter of ball packs and dusty shoes. He busied himself in ritual preparation for the *Match of '96*—the first time he and JT had teamed in more than a year. Both held a competitive edge.

Earlier, the Daniels family had encountered more sinister competition.

Long before JT's birth, Grandfather Brock Daniels smashed a not-so-invisible barrier to Hampshire Club membership. Theoretically, the club offered unrestricted access to applicants who appeared socially respectable and financially successful. In reality, an unwritten code barred admission to all but male Caucasians of the Christian persuasion.

When the Duchess encouraged Brock to join Hampshire, he said "Yes" not because he intended to become an avid golfer or to crash the local scene, but because he felt propelled by the conscience of a trail-blazing social pioneer. Despite his awesome credentials, when Bishop Daniels applied for membership, an ugly undercurrent of preferential treatment and delay kept the application in limbo.

In contrast, the Duchess had inherited her Hampshire Club membership as a birthright of privilege. Behind her back, irreverent detractors labeled the heavy-handed woman, "Iron Pants." Mortals within range of her hard-nosed tactics usually bent to her wishes, knowing resistance was futile. But on the plus side, the Duchess spurned labels, refusing to resort to twisted bias to pigeonhole people. She consistently championed the rights of African-American Magruders

Mill neighbors like the Daniels family, who had earned their spurs as independent farmers since long before the Civil War.

Once Lady Carrington announced her intention to run interference and bulldoze entry into the club for an admired friend, opposition melted. Thus, the Bishop's entry into the club's inner circle of movers and shakers proved a hint of things to come.

Although Grandfather Daniels rarely appeared on the golf course, JT relished the opportunity, earning the plaudits of mentor Lyle Grant, who unabashedly pointed to JT and Josh as "my guys!"

Seeing his pal approaching the range, JT motioned him on, shouting, "I hope you're loaded with cash. I don't take credit." The jest drew a quick retort.

"All I brought is this deposit slip for my account in Silver Spring National Bank. The big-wheel bankers promised your loot would be welcome."

Following the mandatory macho drill of back slapping and high-fives, the two studiously busied themselves whacking strategically placed, 50-yard pitches and sizzling, 225-yard drives.

Each pretended to be woefully out-of-sync. Checking their watches, they sped to the practice putting green across the cart path from the tee-off box of the first hole. They feigned agony at the woeful condition of the putting surface. One suggested that the morning dew slowed the roll of the ball, while the other complained that the surface seemed sunbaked. Despite the hype, both young men mastered finely honed swings.

Exactly five minutes before 10 a.m., Lyle Grant appeared in his monogrammed club cart barking, "Let's go!" The two pals reacted obediently, hopped in a rented vehicle and sped to the tournament-level tee box overlooking a yawning chasm of green foliage.

Once on the course, the pro ordered, "Follow me." Effortlessly Lyle smacked the ball a country mile. The allegedly rusty students followed suit.

By the fourth hole, the composite score included a birdie and a couple of bogies, placing the fast-paced trio hard on the heels of the less-expert foursome they trailed. With no one in sight behind, the pause offered opportunity to reminisce.

Lyle reminded his protégés of their semi-legendary performances on Hampshire's links. "I can see it now," chortled the old pro. He reared back, slapping his knee, savoring the memory. Josh and JT indulged their mentor, listening impassively to well-worn tales as JT attested to the authenticity of the anecdotes. "Our great chief speaks truth—without forked tongue!"

Lyle Grant accepted the left-handed accolade with a grin.

At the seventh hole, the trio again caught up with the lagging foursome and took the occasion to sip the ice water at the tee box. Here, Lyle recalled his favorite golf story.

"I still remember when you guys were twelve or thirteen, earning spending money as caddies. The way Jonathan shut the trap of the big-mouth braggart lives forever as a highlight of club lore."

The teaching pro referred to an occasion when one of Michael Ryan's clients, trying to drum up new business, asked Michael to play host to a notoriously uncouth guest. Michael graciously agreed to humor the client until, too late, he met the ill-mannered pariah. The braggart overindulged in booze, embellished his résumé, and cheated at golf without a qualm.

Michael felt embarrassed to have brought the pompous ne'er-do-well onto the Hampshire links, particularly since he had invited Josh and JT to serve as caddies.

Draped in his mantle of infallibility, Mr. Obnoxious insisted on a wager of $100 per hole, then fraudulently claimed a 22 handicap. A seasoned eight-handicapper, Michael gritted his teeth.

Suspecting the manipulator's calculated tactics, Michael whispered to Josh to be alert and to follow the man's every stroke. JT, caddying for Michael, watched in disbelief as the guest flagrantly nudged his ball to more favorable lies. When Josh inquired if he "meant to do that," the cheater would respond with a profuse apology, claiming an inadvertent mistake.

Thanks to Josh's skill at keeping the fraud within semi-compliance with golf's rules and a near-scratch performance by Michael, the interloper had run up a tab of $500 due his host by the end of the eighth hole. Unable to conceal his distress, Mr. Obnoxious implied with hostility that since he was not accustomed to losing, perhaps the host himself was guilty of fudging his shot count.

Michael Ryan played it cool, refusing to take the bait. Instead he offered the man a sporting opportunity. "Today's your lucky day!" he said. "It happens that I have ten $100 bills in my pocket that I am giving to your caddy to hold. If you care to match it with your $1,000, I'm willing to risk it all on a double-or-nothing proposition, winner take all on the ninth."

The guest hesitated, still debating. While he teetered on the brink of the challenge, Michael threw in the clincher.

"Tell you what! If you'd like, and are willing to forget handicaps and go for gross scores, I'm willing to let my caddy play the ninth hole in my stead—that is, if you're willing to bet against a twelve-year-old kid."

The bait proved irresistible!

The greedy guest drooled with delight, fumbling in his pocket for a wad of $100s and a few $50s to cover the wager. Stuffing the folded bills into the hand of the waiting Josh, he strutted toward Michael, braggadocio restored.

"You've got a deal!"

Then wheeling toward JT with overbearing disdain, he ordered, "Go ahead, kid, you're on!"

Ever the showman, JT added to the ruse with a spin of his own invention. With wide-eyed innocence he inquired, "What club should I use, Mr. Ryan?"

Michael, now playing caddy, responded, "You know, kid, this is a par five. You ought to try the one wood...It's the longest club in the bag." Struggling to keep a straight face, he extracted the club and handed it to the twelve-year-old with the sage wisdom, "It's called a driver. Be sure and keep your backswing slow."

The irrepressible JT chopped some practice swings, displaying uncoordinated form, feigning ignorance about the correct way to hold the club and beguiling uncertainty about where to stand to address the ball.

A smirking Mr. Obnoxious stood by as casual spectator, planning how to spend an easy grand.

Finally, fixed in an awkward stance after much instructional fuss and ado from Michael, JT swung away at an awkward-looking angle. The "lucky shot" sent the ball sizzling into the wild blue yonder, landing squarely in the center of the fairway,195 yards dead ahead.

Turning to no one in particular, a wide-eyed JT queried, "Was that OK?"

The conspirators didn't dare risk looking at each other, or they would have been swept away with laughter.

Mr. Obnoxious gritted his dentures, blaming beginner's luck. His own shot went farther than JT's, then hooked dangerously close to the woods, landing in a snarl of semi-manicured rough.

Moving down the fairway, the charade continued, with JT striking modest hits dead-center. The agitated Mr. Obnoxious fell victim to increasingly errant shots. His third effort threatened a fairway oak, then, like iron filings drawn to a magnet, the ball eventually came to rest, sandwiched cozily among gnarled roots. Still enjoying "beginner's luck," JT's ball bounced to the fringe of the green in three.

The number four shot from Mr. Obnoxious plummeted to earth with a thud, fifty feet off the green. JT nonchalantly dropped a forty-foot putt for a birdie, crowning the performance with the line, "Is it OK, Mr. Ryan, that I sank it in one stroke instead of two?"

That did it!

The now-raging guest unleashed a deep purple vocabulary, wrenched his golf bag away from Josh, and stomped off in a fury. Ever the honorable sportsman, Michael blocked his path long enough to return the guest's entire $1,000, slipping the bills into his palm as he passed. Without so much as an appreciative grunt, the interloper accelerated his retreat, never again to be seen at the Hampshire Country Club.

Lyle Grant finished recounting the adventure, reminding the boys that Michael had used the occasion to treat them to dinner while delivering a lecture on golf courtesy and basic integrity. It went something like "Cheaters never prosper, and real men never cheat!"

Finishing the ninth hole, Lyle broke off his reminiscences, shook hands, and headed back to his teaching responsibilities with a new generation of aspiring teenage golfers. Waving "so long," he urged the two to complete the full eighteen holes. They finished tied at five above scratch.

Reliving old times on the links put the *Monkey II* proceedings on a temporary back-burner. Finally, while heading to their cars, Josh recounted the details of his early-morning confrontation with Dr. Striker. JT didn't bat an eye.

"I'm not surprised, Josh," he said. "Have you thought about joining me in the legal profession?" then quickly added, "You know I'm just kidding. You're destined to be a tiger of a Ph.D., whether or not MAU is dumb enough to let you go. That's their problem, not yours."

Josh acknowledged the bravado of JT's compliment, inquiring pensively, "You think I can still cut it as a witness?"

"Count on it. If you go, I go, too. You've already shown a lot of guts. Now let's enjoy the rest of the ride."

# 10

Josh Ryan spent the balance of Monday afternoon in idle seclusion, alternating between exotic daydreams co-starring Traci Kilburn and cramming for the evening's scheduled show-down. Undisturbed in this joyous reverie, the minutes ebbed away until his grandfather's clock chimed seven times, reminding him that the moment had come to head back to Chamberlain Playhouse.

As he wound up routine chores, Josh checked his E-mail and discovered an unsettling message.

*"Big man on campus. You think you know it all. Keep it up smart boy. You'll get yours—even grandma Iron Pants can't save you. She thinks you're too good for this world anyway!"*

Unsigned, the sinister sentiments baffled more than frightened. Josh's first instinct counseled, "Share this thinly veiled threat with JT."

Not wanting to sound alarmist or to act impulsively, he resisted the impulse, choosing to dismiss the contents as a harmless prank. He crammed a printout of the grim warning into the top drawer of his desk. Unnerved but not intimidated, he headed out the front door of the cottage toward the theater, driven by the curse of inexperienced youth—presumed invincibility.

At the call of Pace Terhune, the players arrived early to update the strategy for the week's entertainment. As Josh approached the MAU campus, he noted scattered crowds milling about in the playhouse lobby, spilling into the manicured courtyard. Striding past the mostly strange faces, Josh encountered occasional sideways glances of recognition. Realizing that the nods were aimed his way, he returned the silent greetings with some misgivings and quickened his stride, seeking the sanctuary of the back-stage entrance.

Inside, cast members appeared fully informed of Josh's private trauma. Most offered reassuring shoulder pats. One bolstering comment caught his attention.

"We're counting on you!" exclaimed thirty-nine-year-old Jessica Saunders as she dropped a motherly pat on his shoulder. Up to now, Josh had viewed her as a convenient prop, albeit an attractive one. He had yet to learn of her impressive credentials as certified public accountant, Olympic sharpshooter, and karate black belt, which belied her diminutive size. He had heard rumblings that she had connections with the Federal Bureau of Investigation. Belatedly, he recognized that the lady exuded presence.

Inspired by the outpouring of support, Josh reaffirmed his determination to testify, come what may, as Judge Stone moved front and center to inform the players of procedural modifications.

"My presence is primarily that of a decorative prop to lend credibility," he acknowledged in resigned good humor, "but I'm going to exert judicial discretion and require that our Darwin be exposed to some unfriendly cross-examination by Ms. Kilburn. Any problem with this?"

When no one objected, he put the crew on notice of a tactical shift. "Last evening, several people called, objecting to a one-sided presentation. I invited two colorful characters to volunteer as witnesses...if that's OK!"

In deference to His Honor, all parties shrugged approval. Pace Terhune beamed, knowing he could not count on a repetition of Karl Striker's fire and brimstone temper tantrum to spice up the remaining five acts.

The amateur cast moved on stage moments before curtain time. Some furniture had been reshuffled. Most notably, the witness stand had been shifted center stage where it attracted the glare of several spotlights. The rising curtain revealed fewer vacant seats than the previous afternoon.

Declaring the court in session, Judge Stone intoned a surprise order: "It has come to my attention that one of the jurors is a longstanding friend and golfing buddy of Joshua Ryan and Jonathan Daniels. While I enjoy absolute confidence in this juror's integrity, I have no choice but to dismiss Mr. Lyle Grant to avoid the appearance of bias. I'm sure you understand the rationale for this action."

A bemused Lyle Grant had expected as much. Reeking authority, Bailiff Terhune strode to the jury box and motioned the ex-juror to follow backstage. The amused audience perceived this as another staged stunt.

While studying the sea of interested faces, Josh did a double-take. Sitting in regal splendor and dripping diamonds, the elegantly attired, 78-year-old Lady Carrington occupied Center Orchestra, Row G, Seat 14. The spot had been vacant the previous day. An avid Redskins fan, cheering for the home team took top priority for the Duchess. Today, and for the balance of the week, Pace Terhune's *Monkey II* pre-empted her dance card.

Left with a jury of seven, the Honorable Judge Stone motioned JT to proceed. He had opted not to write the script for his witness friend. Responding to the judge's earlier instructions, JT chose to lead, inviting a buffet of contentious ideas.

Josh obliged. Not that he felt restrained previously, but this evening he sliced and diced with the delicate skill of a surgeon. He wasted no words dissecting a charade of inaccurate perceptions.

"Darwin's nonsense, garbed in ponderous phraseology and cloaked in more than a century of backtracking advocacy, remains nonsense. Today's unproven conjecture touts repackaged myth. Garbage in, garbage out!"

The audience stirred at the brash jargon. Hardcore evolutionists steamed at the blatant arrogance. Unfazed, Josh continued: "Darwin's recipe portraying a trek from molecule to man remains deafeningly silent about the mechanism spawning the event. Spontaneous generation is blindly medieval, a cruel hoax that sweeps untestable suppositions under carpets of pompous obfuscation."

The audience fell silent. Josh continued, "Did life begin courtesy of random chaos or by intelligent design? The paradox of beginnings puzzles us: Is it rational for us to attribute life's intelligent designs to the whims of chance or luck?"

Surprised by his own eloquence, Josh pressed the attack with show-and-tell grandstanding. "Darwin's intellectual heirs attempt to bridge the quantum gap between inert, inorganic matter and the first-ever organic life by means of a forever missing recipe for *prebiotic soup*."

Scratching his head in pretended perplexity, the aspiring scientist launched a comic quest for the illusive formula. He lobbed one-liners laced with humor.

"Seems as how plain ol' H2O is a rare commodity in our Solar System but manages to inundate 70 percent of the surface of Planet Earth. Good thing, too, since no one expects to find life on Mars unless water exists—no water, no life!"

The audience nodded acknowledgment.

"I guess this means the *soup* must contain water. But what kind of water? Salt water from an ocean? Fresh water from a river? And, how come the difference? Imagine the calamity if that inscrutable prebiotic soup recipe began mixing in some freshwater stream, but eventually reached the ocean and became tainted with salt, scrambling the slimy formula. Then it got recycled by the sun, fell from clouds as drops of rain on some remote stream, there to repeat the journey to biological purgatory!"

Here the witness mounted a vain attempt to mimic a drawling Jimmy Stewart, finishing the water treatment with a splash. "After millions of years of this circular process that always pollutes the mix, even the most optimistic random-chance soup chef might get discouraged and throw in the sponge. Wait a minute now, maybe that's the key: water plus sponge equals *prebiotic soup*?"

Pondering, he let the question hang, clutching his chest in mock shock. "But then, I'm forgetting, the chef didn't have a sponge—spongedom wasn't supposed to debut until millions of years after that first little wiggly something-or-other appeared spontaneously in the soup! And there wasn't supposed to be an intelligent chef either!" The audience clapped as much in appreciation for Josh's feeble attempt to mimic Jimmy Stewart as in reaction to the absurd parody.

Next, shifting to the role of mad scientist, he introduced a motley array of odd-shaped bottles and tubes that fell dismally short of meticulous laboratory practice. In elaborately overstated motions, he made haphazard selections from the elements listed in the Periodic Table of the Atoms and dumped them willy-nilly into an incongruous chemical stew.

The stunt intrigued the jury.

"Starting with *oxygen* and *hydrogen,* wonder of wonders, it appears we have plain old-fashioned water. And of course, since *carbon* is key to organic formulas, it's critical to the recipe. And don't forget a touch of *sulfur* for flavor. Finally, pump in a few bubbles of *nitrogen* and *phosphorous* for flair."

He raised a beaker of bubbling, darkish broth, proclaiming in mock triumph, "There she be!" As to just what it was, no observer had a clue—least of all Josh Ryan. Shrugging, he continued the spoof.

"Now, ladies and gentlemen of the jury, admittedly this basic recipe for *prebiotic soup* is seriously flawed. The percentage of each element may be out of kilter; perhaps a pinch of all ingredients from the Periodic Table of Elements needs to be added for seasoning; and an occasional jolt of lightning or some other energy source; a wet, warm environment, alternating with drying-out periods; and time stretching out to several billion years!

"Since I can't keep you in suspense for that long, you may have to take my word that thanks to random good luck in a sea of chaos, meticulous order prevailed and voilà—the first-ever simple, single-celled life emerged—or so it is claimed."

Like a medieval alchemist, Josh shook the phony concoction vigorously, inviting the audience to scrutinize the ugly brew. "Just think, folks, it may be too early to tell for sure, but if you stretch your imaginations, maybe you can visualize the emergence of that first-ever microscopic cell floating there on the foam of this motley brew. If you believe that, you better hope it doesn't drown, finds something to eat soon, and learns to reproduce—and you better show some respect!

"Theorists conjecture that this could replicate your own common ancestor...the start of your family tree."

JT made no move to interrupt. Traci waited to pounce. Clearing the table of the test tubes, the witness went on: "The brightest scientist has yet to re-create the recipe for prebiotic soup or the simplest living cell from scratch. Prebiotic soup is nothing more than a fancy label for 'I don't know.'

"Alas, the mysterious formula for soup defies detection. Technology unheard of in Darwin's day has yet to solve the riddle as to the *what, when, where,* or *how* of that threshold beginning event.

"Where is the evidence, fossil or living, of a transitional cell evolving from a no-nucleus prokarya en route to becoming a more complex eukarya cell with a nucleus? What does mega-evolution do with studies of microbial genomes that show anomalous genetic sequences that render the roots of the conjectural ancestral tree indecipherable?

"'*Some argue that these events from 3 billion years ago will always be a mystery.*'"[15]

As the witness moved into high gear, the audience listened intently; Producer Terhune stood guard, mastering the guise of neutral, impassive bailiff; Traci

Kilburn, vested with cross-examination authority, had yet to seriously object; and JT's cat-that-swallowed-the-canary expression signaled his reluctance to intrude.

The judge studied audience reactions. No stranger to courtroom antics and manufactured drama, Edward Anthony Stone settled back, patiently monitoring the dog-and-pony show, temporarily setting aside any inclination to intervene. He didn't have long to relax.

Traci Kilburn chose this moment to stir the stew. "Objection, Your Honor! With all due respect to the expertise of the witness and his demands for evidence, he is flaunting personal opinion and staging phony experiments. Where's the hard evidence he demands?"

The judge passed the query to JT Daniels. "What's your position, Counsel?"

"It seems to me, Your Honor, that counsel will have every opportunity to challenge the witness during her cross-examination. The witness is challenging the glaring absence of verifiable empirical evidence that supports the spontaneous generation postulate. Rather than attempting to prove any theory of his own, he is arguing that Darwinian teaching is deficient."

Traci countered, refusing to back off. "With all due respect for Counsel's learned logic, evidence—or lack thereof—is the ultimate test. In the eminently quotable phraseology of this witness: *garbage in, garbage out.*"

The audience loved the exchange. JT fired back: "If the 1925 *Scopes* trial stands as a touchstone, it's relevant that Clarence Darrow built a case primarily on opinion and conjecture. The expert witness in this trial deserves comparable latitude."

Judge Stone, too, savored the stimulating exchange, finally intervening to rule, "Objection sustained."

Traci beamed, barely concealing her satisfaction. JT reacted mechanically, plunging ahead. "You claim no one has the soup recipe, but what hard evidence can you produce to show that the so-called simple cell is anything but simple?"

Josh whipped out a file card with a lengthy quote from microbiologist Michael Denton as deserving "citation in the court record." He dispensed copies to attorneys, judge, and jury.

"*'To grasp the reality of life as it has been revealed by molecular biology, we must magnify a cell a thousand million times until it is 20 kilometers in diameter and resembles a giant airship large enough to cover a great city like London or New York. What we would then see would be an object of unparalleled complexity and adaptive design. On the surface of the cell we would see millions of openings, like the portholes of a vast spaceship, opening and closing to allow a continual stream of materials to flow in and out. If we were to enter one of these openings we would find ourselves in a world of supreme technology and bewildering complexity...The simplest of the functional components of the cell, the protein molecules, were astonishingly, complex pieces of molecular* machinery, each one consisting of about 3,000 atoms...What we would be witnessing would be an object resembling an immense automated factory...larger than any city and carrying out almost as many unique functions as all the manufacturing activities of man on earth...a factory which would have one capacity not equaled in any of our own most advanced machines, for it would be capable of replicating its entire structure within a matter of a few hours.'"[16]

"Bottom line: Simple life forms in molecular systems exist '*irreducibly complex.*'"[17] Dar Ryan quipped his endorsement of the Michael J. Behe catchy phrase, claiming, "The only thing simple or primitive in the world of microscopic organic life is the pointless attempt to impose the prebiotic soup mix as a rationale to explain its beginning."

The assertive witness continued without a pause: "Viewed from space, the still-sterile surface of the moon, hostile to organic life, rotates in stark contrast to the striking blue-and-white marbled colors of Planet Earth. Why *spontaneous generation* on the Earth and not the moon?"

To answer his own rhetorical question, Josh returned to his previous day's reference to the cosmic convergence essential for survival of organic life. He summarized the unlikely probability of finding a galaxy, star, or planet offering a friendly environment to life within demanding parameters, by chance, such as: mass, color, location, and luminosity of stars; Earth's orbit inclination and axis tilt; crust thickness, gravity, magnetic fields, land/water ratio; and ratios of carbon dioxide, oxygen, ozone, and nitrogen in the atmosphere.

"Too much or too little of these delicately balanced environmental factors dooms any prospect of organic life. The probability ratio for the successful confluence of all parameters approaches 10-42. Life-friendly ecology demands flawless equilibrium."[18]

Drawing on authoritative scientific literature, the witness took pains to deliver the evidence demanded by cross-examining counsel.

"Even if all conditions mandatory to sustain life existed in perfect calibration '*To get a cell by chance would require at least one hundred functional proteins to appear simultaneously in one place. That is one hundred simultaneous events each of an independent probability which could hardly be more than 10-20 giving a maximum combined probability of 10-2000.*'[19]

"'*If one were to take the simplest living cell and break every chemical bond within it, the odds that the cell would reassemble under ideal natural conditions (the best possible chemical environment) would be one chance in 10100,000,000,000...If all the matter in the visible universe were converted into the building blocks of life, and if assembly of these building blocks were attempted once a microsecond for the entire age of the universe, then instead of the odds being 1 in 10100,000,000,000, they would be 1 in 1099,999,999,916.*'"[20]

JT invited Josh to comment about life's dependency on the sun's radiant energy.

"Without the protective ozone shield wrapped conveniently around Earth, life could be overpowered by ultraviolet radiation," Josh declared. "Sunlight, radiating ultraviolet and infrared, sustains life in consistently reliable measurements. Instead of overpowering heat or uncontrolled, deadly radiation, energy from the sun comes calibrated in a range to support animal and plant life. If its relationship to the electromagnetic spectrum shifted imperceptibly, the chance of life could vanish.

"Equally stunning is the mutually predictable orbits balancing sun and Earth, suspended in a spatial pattern so precise that the exact proximity can be pinpointed a thousand years past or future.

"A collateral bonus of sunlight is the riot of hues and colors that paint Earth's environment. Restful sky-blue shades, backlighting forests of multi-hued greens, define the landscape. Golden red kaleidoscopes traced by the path of a daily rising and setting sun deliver a rainbow of accents. Combinations conspire to assure psychological peace.

"Too much, too little, too far, too near, too late, too soon—any factor out of kilter and single-celled organic life could not exist. But even that's not enough."

Josh explained that molecular bonding essential for life required the presence of no less than 40 different elements with successful bonding contingent upon the functioning force of electromagnetism operating within a delicately balanced electron-to-proton mass ratio.

Reflecting instincts honed by exposure to gambling odds, he cited the universe's fine-tuning mix of protons and electrons and introduced an independent observer's view that *Unless the number of electrons is equivalent to the number of protons to an accuracy of one part in 1037, or better, electromagnetic forces in the universe would have so overcome gravitational forces that galaxies, stars, and planets never would have been formed.*"[21]

Josh went on to offer illustrations that underscored the outlandish dimension of such odds: "A class-five hurricane can scramble and devour telephone lines in its fickle path of raging fury. No one has ever seen this fearsome force use the random jumble of metal strewn in its twisted wake to create a fully operational TV set or personal computer—much less manufacture self-replicating models!

"But don't rush pell-mell to judgment! Accidental creation of an operational TV by a hurricane's random havoc seems less than farfetched compared to the chance of a living cell spawning itself from slimy chaos thanks to *spontaneous generation*!

"The magic elixir powering unproved and unprovable *spontaneous generation*? Random-chance advocates think if they allow enough time, anything can happen! Hello? Anyone home on the range?"

To clinch the mathematical impossibility, Josh switched to a different illustration.

"'*Cover the entire North American continent in dimes all the way up to the moon, a height of about 239 thousand miles...Next, pile dimes from here to the moon on a billion other continents the same size as North America. Paint one dime red and mix it into the billion piles of dimes. Blindfold a friend and ask him to pick out one dime. The odds that he will pick the red dime are one in 1037. And this is only* one *of the parameters that is so delicately balanced to allow life to form.*'"[22]

"You're saying you're not placing any bets on the red dime?" interrupted JT.

Josh rolled his eyes heavenward. "Not one red cent, Counsel!" The pun went over the heads of most listeners but drew a chuckle from Judge Stone.

"But the red dime odds may look like a winner when compared to the one chance in 10161 that a single useable protein might just happen '*even if all the atoms on the earth's surface, including water, air, and the crust of the earth were made into conveniently available amino acids and four to five billions of years were involved.*'[23]

"Speaking of impossible odds, there's one calculation in play that suggests the random chance '*for producing the necessary molecules, amino acids, proteins...for a cell one-tenth the size of the smallest known to man is less than one in...10 with 340 million zeros after it.*'[24]

"Counting on *spontaneous generation* to evolve a single cell from *prebiotic soup* that becomes the common ancestor of the human race is more foolhardy than attempting a space walk to Jupiter—unaided by a space ship or other mechanical means."

Josh hurried on, intent on impressing listeners with his next bombshell.

"The *fully oxidizing* nature characteristic of Earth's prehistoric atmosphere is hostile to the assembly of new life, creating a chemical dilemma. Atoms and molecules tend to bond with oxygen atoms. Free oxygen inclines to oxidize organic compounds. Oxidizing conditions are reportedly thirty million times less efficient than reducing conditions, decidedly unfriendly to the processes producing amino acids and nucleotides.[25]

"Life demands a reducing atmosphere, where atoms and molecules bond with hydrogen rather than oxygen. Simply put, the amino acids essential to mega-evolution's formula for generating organic life requires an oxygen-free, reducing environment—a condition missing when *spontaneous generation* was supposedly doing its magic!

"Oxidation renders random chance generation of life impossible '*Oxygen destroys the chemical building blocks of life...*' This creates a classic catch-22. If in fact the '*early atmosphere was oxygen-free,...then there would have been no protective ozone layer. Any DNA and RNA bonds would be destroyed by UV radiation...Either way, oxygen is a major problem.*[26]

"Oxidizing atmosphere and *spontaneous generation* cannot coexist!" Glancing sideways at Traci, he noted triumphantly, "By the way, we are talking hardcore evidence here, not witness opinion."

Judge Stone tapped the gavel, gently admonishing, "The record will speak for itself, Mr. Ryan—without unnecessary reference to Counsel's objections."

Duly chastised, Josh pushed forward with another perceived watershed disclosure. "In an oxidizing environment *'the minute amino acid production would almost entirely be composed of the simple acid glycene. The more complex acids that are also needed would be virtually missing.*[27]

"The amino acid barrier to *spontaneous generation* is another nail sealing the fate of Darwinian myth. Artificially created amino acids manufactured from non-biological systems produce two forms: a mirror image *left* and *right* in essentially equal amounts."

Josh reiterated the message like a patient professor introducing college freshmen to Science 101.

*"'Amino acid, when found in nonliving material...comes in two chemically equivalent forms. Half are right-handed and half are left-handed—mirror images of each other. However, the amino acids in life, including plants, animals, bacteria, molds, and even viruses, are essentially all left-handed. No known natural process can isolate either the left-handed or the right-handed variety.*

*The mathematical probability that chance processes could produce merely one tiny protein molecule with only left-handed amino acids is virtually zero.'"*[28]

Josh addressed the jury earnestly. All seven citizens in the jury box listened with fascinated attention.

"This left/right conundrum is no incidental footnote relating to flawed theories of organic life origin. It's landmark stuff! For reasons unknown and perhaps unknowable to sophisticated scientists, amino acids never appear ambidextrous in living systems! It's always either/or and never a mirror image containing both left and right! That's a resounding *never!*"

Listeners grasped his passion, unaware that the witness was struggling under the cloud of excommunication from a doctoral candidacy. Josh topped his virtuoso performance with the challenge:

*"'The tiniest bacterial cells are incredibly small, weighing less than 10-12 gms; each is in effect a microminiaturized factory containing thousands of exquisitely designed pieces of intricate molecular machinery, made up altogether of one hundred thousand million atoms, far more complicated than any machine built by man...The size, structure, and component design of the protein synthetic machinery is practically the same in all living cells...no living system can be thought of as being primitive or ancestral with respect to any other system, nor is*

*there the slightest empirical hint of an evolutionary sequence among all the incredibly diverse cells on earth.'*[29]

"The most sophisticated laboratories consistently fail to manufacture the simplest living cell exclusively from inorganic matter. It has *never* happened—ever! Several billion years is not enough time to compensate for impossible odds. The cornerstone essential to making evolutionary theory plausible does not rest with random chaos. Spontaneous generation consists of misleading conjecture presently unproved, forever unprovable."

Seeing the witness reach an emotional peak, Judge Stone ordered a recess with all the stern decorum characteristic of his real-life courtrooms.

# 11

Refocusing backstage, Josh glimpsed the edge of an envelope poking under his dressing-room door. Curious, he picked it up, fingers gently tracing the embossed family coat of arms emblazoned with the name *Morgan*. It was the first written message he could remember receiving from his grandmother, Lady Regina Ann Morgan-Carrington.

Formal but friendly, the handwritten message summoned "Mr. Joshua Chamberlain Ryan" to travel to Morgan Manor, 10 a.m. Tuesday, September 3, 1996, for a first-ever, personal visit. The gold ink signature sparkled, "Affectionately, Regina Ann Carrington."

Slipping the watershed communication into his jacket, the recipient felt only mild surprise. He couldn't begin to fathom the plot or the motivation behind Her Majesty's decree announcing a command performance. He had another twelve hours to cogitate about what the Duchess had in mind. Postponing conjecture for the moment, he responded to another curtain call. Whispered chatter from the sea of observers faded as the curtain rose for the last act of *Monkey II's* second day. Judge Stone dutifully gaveled the scene to order. Josh noted that the Row G orchestra seat, earlier occupied by the beaming Duchess, had been vacated.

Shifting his attention to the impending cross-examination, he caught a glimpse of a bespectacled, vaguely familiar face. The visitor sat hunched over, busily jotting notes. By Tuesday evening, Josh and the stranger would cross paths; by Thursday evening, they would cross swords.

Traci Kilburn's entrance, as she strode to a spot adjacent to the jury box, distracted him from the curious scribbler. She sought out a site where she could simultaneously eyeball the witness and assess jury reaction. The instant she crossed the threshold of the make-believe courtroom, the stiletto directness of her inquiry warned that on stage, friendship took a back seat to professionalism.

"Is it not true, Mr. Ryan, that you have testified to this court that empirical evidence is the fundamental foundation to scientific fact?"

"Yes, yes...of course!"

The witness struggled to concentrate, given the beauty and grace of the interrogator and the fragrant scent of her intoxicating presence. She stood impeccably clad in a form-fitting but conservative navy-blue suit with long sleeves, softened by white collar and trim. Though for months, Josh had been aware of her beauty, he felt overwhelmed. He struggled to avoid coming across as a blathering, tongue-tied putty man.

Her rapid-fire questioning banished abstract reveries, returning him to the reality of his role. "Still, Mr. Ryan, is it not true that much of your testimony has consisted of your personal opinion?"

"Yes, but—"

She cut him off.

"Do you expect this jury to value your opinion any more than Darwin's?"

"Why not? Who wants to be a monkey's uncle?"

The instant he resorted to flip bravado, he regretted the sarcastic tone. Regaining his composure, Josh felt his intellectual sea legs returning. He chose his words with care. The audience loved the verbal jousting. No one other than JT, and possibly Judge Stone, had the remotest clue that the duelers were exceptionally good friends.

Neither blinked, neither retreated.

Both earned their spurs, basking in the glow of an animated audience. She taunted and teased, suggesting maybe he lacked the daring to be a successful riverboat gambler, with all his talk of impossible odds.

Ignoring the bait, he responded, "Thanks for the compliment!"

She reminded him that a chorus of big-name scientists representing major academic disciplines touted evolution as *fact*. He countered that a similar audience of "experts" once scoffed at Copernicus' claims that the Earth orbited the sun rather than the reverse.

"Majority vote works in politics but is a poor substitute for empirical evidence in science."

At this point Traci launched into a demanding examination of Josh's negative reaction to the basis of Darwinian dogma. "Do you categorically reject the notion that today's complex life forms emerged from a single, simple-cell origin?"

"I reject banal generalities masked with flawed ambiguity. Microscopic size doesn't equate to simplicity. Darwin's fairy tale conveniently omits an introductory chapter explaining—with empirical evidence—the *how, when, what, where*, and *who* of that alleged common ancestor.

"Where's the proof that cells with an organized nucleus evolved from a no-nucleus bacteria, or something similar?"

He spread his arms wide, palms open, acknowledging his own ignorance of the unknown. "Darwin suspended his academic roof in intellectual space without visible foundation! It's like saying, 'I'm planning to live forever...so far, so good!'"

"Surely you recognize that modifications can occur, given enough time?"

"Of course! Micro-variations flourish, exploiting the genetic potential inherent within each prototype. Darwin never introduced evidence proving that a bacterium, or other single cell, is the common ancestor for radically different taxons—for example, a several thousand pound pachyderm or a Monarch butterfly weighing in at 1/100th of an ounce!"

"But wouldn't you admit that *survival of the fittest* is a fact of life? Don't the healthiest and strongest prevail?"

"First, define *fittest*. Muscular strength? Biggest? Smartest? Best-looking? Most efficient reproductive capacity? If it's repro, bacteria win, hands down. Whatever the definition, *survival of the fittest* does not extend to producing an entirely new prototype."

Abruptly departing from this line of questioning, Traci launched a tongue-in-cheek assault designed to expose witness fallibility.

"You've made a big point about impossible mathematical odds, so I assume you know something about mathematics?"

"It's not my major, but sure, I know the answer to 2+2!"

Again he regretted his glib impetuosity. He worried that Traci might view him as slick, but she ignored the sass.

"Do you put these odds-making skills to work to your financial advantage?"

"Do you mean do I ever bet real bucks? Yep. Guilty as charged. Since coming of age, I've been known to place a few bets at Pimlico racetrack."

"Is that all?"

"Sometimes I try to outguess the odds at a Vegas 21 table."

"And no doubt your considerable mathematical skills have made you filthy rich?"

Josh scoffed, appreciating the spoof. "Don't I wish! I've been as good as the next guy at losing. While I admit to feeling pumped by the excitement, I've never struck it big like my legendary great-great grandfather, Sam Houston Morgan."

"Hmmm...Let's see, now. Fallible at gaming but infallible in the cosmic realm?"

The audience tittered. The witness fumed.

"I have never claimed infallibility!"

The contrite retort came with a flush.

"So, would you say that the odds you quoted against *spontaneous generation* are worth a \$2 bet?"

Josh's audacious determination returned. "If anything, the odds quoted may be conservative. So outrageously remote and impossible, not even an inebriated amateur gambler in Vegas could be persuaded to drop a quarter in the slot of a slot machine offering no chance to win. The insurmountable problem for mega-evolutionists is this: Spontaneous generation is unverified myth! Evidence confirms *life from life, like from like*."

Josh shifted to the philosophical. "Conventional scientific thought proposes that an inorganic system comprised of Earth, sun, and moon originated four-and-one-half billion years ago. They offer no grain-by-grain, atom-by-atom gradual process that takes billions of years, but an *event!* If you pinpoint a date, it sounds more like a happening than a process!

"Darwinists and interventionists can agree on one thing—there was a *beginning* event! It's the *how* and the *when* that fires the dispute. The corollary issue asks whether the first-ever organic life on Planet Earth could also have been a *sudden event* rather than an ill-defined, protracted process. After all *'If life were spontaneously generating 3.5 billion years ago, we could expect to see it doing so today.'*[30]

"Do you doubt the 4.55 billion year age of inorganic Earth?"

"My opinion is that inorganic matter is very ancient and that organic life appeared after-the-fact, relatively recently!"

"And you fault Darwin as a *master wriggler*? Get real, Mr. Ryan!"

Wordlessly, he admired her spunk. She continued: "Rather than beat around the bush with high-sounding speculation, can't you stick your neck out for once and try being specific for the jury? Do you or do you not believe in the authenticity of radiometric dating?"

"The dating of rocks is not an exact science!"

"Is it too much to ask for proof, Mr. Expert?"

"I'm no expert, but no less an authority than Gunter Faure warns of *discordance* in isotope geology.

"*'Unquestionably, 'discordance' of mineral dates is more common than 'concordance'...the mineral dates generally are not reliable indicators of the age of the rock...Although examples of nearly concordant U, Th-Pb dates can be found in the literature...in most cases U-and Th-bearing significance is questionable.'*"[31]

"Can you answer without trying to impress us with high-falutin' language?" Traci challenged.

"Geochronology requires more than a conjectured beginning fashioned from unproven assumption. Mineral dates are not reliable if the radiometric time clock did not reset to zero when the sedimentary deposit occurred!"

"*'Radioactive decay...is badly compromised as a historical timekeeper, because it is not the rate of decay that is being measured by the amount of decay products left. For this reason, all radioactive methods of geochronometry are deeply flawed...Accuracy in the measurement of elapsed time requires that the process does in fact remain constant...the starting value of the clock...(how tall your candle was before it was lighted)...and certainty...that some external factor cannot interfere with the process.'*"[32]

"Do you have any *external factors* in mind?"

"Could there once have been more oxygen in the atmosphere with some kind of vapor envelope cutting the intensity of ultraviolet rays reaching the Earth? Some claim that such an environment would make organic life grow larger in size and last longer in time! Could that also interfere with the process of radiometric measurement?"

"Mr. Ryan, where are the *badly compromised* dates you glibly allude to?"

"In 1973, Richard Leakey discovered what he believed to be a human skull in Kenya below rock *securely dated* at 2.6 million years BP (before the present). Radiometric dating of the KBS Tuff site has ranged discordantly from.52 million BP to 17.5 million BP[33]—not exactly a vote of confidence for measurement precision."

"Is it asking too much for you to cite more examples of discordant dates?"

"Ask away! Washington's Mount St. Helens blew its top in 1980. Preliminary attempts at radiometric dating indicate an ancient lava dome—with time measurements ranging from 340,000 to 2,800,000 years in the past![34] That's quite a discordant stretch for a mere 16-year-old event—don't you think, Ms. Kilburn?"

"So what's your point?"

"The point, Counselor, is discordance! What age do you prefer? Sixteen years? 340,000 years? 2,800,00 years? Since the newspapers reported the event in my early lifetime, my money rides on the 1980 date."

"Regardless of discordant results, surely you don't quibble with radiometric decay rates?"

"I quibble with discordant radiometric reads on rock!"

"You're so glib—surely you can be more specific!" Traci challenged.

"*If we take the measure amount of helium 4 in the atmosphere and apply the radioactive dating technique to it...we find that the calculation yields an age for the Earth of around 175,000 years.*"[35]

"Are you trying to say Earth is less than 200,000 years old?"

"I'm trying to say the jury is still out as to radiometric dating: *'using [Willard] Libby's own data, the age of the atmosphere is around 10,000 years!'*[36] This could mean Planet Earth originated before the atmosphere necessary to sustain organic life appeared!"

"Are you claiming Earth is only 10,000 years old?"

"How could I know...I'm only twenty-three, not 10,000 years old! I'm saying life on earth without an atmosphere would be short and not so sweet!"

"Surely you don't expect this jury to believe that occasional isolated anomalies prove discordance?"

"These are not isolated anomalies!"

"For example?"

"Earth's decaying magnetic field, for one! The data doesn't jibe with projections of organic life on the planet going back millions of years."

"You'd better enlighten us, Mr. Ryan."

"If the half-life of the magnetic decay rate is 1400 years and heat dissipates from '*circulating electric currents,*' then '*only 50,000 years in the past, the heat generated in the core...would have been too great for life to have been possible on the surface'*...Consequently *'the decay is not likely to have begun more than 10,000 years ago.'"*[37]

"Is that all?"

"Coloration imprints of radioactivity known to have a fleeting existence can be seen when thin slices of some granites, common to rocks in the earth's crust, are viewed under a microscope. For the imprints to form required those granites worldwide to crystallize almost instantly at the precise moment the tiny radioactive specks were sprinkled throughout their massive volumes.

"Radio-halos, *'the signatures...of the Uranium 238 series'* inhabit igneous granite. Polonium 218 *'occurs about midway through the overall uranium 238 decay process and has a half-life of only 3.05 minutes.'* Searching with an ion micro-probe, Robert V. Gentry spotted *'daughter elements without a trace of the parent...'* leading to the conclusion that *'the decay process began with polonium 218.'"*[38]

"Please interpret for those of us inhabiting Earth's surface!"

"With such a short half-life, the decaying process of polonium 218 could not form these radio-halos unless the rock was already in a solid state. Granites and imprints both had to form suddenly and simultaneously!

"Mega-evolution's uniformitarianism cannot co-exist with igneous granite originating in the solid state. Since it's an either/or, mutually exclusive scenario, scoffers snub Gentry's idea as flawed—on the rationale that the radioactive precursor of polonium did not leave behind any traces of its alpha particles."

"Would you mind speaking English?"

"I'm not an English major, Ms. Kilburn, but here goes: Unless Gentry's findings can be demolished, the liquid base of uniformitarian's gradualism theology evaporates!"

"Intriguing! Now how about chapter-and-verse specifics?"

"You lawyers talk of the exceptions that eat up the rule! Radiometric dating is loaded with discordant results that scramble the rule."

Josh lapsed into his lecture mode.

"*'Illustrating the difficulties inherent in the potassium-argon method, scientists have obtained ages ranging from 160 million to 2.96 billion years for Hawaiian lava flows that occurred in the year 1800[1].'*[39] A cross-section of lava specimens taken from New Zealand's Mt. Ngauruhoe volcanic eruptions in 1949, 1954, and 1975 show potassium-argon dates ranging from 270,000 years to an ancient 3,500,000 years before the present.[40]

"Even if the radiometric date proves as accurate as my Timex, does a mix of sediment from diverse sources, with multiple ages, provide a reliable clue as to the actual time the mix occurred? If a cataclysm buried a saber-toothed tiger, does the age of the fossil tiger reflect the age of the catastrophic event or the age of the inorganic matter that engulfed it?"

"The question remains: *Does the rock creating the fossil cemetery necessarily represent the actual date of the event that buried the fossil?*"

"Presumably you can answer the question, Mr. Ryan?"

Attempting a show of sophistication, he prefaced his response with a broken-accented, "Con mucho gusto!"

"Translated any way you choose, the answer remains, *'No!'* A creature trapped on the slopes of Mount St. Helens by 1980's torrent of lava could hardly qualify as a 2.8 million-year-old fossil!

"The volcanic island of Surtsey, just south of Iceland, first surfaced 33 years ago in the Atlantic—November 14, 1963. Less than four years after its birth, a visitor flew in to inspect the pristine scene. What he observed astounded him *almost beyond belief.*

"The brand-new piece of geography appeared aged and weatherworn. Already the sea had ground out black sand beaches. Multi-layered cliffs composed of a series of lava flows guarded the more than four-mile coastline, carved by the Atlantic's tides. Foot-long stalactites hung from the ceilings of caves. Basalt blocks appeared as rounded boulders, chiseled by the elements. Sea gulls, insects, and three species of plants called Surtsey home.[41]

"That describes the scene glimpsed by a visiting paleontologist thirty years ago. A casual visitor today might inadvertently presume the island to be several thousand years old. The question remains: *Does Surtsey's source material reflect a radiometric date that coincides with the 33-year history of the event?*"

"Well?"

"Well, what?"

"What's the answer, Mr. Ryan? Let's not play games! You're the so-called expert!"

"The answer is, 'I don't know.' But it's a doozy of a question! If the date of the *material* doesn't match the date of the *event*, this wreaks havoc with the assumption that the radiometric date of the material that buried a fossil represents the date of a more recent burial event!"

"But surely you must admit, Mr. Ryan, that fossils in the geologic column substantiate conventional radiometric dating results?"

"You've just described a merry-go-round of circular reasoning foisted on geochronology. It's not exactly front-page news that dates attributed to fossils sometimes represent guesstimates: strata dates contingent upon fossil content or fossil dates contingent upon time allocated to a slice of the geologic column. Sometimes the presence of a fossil is used to date strata; on other occasions, the date assigned strata is used to date a fossil. Either way, unproved assumptions assure flawed results.

"Surely you respect the accuracy of Carbon-14 dating of organic matter?"

"I'm curious as to the apparent conflict between Carbon-14 dating and Accelerator Mass Spectrometry methodology. For example, AMS dates the Los Angeles Man fossil at 3,560 years in the past while the Carbon-14 method produces a date of 23,600 years before the present.[42]

"Does that prove the Carbon-14 date is wrong? Did it never cross your much-touted scientific mind that maybe the AMS date is flawed?"

"I'm just raising the issue, Counsel. Other anomalous readings raise an eyebrow or two. '*Living mollusk shells have been dated by the C-14 method at up to 2,300 years...a freshly killed seal at 1,300 years, and wood from a growing tree at 10,000 years.*'"[43]

"And your wise insight can unscramble this confusion?"

"Ask the mega-evolutionist! I plead major-league ignorance about any methodology loaded with discordancy."

Traci appeared to ease up before sending a zinger.

"You categorically reject Darwinian thought as *fact* or *testable science*. Perhaps you're being as opinionated as the philosopher/naturalist you accuse?"

Josh responded without a hint of arrogance.

"'*The evidence should be so compelling that it convinces even the most serious sc[k]eptic...science must admit what it does not or cannot know.*'[44] Count me a serious skeptic."

Pretending exasperation, Traci interrupted him. "Aren't you willing to credit the long-gone Darwin with anything positive?"

"Sure! His *it may be inferred* opinions riveted world attention on pre-history and accelerated serious inquiry into life's beginning. But a darker downside outweighed the plus. Meandering assumptions planted the seeds of elitism and social Darwinism.

"Remarkably, evolutionists exist who do not worship at the shrine of Charles Darwin. Søren Løvtrup pulls no punches: '*I believe that one day the Darwinian myth will be ranked the greatest deceit in the history of science.*'[45]

"In my book, that *day* has arrived! A growing cadre of young scientists will step into the next millennium, prepared to call mega-evolution's bluff!"

Abruptly, Traci announced, "No further questions."

The judge and JT exchanged glances, both marveling at the powder-puff treatment granted the witness on cross-examination. For JT, the truth behind the soft-touch sparkled bright as the morning dew. Judge Stone surmised as much. Shades of a made-for-TV soap opera!

Could the glamorous lawyer be as romantically smitten as the witness? Would Traci join Josh's uncharted trek to a happily-ever-after promised land? The savvy Edward Anthony Stone recalled the counsel given to him in his youth by a worldly wise uncle: "Never underestimate the power of a woman!"

Interrupting his own reverie, the judge inquired, "Mr. Daniels, do you have further questions?"

JT stood courteously to offer his "Not at this time, Your Honor."

With that, Judge Stone rapped his gavel, reminding observers, "This court will reconvene at 8 p.m. tomorrow, Tuesday, September 3."

The curtain dropped.

# 12

The jury departed immediately. Pace Terhune called the other members of the cast together for a back-stage confab. "During the intermission, I received a telephone call from none other than Dr. Karl Striker. He advised me that this *sham* has gone too far, even for entertainment value."

Smiles greeted the unsurprising announcement.

"He also assured me that he would stop the show if he had the authority, but lacking that, he has issued a challenge. Rather than the *three musketeer* student performers he sees making a mockery of science and the legal process, he proposed, at his expense, to invite one of the nation's pre-eminent criminal defense lawyers to headline the act."

"Who?" asked Edward Anthony Stone.

"Marcus Brogan."

"I could have guessed as much," observed the judge. Then, in an aside to Josh, he added, "After the intermission, I noticed you looking at the gentleman taking notes in the audience. You've now seen Attorney Brogan in action—in case you were wondering."

"So what do you all think?" Pace sought consensus.

Traci and JT shrugged, revealing no sentiments pro or con. Jessica Saunders, the quiet court clerk, brightened.

"The audience should welcome Brogan's appearance. The media will jump at news of a national name. After all, maybe Dr. Striker's bark is worse than his bite? We do work for the same university, and I have no objection to assuaging the dear man."

Gazing levelly at the student witness, Jessica inquired, "How does it strike you, Josh?"

Whether or not he shared the *Monkey II* spotlight with the notorious advocate mattered not a whit to Josh. Last spring, he had committed to play the role of expert witness, and he intended to hang tough, in spite of the fiery verbal darts likely to be hurled by the tiger litigator.

"Does this mean we're now ready for prime-time TV?" he joked lamely. When the comment drew no immediate laughs, he quipped, "In the words of my Grandfather Ryan, 'I might as well be hanged for a goat as a lamb!'"

Smiles broke into relieved laughter. Most welcomed the rare opportunity to occupy the same courtroom space with the flamboyant Brogan.

Eyeing the two future lawyers, Judge Stone advised, "Exposure to Attorney Brogan's style may be worth the equivalent of a law school lecture series." JT and Traci nodded agreement. The judge shifted his attention back to Josh.

"As for you, young man, remember not to get riled or to take insults personally. It's just a mind game, played by a pro!"

Edward Anthony Stone concluded with a nod to reality. "And remember, at 500 big ones per-hour, the crafty counselor will dance more than an Irish jig."

Pace tasted elation. "With four days to go, the remaining issue is *when?*"

Judge Stone's calm demeanor emerged again.

"Tomorrow is too soon, and you certainly won't want to give him the last word on Friday. That leaves Wednesday and Thursday. Unless you want to see the entire scenario turn into preposterous hyperbole, you should offer him a one-day-only appearance—take it or leave it."

Pace Terhune signed on to the strategy.

"Dr. Striker's demands should be met with a single appearance...The publicity value will bolster public interest for the balance of the week!"

The win-win news seemed to tickle the funny bone of the sly promoter. With the die cast, Josh weighed in with his own wish list. "If it's a single day, I'd prefer Thursday. That's when I plan to hype my favorite subject—fossils!"

All heads nodded consent. Marcus Brogan's appearance could provide some spice for the audience, and the appeased Striker should jump for joy. As the cast adjourned, Terhune motioned them back into the huddle, preparing to share a more distasteful footnote to the agenda.

"Oh yes, I'm sorry to keep you, but there is a P.S. to the Striker request. He insists that third-year Science Division grad student Slade Lassiter be given a day in the sun as an additional witness—to, as he put it, *provide desperately needed balance to the circus.* Specifically, he demands that Lassiter share the stage Wednesday evening—in the lingo of his own indelicate phraseology, *before the Mid-Atlantic University science program becomes irretrievably discredited by this entertainment travesty.*"

Pace accompanied news of the demand with another quest for consensus. "What do you all think?"

All eyes focused on the primary target of the disgruntled professor. Josh didn't flinch. "Bring him on. Makes no difference to me if a Striker sycophant is willing to abandon his personal dignity and risk exposing the shallow veneer of his book larnin'—providing, of course, the about-to-be-legendary Jonathan Thomas Daniels gets to cross-examine."

"I predicted to Terhune that you would not object, young man. As to the possibility of cross-examining a potentially hostile witness, I have no doubt Mr. Daniels will be more than up to the task."

Peering at JT and Traci over the top of half-lens horn-rims, Judge Stone invited their reaction.

"How do you two young lawyer types feel about sharing the stage with this unpredictable witness? Understand, it would be Wednesday evening only."

Traci said: "In the words of Mr. Ryan, *makes no difference to me.*"

JT's comment predicted everything short of the rockets' red glare. "I can hardly wait to meet the esteemed Mr. Lassiter. In the words of Marshal Wyatt Earp, *Welcome to the OK Corral.*"

The unanimous support pleased Pace Terhune. "Heightened suspense can't hurt the box office. You people are a class act. My thanks to each of you. Sleep well, my friends."

Pace darted out the door, bound for a desktop computer and a suggested headline for the *News Press* Tuesday edition: *Tiger Brogan Joins Monkey II Battle.*

As the informal consultation dissolved, the maintenance crew moved in. Before Josh could break away, the petite Jessica Saunders tapped him on the shoulder.

"If you don't mind, Josh, I have a favor to ask."

"Sure."

"A major donor may deliver big bucks to Mid-Atlantic University in the reasonably near future. A behind-the-scenes debate rages as to whether this anonymous benefactor would be happier with a building honoring the memory of an unnamed friend or underwriting a new science foundation entity to fund Dr. Striker's research programs. As we speak, the gift teeters precariously in the negotiation stage. I trust you'll keep the scoop under your hat."

"I don't see why you need me, since my relationship with King Karl wallows in big-time disarray."

"I know; I've heard. But the donor requires that gift discussions include input from at least two grad students representing Dr. Striker's domain. Something about diverse objectivity, or so I've been told."

"Now doesn't that beat all!" Josh exclaimed.

"If the unnamed donor is exposed to my thinking, MAU might well be breaking ground for the memorial science building by early next week."

After shooting from the hip, Josh's outburst came capped with a more reasoned afterthought: "Although he is a major-league jerk, I wouldn't sandbag the guy...but then again..."

Giving in to human emotions, his voice drifted in mid-sentence.

"Feel free to say what you want. I've set up a meeting at 2:30 p.m. tomorrow in the Science Research Foundation's conference room. Can you make it?" she asked.

"Count on it! Wouldn't miss the chance!"

"As you can imagine, you were not Dr. Striker's first choice, but the potential donor demanded that the grad student with the highest GPA be present. If you had been Dr. Striker's first draft pick, bully for him! Since you didn't make the cut, he gets a bonus choice to balance input. You may not admire Dr. Striker, but I'm delighted that you will show for the good of the university."

"I'll show for the good of my curiosity. How about some insurance for this extra-hazardous duty?" Josh retorted.

"Not on your life, Mr. Ryan. Your reward is the joy of living on the edge, a niche which seems to fit you comfortably. See you mañana; 2:30 p.m. sharp."

Mission accomplished, she fled into the darkness.

Driving at a leisurely pace, Josh breezed homeward. Surveying the shambles of his hoped-for career, the weird chain of the week's events paraded through his thoughts, then churned in a pointless jumble. He wondered, "How come if experience is the best teacher, it doesn't arrive until after graduation?" Nothing made sense.

Trying to look on the bright side, he ultimately concluded that any price paid to share the company of the magnetic Traci Kilburn represented Bargain City. He began to feel downright philanthropic, recognizing his potential role in generating a major gift to a university unlikely to lift a finger to sponsor his doctorate.

Despite the emotional turmoil, he felt unexpectedly buoyant.

Reaching home around 11 p.m., he pushed open the entry door to the phone's incessant ring. Answering, he heard a cheery voice. Good old Gus Morley had called to commiserate. "Josh? I'm relieved I was able to finally catch you. I may have some encouraging news. Since it could make you happy, I didn't think it could wait till morning."

"Would you care to define *encouraging*?" Josh wasn't interested in midnight telephone games.

"No, really! I've been speaking to Dr. Striker, just as I promised. If I read him correctly, he may be willing to forgive and forget, recognizing that you are one of his all-time brightest Ph.D. candidates!"

"And the conditions that inspire this extravagant magnanimity?"

"Well, of course, he values team play, as you know. You'll have to admit, Josh, your team play has been less than stellar in recent times!"

"That's what I like about science: its great credo of *team play*. Amazing, isn't it, how geniuses like Isaac Newton and Louis Pasteur are lauded without reference to *team play*? I doubt the résumé of Charles Robert Darwin cited *team play*!"

Gus Morley ignored the sarcasm. "I'm not proposing you sacrifice the principles we all respect. Only that you act with honor and fairness when you meet with Ms. Saunders tomorrow to discuss benefits to this university's Science Division."

"I hear you. Of course, this is not a *deal*. Rather, just a few nice words from me endorsing research funding rather than capital construction! Then hocus-pocus, that doctorate may suddenly reappear on the computer screen, right?"

"Exactly!"

"You've always appeared to be generous in your concern for students and my professional well-being, Dr. Morley. I do appreciate that. But the only promise I can make tonight is that come tomorrow afternoon, I will say what I think."

"Meaning?"

"Meaning there is a long Ryan history demanding truth—and letting the chips fall wherever!"

"You can't know how sorry I am that I've been unable to convey to you the seriousness of your situation. Good night, Mr. Ryan."

The line went dead.

Josh felt drained. Stripping for a shower, he sought refuge in the cascade of soothing warm water. He wondered how he managed to meander into this minefield of academic muck. He had heard of people with book larnin' smarts who were dumb as dodos in the real world. His education seemed careening at breakneck speed in that direction.

Within minutes, he hit the sack with a sigh, and fell asleep, sedated by dreams of Traci Kilburn.

He hadn't set the alarm clock; why was it ringing?

Groggily, he reached to shut off the annoyance when he realized it wasn't the alarm that had invaded his space.

In the pitch black, he stumbled clumsily toward the only phone in the house, mounted on the hallway wall. Disoriented and bleary-eyed, he stubbed his toe en route, the jolt of pain drawing a squawk. Grasping for a light switch, he squinted at the clock, confirming the time: 2 a.m.! His adrenalin sparked, he wondered about the safety of his father, heading home to Maryland and now airborne over the Pacific.

At last, lifting the receiver, he grumped a grave, "Hello."

The greeting drew silence.

Clearing his throat, he tried again, blaming a bad connection.

"Hello?"

Nothing. Finally, with the full benefit of his now-awakened faculties, Josh barked a vigorous, "Hello!"

Still no response.

Abruptly, the caller hung up. The only sound came from the eerie humming of a vacant line punctuated by Josh's exasperated breathing.

Surreal! Just like the movies! Only he didn't have the remotest clue about the plot.

At last, fully aroused, he realized that the caller intended harassment, compliments of Ma Bell! Perhaps the implied threat earlier in the day amounted to more than a collegiate prank.

Another set of rings accompanied by silent lines at 3 a.m. and again at 4 a.m. confirmed his suspicion that the intrusions represented more than Darwin's random coincidence. He vowed to order caller ID first thing in the morning.

Somebody out there did not like him!

# III
## Day Three: Tuesday

# *It's All in the Genes*

# 13

Piercing sun's rays stabbed at his eyelids, intrusively assaulting restless sleep. By the time Josh surrendered to mother nature, the grandfather's clock chimed eight times. Fumbling clumsily, tired fingers eased to his face to roll back the cobwebs spun during the fitful night. The last harassing phone call shattered the night's peace only four hours earlier.

Rubbing his eyes, attempting to yawn himself awake, Josh jerked upright. The Duchess demanded his Morgan Manor presence sharply at ten. Swinging ungracefully out of bed, he stumbled to the ritual relief of a morning shower. The steamy drops caressed his sleep-deprived frame in balmy embrace. The morning fog shrouding his brain lifted slowly as he reassembled command of his senses.

Try as he would, the rationale behind the disturbing phone calls, much less the identity of the mysterious caller, eluded. He didn't have a clue as to who; nor could he make sense of the why.

For one brief instant a vision of Dr. Karl Striker flashed, but the idea faded fast—this wasn't Striker's style. The professor enjoyed demolishing his prey eyeball-to-eyeball, not through anonymous harassment.

Stepping from the shower, revived and refreshed, Josh began pondering the unprecedented behavior of Lady Carrington and his first ever-invitation to visit the Morgan Manor homestead.

Except for pulsating curiosity, he felt nothing. The Duchess reigned uncontested, ever the enigma. For a quarter-century she had vented back-handed contempt for Michael Joseph Ryan, the unwanted son-in-law. Sole heir, Joshua Chamberlain Ryan, fared no better.

"Let's see," reasoned Josh, "If *Iron Pants* invites me every twenty-three-years, I'd better show up for this one! It's not likely she'll be around to try again the next millennium." He chuckled at the irreverent impudence.

Washing down a sticky bun with a cup of aromatic Colombian, he topped the continental interlude with a glass of "not from concentrate" Florida sunshine. Devouring this luxurious leisure, the student browsed the headlines of the Tuesday, September 3, 1996 *Washington News Press* spread out on the cluttered breakfast table.

Trouble brewed in Iraq; the Redskins still licked their wounds from Sunday's loss to the Eagles; and the lethargic presidential campaign pitting incumbent Clinton against a dutiful Senator Dole, drew monotonous, boiler-plate reporting. The aspiring fossil hunter had guzzled his way into a second cup of coffee by the time he reached the newspaper's Science Section.

Startled, he spotted an editorial reference to *Monkey II.* Surprise shifted to chagrin when he discovered the author of the piece, an editorial writer he surmised to be devoid of scientific credentials, had unleashed a scathing attack on *Monkey II* in general and its student star in particular. Unprepared for his 15-minutes of fame, Josh posed an inviting target, naively oblivious to the potential for public notoriety.

But here it came, ready or not! Three days earlier, an identity virtually anonymous except for family and intimate friends, attracted little attention inside or outside the Beltway. Now, he was smeared by a muck-raking journalist who, taking careful aim, had depicted Josh as a villain lacking academic credentials, afflicted with *minimal intellectual integrity* to match.

Although the writer never ventured to excavate a fossil or to scrutinize life under a microscope, he didn't lack for skill in expressing malevolent scorn for anyone daring to question the inerrancy of conventional evolutionary dogma. The blistering tone savaged the student's testimony reminiscent of Baltimore's own H. L. Mencken reacting to the 1925 *Scopes* trial!!!

> *Mid-Atlantic University's Chamberlain Playhouse is the unlikely host for a week-long tedious pseudo-legal drama featuring a tawdry attack on a scientific theme. This atrocious mutation presented to the public under the guise of educational entertainment casts a dark shadow, certain to be an embarrassment to its University sponsor as well as to the community it victimizes.*
>
> *Headlined by a cast of amateurs (except for the distinguished Judge Edward Anthony Stone who, for dubious reasons, chose to put his reputation on the line), the production scandalously, if not maliciously, attacks treasured traditions of the science of evolution as articulated by Charles Darwin and espoused and defended by the MAU science faculty and the scientific community.*

The pointed rhetoric struck a nerve, a hint of things to come!

> *The majority of intellectuals who hold hard-earned credentials in scientific disciplines must cringe at the thought of an inexperienced graduate student perhaps motivated by some combination of fame, fortune or mischief, undertaking to single-handedly sabotage sacred scientific tradition, while simultaneously thumbing his nose at world-renowned scholars like Dr. Karl Striker who have contributed unselfishly to science education.*

*While I believe in the First Amendment's guarantee of free
speech, I would personally applaud the courage of the university
administration should it order the immediate closure of the
tragi-farce dubbed Monkey II.*

Sliced, diced, and filleted, Josh absorbed the center-forehead hatchet-job. Stung by the ferocity of the broadside, he slouched to a chair, trying to digest the blow. Previously unexposed to the scathing scrutiny of spin doctors posing as wordsmiths, nausea twisted his stomach.

He felt heat in his throat before realizing he was gagging on too-hot coffee. He cringed at his role as designated target, walking unarmed into the cross-hairs of a sharp-shooting journalist. Angrily he dismissed the poison-pen author as a "gutless wonder."

Rallying his senses, he thought, "The dude could at least have called for an interview." Then he rationalized, "Like it or not, press freedom is protected by the First Amendment." Scorning this logic, he bemoaned, "Too bad freedom of speech for MAU students doesn't rate the same privilege!"

By this time the simmering coffee brew cooled, and so had Josh.

Realistically, Pace Terhune would ignore the searing attack—seeing it as gratuitous publicity for the university's centennial. Josh, reconciled to reality but still seething, impulsively grabbed the phone, and dialed the *News Press* editorial offices.

Gritting his teeth, he determined to speak responsibly. Specifically, he wondered about the scientific credentials of his critic, reasoning that justice awarded him the right to know. Expecting to direct the question to some disinterested office underling, he was surprised to hear the author of the attack pick-up in person. Identifying himself honestly, Josh tried for an air of detachment.

"I read your editorial attacking *Monkey II's* challenge to your so-called *universally accepted traditions of the science of evolution.* I felt curious about the academic credentials of the author. Per chance Sir, are you trained in scientific disciplines—or just a journalist skilled at generating headlines?"

Josh tried to be courteous but his detractor proved thin-skinned. The telephone hotline exploded with expletives that turned the air blue. Josh Ryan had inadvertently stumbled into a quagmire of confrontation.

"You medieval minded religious bigots are all alike. You wouldn't know science if it hit you in the head with a baseball bat. You and your kind have been clogging this line all morning. I've heard enough trash for one day.

"Believe it or not, Mr. Ryan, the world is not flat! Nor is every newspaper editor a scientist. But for your information, I've attended seminars conducted by splendid scholars such as Dr. Karl Striker. Also, I possess the intellectual equipment to cope with the coming twenty-first century—unburdened by horse-and-buggy mentality."

The tirade came across as over-kill—the reaction of a wounded ego. Josh responded carefully, determined to side-step an unpalatable brawl in the tar pit of bruised personality. At least the answer to the initial question seemed clear—Mr. Editor lacked scientific training, compensating with a bow to the bias of old-fashioned, yellow journalism.

While the editor paused to catch his breath, Josh searched for a disarming response. He ventured carefully, "Thank you for the candid admission. I respect your right to vent, under cover of the First Amendment. The same goes for religion. Last time I checked, the Bill of Rights applied to all citizens. But for the record, I've never attended church, as you presume!"

Pumped by the verbal jousting, rivulets of perspiration raced down the student's ribs, soaking his shirt. The offended journalist, refused to be assuaged.

"While you wasted Monday evening scoffing at *spontaneous generation*, it's apparent you don't know about research taking place in a Montgomery County lab. Word is they've already stripped 170 superfluous genes from the 470 gene parasite, *Mycoplasma genitalium*—and the bug still lives! They want to reverse the process, string genes together, one at a time, and create life!"[46]

"You're wrong, Sir! I'm aware of the process and respect the lab," Josh countered. "However, the claim of *spontaneous generation*, raises two questions! Does the experiment rely exclusively on inorganic, inert matter or does it require use of pre-existing genes? Is the process designed and engineered by supervising intelligence or would it qualify as a miracle, attributed to random chance???"

Aware that his questions had prompted only a muffled grunt from the editor, Josh reversed course and mounted an offensive designed to probe his tormenter's sophistication.

"No doubt you're familiar with the works of Charles Robert Darwin? *Origin of Species? The Descent of Man?*"

The naked challenge drew an agitated reaction.

"Yes, yes, of course! What do you take me for...as blind and biased as yourself?"

"Have you read every word?"

"Enough to be fully informed!"

The editor's impatient answers exuded confidence.

Josh played cat and mouse, adopting the feline's role. "And you agree with all the philosopher's assertions?"

"Certainly I agree! Let's not get redundant!"

"Do you seriously contend '...*that the five great vertebrate classes, namely mammals, birds, reptiles, amphibians, and fishes, are all descended from some one prototype...?*'"[47]

"Must be true...if Darwin said it!!"

"And you're satisfied that '*all the members of the vertebrate kingdom are derived from some fish-like animal...?*'[48] Even though the hocus pocus description of a '*fish-like animal*' comes without fossil evidence of an identifiable pedigree?"

"If the guy said it...I believe!"

"Then you're comfortable believing that '*animals so distinct as a monkey or elephant and a humming-bird...could all have sprung from the same parents...?*'[49] For example a '*fish-like animal?*'"

"Why not? Look kid, your wasting my time!"

Josh ignored the rebuff, refusing to back off. "It doesn't bother you that your esteemed guru of science asserted without proof that '*Man, the wonder and glory of the Universe, proceeded...from Old World monkeys...at a remote period?*'[50] Then, thanks to Darwin, you can boast to your readers that some '*Old World monkey*' founded your family tree!"

Josh sensed a slight crack in the columnist's confidence, concluding the man had never heard Darwin's speculative words in raw format. Shuffling papers, the editor muttered softly garbled phrases sounding something like, "I'd have to review those quotes before further comment."

Encouraged by the response, Josh sprang his *pièce-de-résistance*.

"You'll be relieved to know Darwin favored you with his personal vision of the '*early progenitors of man,*' allegedly your ancestors—'*No doubt covered with hair, both sexes having beards...ears pointed...a tail having the proper muscles...great canine teeth...formidable weapons...a third eyelid...lungs...modified swim bladder...heart existed as a simple pulsating vessel...*'"[51]

Getting no audible reaction to this revelation, Josh chose to poke some fun at the editor's expense.

"Oh, by the way, Charles Robert Darwin proposed '*some extremely remote progenitor*' of yours was a single sex creature—neither male nor female, but a '*hermaphrodite!*'[52] No chance of gender discrimination here! And no problem with a gender specific name in the family tree: Robert or Roberta—take your pick! Just think, all this certitude without a fragment of corroborating evidence!"

"That's quite enough, Mr. Ryan. You've had your fun. I work for a living, so now if you will excuse me...."

Josh seized the belated attempt at withdrawal from the conversation as a last chance at editorial reason. "Before signing off, I have a couple of questions...if you don't mind."

"I do mind, Mr. Ryan, but get on with it...Get it off your chest."

The steam had evaporated but the words still carried belligerent edge.

"Thank you, Sir! This will be the-30-sign-off!

"Remember those hundreds of ancient terra cotta soldiers excavated recently in China? Every one different; every one uniquely sculpted. Any archeologist suggesting those works of art to be the product of order out of random chaos thanks solely to millions of years of fortuitous blowing winds and bubbling waters would be scorned as an incompetent charlatan.

"But in another part of China, when someone discovered fossil bones that were quite possibly the remains of once living and reproducing human models for such terra-cotta figures, the Charles Darwins of the world credited *their* design to billions of lucky coincidences over millions of years—descending from a fish no less!

"Tell me Sir, what happened to proof and rational reason in this vacuous scenario? Is mega-evolution's faith built on science or philosophical flimflam? I call the idea *trash bait*...with a hook for the gullible willing to adopt a grandpa fish into the family pedigree."

When silence greeted the spiel Josh tossed a last pitch.

"Darwin claimed all life evolved from the simplest original form. Right?"

The journalist mumbled affirmation.

"Do we agree that since a virus parasite is simpler than a bacteria, it should predate bacteria—if it conforms to mega-evolution's mandated simple-to-complex sequence?"

The "Of course!" answer exposed the Journalist's agitation.

"Since the virus is a parasite that can't survive without a host cell, how could a virus evolve if the cell it needs to feed and reproduce hadn't yet evolved? How could evolution have occurred if the alleged first ever ancestor appeared before the arrival of its more complex food source? After struggling to emerge through *spontaneous generation*, the virus would finally debut, only to drop dead, DOA!!!"

"You're an interesting kid, Mr. Ryan...I'll give you that."

The editor cut the conversation with a retreat to civility. Formal "good-byes" followed courteous "thank-yous" for the informative dialogue. Weeks later, Josh learned coincidentally the hard-edged writer spent four college years as a Karl Striker fraternity pal.

Satisfied he had acquitted himself effectively in a difficult situation, the scholar sought respite reading the rest of the *News Press* until rousing thoughts of Traci Kilburn stifled his concentration.

*******

102

Despite having to forgo the insights gained through the coaching of a sister or a mother, Josh had enjoyed an average social scene during high school and college years. Though intrigued he had never been swept off his feet by the opposite sex. While enjoying the routine social experiences that came his way, he had studiously avoided entangling commitments. He seemed to have no clue that women considered him a "hunk."

But once Traci Kilburn appeared, he spurned casual relationships! The unprecedented obsession surprised and overwhelmed him: He was surprised at the preemption of his faculties and overwhelmed by giddy ecstasy riding surprise's coat-tails.

Meetings allegedly related exclusively to *Monkey II* assumed not-so-subtle social implications. Just the previous week, he had invited her to dinner at Baltimore's Spike and Charlie's. He topped the sultry evening with an orchestra seat at the Myerhoff for a Baltimore Symphony Orchestra pop concert featuring the magic of guest trumpet Walter White. As she slipped her hand through his arm while entering the elegance of the music auditorium, Josh glowed. And when he dropped her at her Columbia home, he had impetuously kissed her good-night.

The episode marked a threshold of no-return! Homeward bound, Josh sensed that he had wandered boldly into unexplored domain. Sure, he had kissed girls good-night, but never with the wave of euphoria now bursting over him. He groped for answers, wondering whether any human, from prehistory to the present, had weathered such an emotional roller-coaster.

Musing absently, he wondered, "Could Dad have felt this way about Mom?" The newspaper dropped from his hands, in the middle of the comics, jolting him awake. His usually reliable rationality faltered, then fled as his mind reverted to the mental equivalent of atrial fibrillation. This September morn, twenty-three years of focused mental skills surrendered to romantic daydreams of Traci Kilburn.

Josh sat bolt upright, diagnosing his pitiful plight. "I must be in love!"

Five minutes before heading out to Morgan Manor, an icy blast from the emotional Arctic chilled the epiphany. "Here I've been watching my career go up in smoke, imagining it must be worth it because I've met Traci. How could I be so incredibly stupid and presumptuous? Traci radiates friendship to everyone—even a klutz from Magruders Mill. Would one semi-passionate good-night kiss sweep a liberated lady off her feet?"

The introspection produced no comforting reassurance. Perhaps Traci eyed him with amusement? Did she picture him as a spineless boy toy? The thought sent shivers. Misgivings aroused, he recalled her casual cell phone call from the Tennessee trenches the previous April, "The eagle has been landed!" Guessing erratically, he filled in blanks he didn't like.

"And the gullible eagle falls prey to her charms—more like a clay pigeon, or sitting duck—end of sad story—Yadah, yadah, yuk! When the curtain falls on *Monkey II* for the last time come Friday, Ryan daydreams could be as extinct as a fossil skeleton. Joshua Chamberlain Ryan will no longer be on display as lead witness. My career as a geologist looks to be on hold, at best…more likely defunct. As for romance, this script could end up grist for some wanna-be Nashville songwriter."

Josh wound up the soliloquy thrashing out an off-key musical note.

"I can see it now. I can learn to strum the guitar, put on my Dockers, boots, and Stetson and pick-out mournful sounds with lyrics like, 'My love is as dead as that old fossil.' If I'm lucky, I'll find a lonesome bread basket to carry the tune!"

Abandoning the absurd satire, he rejected self-pity in favor of definitive action. He would kick into high gear and find a remedy. Feeling inspired, he refused to mope or to sit on his hands, hoping for random chance good fortune. Romance deserved to be an *event* rather than evolution's snail-paced reward—and he would make it happen. Hypnotic love demanded no less.

The time had arrived to offer Traci fodder for her daydreams!

He'd stuck out his neck a country mile already—another inch or two wouldn't matter. If the lady had no long-term interest in him, the time had arrived to find out—before *Monkey II's* final scene. As he tried to evaluate the odds of her cherishing reciprocal feelings, his mathematical mind drifted in fog. If he could have loaded the dice in his favor, he would have been sorely tempted.

Racing to the phone, he picked up the receiver. It took only one touch of the finger to ring the Kilburn residence—he had long since put her number on speed dial. He jotted an outline for reference—just in case he fumbled.

The musical voice of Traci Kilburn responded via answering machine. "Hi. If you're someone I wish to speak to, thank you for calling. Since I'm not available at the moment, if you leave your name and number I'll be happy to return your call. If you're selling something, try the Yellow Pages."

Josh got right to the point, before his eroding nerve abandoned ship in mid-sentence.

"This is expert witness Ryan checking in. I don't think the Yellow Pages are interested in what I'm selling. I hope you are! This call has nothing to do with fossils. I want you to be my guest at the Hampshire Club aquatic center tomorrow afternoon around 2:30 p.m. I hope you can make it. I guarantee the water will be Laguna Beach ready."

As an afterthought, he blurted, "By the way, Dr. Striker is encouraging me to pursue a new career. Right now I'm thinking about country music if I can find a crash course in guitar strumming."

Signing off to a fate unknown, he reasoned, "Let the chips fall!"

A crimson flush painted Dar's face as he hung up the phone. In this discombobulated state, the Ryan scion rushed out the front door to his know-not-what rendezvous with grandmother Regina Ann Morgan-Carrington.

# 14

The distance from Fort Ryan to Morgan Manor was a stone's throw as the crow flies. But travel by way of circuitous country lanes, skirting patches of meandering woods and lush stretches of carefully manicured farmland, consumed a good fifteen to twenty minutes.

Not that Josh Ryan's rust bucket was up to 1996 standards.

The exterior of his road-weary '65 Mustang was an indicator of its overall ramshackle condition. Josh had bought it with his own cash the week he turned sixteen and earned a driver's license. The decrepit heap matched the reach of a thin wallet. Josh's dad had joined him on this quest for first wheels, counseling that the tired old hunk of metal could be a buy as long as Josh managed do-it-yourself repairs.

The bargain-basement sale moved as smoothly as soap down a water slide. All smiles, the new owner sped away, guiding the sputtering relic to a parking slot under the protective overhang of a Fort Ryan barn. The teenage craftsman committed his creative energies to restoring the classic collector's model engine to pristine perfection. Josh looked past the rusting hulk to visualize a born-again vehicle that would someday sparkle with chrome accents and a fire-engine-red baked-enamel finish.

Josh took to the challenge of dismantling, replacing, and rebuilding worn-out parts with more zeal than expertise. When complete, the engine purred with a soft hum reminiscent of its sixties-era debut.

No such luck awaited the battered remnants of the car's body shell. Having long since run out of funds, Josh put off work on the exterior until he enjoyed the financial rewards of a real job. Meanwhile, the semi-salvaged clunker performed yeoman's service in its newfound life, dutifully ferrying friends and fossils. This day, Josh headed down the familiar route leading to the Mid-Atlantic University campus and the adjacent Morgan Manor mansion.

He imagined what the Duchess would say when he drove up. "Never in a million years has a disgraceful bucket of bolts like this cluttered the driveway to the Carrington estate." His face split in a wicked grin. Since he didn't really know the Duchess, he hadn't the remotest idea as to her reaction when she saw his dilapidated eyesore. What's more, he didn't much care.

Lady Carrington had check-mated more than one erstwhile mighty man. The victims of her occasional displeasure considered themselves fortunate to escape with quaking calves and sweaty palms. When her laser-like eyes perceived less than honorable conduct, retaliation followed as surely as night follows day.

Though burdened with the accouterments of wealth and privilege, the Duchess proved to be anything but uppity. But despite that saving grace, her claim to notoriety grew out of her judgmental emotionalism. Pity the targets of Iron Pants' unsanctified wrath. Lady Carrington viewed the punishment she inflicted as moral retribution.

Kinship offered no refuge. For a quarter-century she had stockpiled an overabundance of frustration and disappointment. In reaction, she unleashed her cold fury on her grandson, Josh Ryan, and his father, Michael. In her own mind, she defended her vindictiveness as upholding principle.

Until his flaming friendship with Traci, the grandson of the mistress of Morgan Manor had rarely experienced trepidation. Michael had sought to fashion his son's character to reflect "the heart of a warrior, and the soul of an artist." Josh wasn't sure exactly what that meant, but he recognized that standing tall for an idea or a principle was as much a part of his Ryan heritage as the blue-gray of his eyes. And so, at the prospect of visiting this enigma of a grandmother, one-on-one for the first time, he felt no emotion except a faint glimmer of curiosity.

No love, or hate, or fear, nothing...just lingering puzzlement! "Who is this person—and what does she want with me now?"

Cruising through Maryland woodlands, Josh inhaled the damp, earthy aromas that saturated the forested glens. He felt exhilarated by the approach of autumn with its earth-tone hues—it was his favorite season, and more so this year because he had fallen hopelessly in love!

Rounding the last bend of the road, Josh spied his destination dead ahead. The pearl-white Victorian mansion perched astride the highest hill, anchoring a Norman Rockwell setting. The manor enjoyed a commanding view of two sister summits marking the sites of Mid-Atlantic University to the north and Hampshire Country Club to the south.

An expanse of decorative, cast-iron gates bridged the common entry linking the university campus to the Morgan estate. A hammered-brass filigree outline of a shield with a silhouette of a rearing horse carried the caption "Morgan" in a banner unfurled at its base. Folklore reported that Sam Houston Morgan had designed the family coat-of-arms without resort to European peerage records.

Hung on a brick archway with Victorian frills, the gates cranked open by hand. Left ajar from before dawn till midnight each day, the creaky iron shouted welcome with a nineteenth-century accent. Once inside the gates, the road split into a double drive divided by a brick and granite monument that announced in chiseled bronze that Mid-Atlantic University lay directly ahead.

Across the crest of the ridge rose an array of strategically positioned, meticulously crafted brick buildings with massive white columns suggesting Ivy League stability and academic longevity. Front and center of this impressive architectural display stood Chamberlain Playhouse, honoring the memory of the intrepid Civil War hero.

A left turn away from the campus led to another smaller, gated entrance opening to a curving, single lane. The asphalt ribbon climbed up a sloping hillside. Eventually it opened to a circular cobblestone drive at the Morgan mansion perimeter. The highest elevation within five miles, the manor enjoyed an unobstructed 360-degree view. The family had retained the core fifty acres of sylvan beauty to preserve a touch of privacy and to shelter a modest grazing area for the thoroughbreds that resided in pampered comfort at the Morgan Manor Stables.

Entering the mansion grounds for the first time, Josh admired their immaculate condition and the imposing stature of two towering Colorado blue spruces on either side of the drive. Arriving at the top of the rise and continuing around the subtly arcing road, Josh approached the native stone entryway to the house, satisfied to have arrived moments before the commanded time of arrival.

As he eased to a stop, a hissing rush of steam erupted from the jalopy's engine. It was too late to do anything but pop the hood and watch. Ugly spurts of rusty liquid gushed from the radiator through a ruptured rubber hose. As if protesting the unspoiled elegance of the setting, the stain spewed indiscriminately in all directions, seeping across the immaculate pavement—directly in front of the manor's pedestrian walkway.

Josh felt mortified at the dismal sight. Transfixed, he stood by helplessly while bubbling hot liquid spread mindlessly in an array of colors mimicking autumn's orange, red, and brown hues. As he retrieved a spare hose from the trunk, a mellow feminine voice interrupted.

"I'm pleased you were able to make it." The Duchess herself had arrived at the curb.

Trying to be cool, Josh resurrected his Tennessee greeting to Traci, stammering, "We'll have to stop meeting like this."

"But let's not stop meeting," she shot back with an inviting twinkle. The strong, musical tone in the voice belied its seventy-eight years.

Then they both laughed at the preposterous circumstances overwhelming their first-ever meeting. On the plus side, the heir's inauspicious arrival had broken the ice and swept away any reserve. The Duchess spoke with a jovial lilt. "I see you brought a spare hose. Good for you, young man! Always prepared, just like a Morgan."

Beckoning to a handyman, she assured him, "You won't need to impress me with your mechanical skills. My man will take care of this while we visit." Then as afterthought, "And don't worry about the driveway. It's already scheduled for rehab."

As they walked toward the lattice-framed veranda, Josh fell captive to her spell. She didn't look as tall as he had expected—a fraction under 5 feet, 5 inches. Straight-backed, she walked with a confident grace that emanated power.

Her sparkling spirit masked an aging body and bravado concealed her sensitive fragility. A powder-blue pantsuit complimented the sapphire-blue eyes. The matching fire-and-ice necklace and bracelet, along with diamond earrings and three-carat chunk of ice on her ring finger broadcast affluence.

Josh thought, "The woman is gorgeous, with or without the gems."

With a polish that disarmed all comers, Grandmother Carrington placed her right hand under his left elbow, steering him toward a wide-open front door.

"You probably have a million questions as to why I invited you here...I don't blame you if some of those questions are tough."

At ease but cautious, Josh acknowledged bluntly, "Years ago I pestered Dad with questions...but I've forgotten most of them."

The Duchess winced. "I've not been much of a grandmother. A long time ago I learned to regret some terrible mistakes. But proud and obstinate, I lacked the courage to admit my error and lacked the foggiest notion how to right a wrong."

Sighing audibly, she looked away. "I guess when I saw you dueling the likes of Dr. Striker, the truth dawned—it was now or never. Pig-headed stupidity describes my self-imposed isolation in refusing contact with you and your father."

She added with a wan smile, "There's no way I can retrieve the lost years. But there's nothing to stop an old lady from admitting she has been terribly wrong in the past."

Despite her hunger for approbation, Josh made no move to respond. What could he say?

Introductory small talk over, she ushered him into the glittering entry hall. Beveled glass windows bent the sun's rays into rainbow sparkles. From here, the Duchess motioned him into a sitting room. Two plush, red-velvet chairs faced a fireplace trimmed with white marble.

Though notoriously opinionated, the lady seemed blessed with outspoken integrity, akin to his own. How could she have become so embittered, rejecting her only child? How could she have callously buried her grief at her own daughter's funeral? Why did she boycott son-in-law Michael Joseph Ryan? How could she have ignored her grandson and only heir?

As if reading his mind, the Duchess confessed. "I have absolutely no excuse. Perhaps I never fully understood the meaning of love, since your Grandfather Carrington and I were never that close. Not even best friends...maybe cordial co-existers...when he was sober."

Glancing furtively at Josh, and hearing no response, she continued, "You must have heard terrible things about me from your father." She craved some kind of reassuring response.

"Never happened. Hostility never surfaced. If it had, I probably wouldn't have shown up today. Dad answered my questions as best he could. Sometimes his comments came across as vague. I'm not sure he ever understood your thinking. But no, my dad has never bad-mouthed family—not the Ryans, Magruders, Morgans...or the Carringtons!"

Hesitating, Josh ambled on. "I was blessed with a storybook father. You've missed a great deal of happiness by shutting Michael Joseph Ryan out of your life."

The blunt assessment bordered on chastisement. Eyes glistening, the Duchess nodded, meekly accepting the mild censure. She chose the moment to hint at a partial purpose for the invite.

"Now with your permission, Mr. Ryan, I want to share some Morgan family history, seeing as how you are the only heir to the throne, so to speak."

Watching his face for emotional clues, she inquired, "Are you comfortable with that? Do you have questions before I continue?"

"As a matter of fact, there's one question that baffles me."

He tried, but failed, to frame it tactfully. "I've read of your faithful attendance and generous support of a church loaded with the rich and powerful. Pardon me for wondering how, after a lifetime of allegiance, someone who claims to be a devout believer could miss the message of compassion touted as a core belief...*turn the other cheek...forgive seventy times seven...do unto others...*"

Sensing her discomfort, he avoided her eyes.

"Are all those sweet-sounding phrases just sloganeering clichés?"

The brutal honesty struck a nerve, but the Duchess didn't flinch. She paid stoic attention as twenty-three years of smoldering resentment erupted in a verbal hammer-blow. "My father is a good, even a great man. He says my mother was an angel. If your religion was worth a ten-cent movie, why couldn't *compassion* rather than *ostracism* have overcome animosity?

"Forgive me, Ma'am...I guess I'm too dense to understand religion!"

The words stung Lady Carrington. But the resilient dowager never batted an eyelash. Her proud facade neither cracked nor crumbled!

He marveled inwardly, "The woman matches her *Iron Pants* nickname—a tough old broad, for sure!" He found himself admiring her gumption.

"You're terribly right, my boy. There's something in Scripture about a worthless religion that boasts a form of faith but denies its power. As I recall, that kind of counterfeit is exposed as a *sounding brass or a tinkling cymbal*. I'm not proud of it, but for most of my life, church amounted to little more than the socially correct thing."

Lady Carrington's spunky honesty revealed her genetic linkage with Josh Ryan's in-your-face candor. She recounted her privileged life: Pampered unmercifully, she had commanded the world at the snap of a finger. She insisted that her daughter follow in her footsteps and dance to her tune. So when Leslie Anne defied matronly counsel and eloped with Michael Joseph Ryan, the mother's heart hardened with vitriolic anger.

Lady Carrington had concocted a potential match with an MAU grad student she perceived as a "catch." Devoid of charisma, the would-be suitor embellished his potential while projecting romantic interest in Leslie Anne. To placate her mother, the reluctant daughter occasionally dated the pursuing scholar—but displayed lukewarm interest.

While bogged down in this contrived relationship, the strong-minded Leslie Anne met Annapolis Midshipman Michael Ryan. It was love at first sight! The widowed socialite fumed, warning her recalcitrant daughter of the pitfalls of marriage to a "restless sailor." But the more she cajoled, the stronger the bond between Leslie Anne and her "no-account sailor."

The Duchess tried strong-arming Leslie Anne into one last encounter with her preferred choice for a mate. The confrontation left both mother and daughter in hysterics. A day after this clash of strong wills, Leslie Anne eloped with Midshipman Ryan, dashing all aspirations of a mother who never experienced a heart-throbbing romance of her own.

"In a fit of temper, I got even by cutting Leslie Anne out of my will. I even rewrote the trust her father and I created for her at birth, excluding her and her heirs forever. Idiot that I was, I never returned Michael's and Leslie Anne's phone calls nor responded to their letters. When your mother died shortly after your birth, I blamed your father for her death and I rejected you as a bitter reminder of unhappy times."

The Duchess didn't plead for forgiveness. But her eyes misted as she admitted, "I was terribly, terribly wrong. I've wanted to share this with you for many a day. Today seemed the time."

Curious to learn more, the grandson probed her guilt. "What made you change your mind?"

"About a year after Leslie Anne died, I discovered her diary in a hiding place in the tack room she maintained in the stable. Years passed before I could turn the pages. Too late I discovered the hard-nosed fool I had become. Since I could never undo the damage, I kept trying to forget—until I saw you performing this week at the Chamberlain Playhouse.

"While I knew I had no right to expect you and your father to forgive the ways of a wicked witch, I finally summoned the courage to meet you in person and try to erase a lifetime of regret with a few minutes of sharing." She exhaled slowly, faint optimism peeking from behind her words. "I want to walk you to the stable and give you your mother's diary."

Josh wondered at the secrets in the diary, which might explain so much about his life. Studiously polite, he attempted mature insight. "I appreciate your

111

courage. Seems to me we're all human and tend to zero in on faults in others long before we look at ourselves in the mirror."

Gamely trying to lighten her load, he took a stab at the genetic link. "Those Morgan genes tend to stir up action in every generation." She brightened at the opening.

"That's something else I want to do—tell you about those dashing Morgans who contributed to your chromosomes."

She took him by the hand, ushering him to an expansive library dominated by four oil portraits. Standing by each in turn, she offered riveting details of the Morgan gentry who had gone before.

"Sam Houston Morgan, the Civil War survivor who designed and built the original Morgan Manor and founded Mid-Atlantic University. Honest but crafty, his gambling habits funded the Morgan fortune and underwrote wildcatting adventures for oil in Texas and Oklahoma."

Nudging Josh with a brush of her elbow, the Duchess teased. "Hearsay has it that you may have fallen heir to your great-great grandfather's gambling instincts."

Josh gulped as she led him to the next picture.

"It was your great-grandfather, Joshua Chamberlain Morgan, who mustered the corporate skills to blend the family's petroleum holdings into Pan Oceanic Petroleum, the international corporate conglomerate. The advent of the automobile on American highways powered the pumps of Pan Oceanic. Even that junk heap of yours may have inhaled POP's fumes."

Her sense of humor touched his funny bone.

"My dad created the Morgan Manor Stable of thoroughbreds, led the Hampshire Hunt, and founded the Hampshire Country Club."

The third painting revealed a smashingly beautiful young woman. "Now, this is your grandmother without wrinkles!"

The grandson grinned, recognizing stunning beauty when he saw it.

"I thrived inside fortune's cocoon. By the age of ten, I rode to the hounds in the hunts with the best of 'em. At eighteen, I danced every dance at the hunt balls, flirting shamelessly with every eligible bachelor who could cut a rug.

"That's where I met your grandfather Carrington." She sighed with regret, rationalizing a love that never was. "He appeared every inch a dashing, socially correct gentleman. Scion of a titled English family; dazzled at parties; all that malarkey. Debutantes of the day marched in lock-step: Meet eligible charmer; get married with an extravagant splash; and wish in vain for happiness-ever-after. Except for the birth of your mother, the 'happy' and the 'ever-after' never showed up.

"Playboy Phillip Carrington, always a hail-fellow-well-met hit at parties, could down considerably more than his fill of the bubbly. But his skill and ambition hit the wall beyond that. Thanks to Morgan money, he never worked a real job. Instead, he focused on his forte—partying and drinking. He flunked a World War II British Army physical and died of cirrhosis of the liver in 1962.

"The fraternal parent of Leslie Anne deserves some credit for creating a 1948 trust fund for her benefit. Fortunately, he never raided the fund—with the result that the money grew geometrically without serious oversight."

A pensive Duchess lapsed into silence. Glancing sidewise, Josh caught her in a half-smile, staring at his profile. His inquiring "Yes?" brought her back to the present.

"You'll have to excuse an old grandmother, Josh...For a moment I caught a glimpse of your mother's features in your profile!"

"She looks gorgeous in photos...I guess I can take that as a compliment!"

"And well deserved, I might add! Coeds in my day would have described you as an Adonis!" The extravagant flattery turned his face a self-conscious crimson. The doting hostess savored the next surprise.

"Now's as good a time as any to share a touch of irony. You would have no way of knowing that your Granddad Carrington claimed a shirttail relationship with the Darwins and the Wedgwoods!"

She slapped both knees in glee. "Doesn't that take the cake? Poor old Charles Robert would be apoplectic if he could see how the family genes have gone so far astray as to evolve the likes of Joshua Chamberlain Ryan! Maybe the old boy would even abandon his *progress toward perfection* mantra if he understood the sad state of affairs?"

Grandson joined the tale-bearing grandmother in cracking up at the improbable paradox. She wondered rhetorically, "Now wouldn't the press love to get a hold of that tidbit?"

Josh relished the mirth, pleased at Lady Carrington's zany relevance.

A portrait of a young woman whose dazzling beauty matched that of Traci Kilburn provided the next stop on the guided tour of family history. The bronze caption read: "Leslie Anne Carrington, 1948-1973."

Josh responded point-blank, "I wish I could have known her."

A vulnerable Duchess choked, but said nothing. She tugged at his hand, led him outside toward the south veranda, and pointed to a stand of four silver maples. A gap silhouetting the horizon suggested a time when five trees stood side by side in an unbroken row.

"I want you to see these trees, Josh." A large brass plaque memorialized the site of each maple. The oldest, now crowding the house, appeared gnarled but very much alive. The marker read: "Sam Houston Morgan."

"This giant swayed in Maryland breezes as your great-great grandfather built this house," explained the Duchess.

Josh got the picture. The second tree honored the memory of Joshua Chamberlain Morgan; while the third carried the two-word plaque: "The Duchess." Yawning emptiness opened where a fourth tree once flourished. There it was, a space in time identified by the weathered words, "Leslie Anne."

"I planted that tree when your mother was born in 1948. The year following her death, this tree up and died of a broken heart."

Lost in thought, the two strode on to the fifth and last tree, isolated and standing alone. Obviously the youngest, It bore no identifying plaque. Josh looked at the Duchess.

Staring at the tree but not at him, she declared, "A few months after your mother died, I planted this tree for you, Josh. I waited far too long to order a marker with your name. Little did I realize in 1973 how bitterness feeds procrastination, postponing happy moments."

Quickly she looked away, as if searching for some faraway object. Josh pondered, at a loss. Fumbling, he finally spoke, brushing her arm with his fingers. "I understand. It's OK to put my name there...anytime you choose."

Turning, she smiled up at him, grateful for the vote of confidence. Then grasping his hand, she pulled him in the direction of the stable to display another family legacy.

As skilled in horsemanship as Lady Regina Carrington and the other ancestral Morgans, Leslie Anne had insisted on training every horse she owned. She "broke" horses with murmured whispers rather than force and cunning. Her gentle touch produced prompt adjustment to the saddle, without bucks or kicks.

With stalls for 24 horses crafted in finely finished oak trimmed in solid brass, the mansion's stable accommodated thoroughbreds that dominated race cards from Pimlico to Charlestown. By comparison, the Fort Ryan barn ranked as a shed.

Unlocking the door to the tack room, the Duchess invited access to Leslie Anne's private shrine. A cowboy-style bunk bed had enabled her to nurse a sick horse round-the-clock when necessary. Guarding her cherished memories, the Duchess had left the space untouched. Secured under lock and key, the shrine remained off-limits to all but her—until this moment.

Inside, ribbons cascaded from the walls. Victory trophies vied for space on sagging shelves. Morgan horses under Leslie Anne's tutelage had been consistent winners at races and horse shows.

Hanging on the wall at the room's center, facing a dusty writing desk, a small portrait of an exuberant Leslie Anne graced the decor. Rather than the formal oil painting in the manor house library, the artist had captured a breathless rider in action, sailing recklessly over a fence in all-out hunt mode. Framed in gold leaf, it hung slightly askew.

114

"It's yours!" the Duchess said, reaching up to lift the painting from the wall and deliver it to a stunned grandson. Hands raised in protest, he retreated. Trying to be polite he stammered, "No, really I couldn't, really..."

She responded with fierce authority, grabbing one of his hands, pulling it toward her while pushing the picture at the reluctant beneficiary.

"Nonsense! Here! It's a done deal! For heaven's sakes, haven't you ever heard the expression 'Don't look a gift horse in the mouth'?"

They both snickered at the pun. Thoroughly nonplussed, Josh accepted the prized offering, mumbling something like, "Thanks. Really, thanks a lot."

The magnificent generosity of the gift astounded him. The benefactor went on, "By the way, the horse in the picture is the sire of Turbo. Your mother loved Arabians, too!"

Overwhelmed, Josh moved to leave the room. But the Duchess blocked his path, tugging at his sleeve. No one said "No" to Lady Carrington.

"Don't be in such an all-fired rush! There's more show-and-tell!" With this, she handed him two candlesticks hand-carved from maple.

"When I lost your mother's silver maple, I commissioned the best artisan I could find to create this pair of candlesticks from the heart of the tree. When I'm in the mood, I take them to the manor house, light a couple of candles and play happy memory games. It's time you played, too!"

With a flourish, Grandmother Carrington ceremoniously shifted ownership of the keepsake to the heir apparent. Tongue-tied, Josh gawked in amazement.

"Oh, one last thing,...just for you." She stuffed a gold-embossed leather volume into the side pocket of his jacket. With his hands already clutching the oil portrait and the candlesticks, he made no effort to resist this outpouring of belated generosity.

"That's your mother's diary. It's important for you to have it. You must promise you won't read it until the *Monkey II* hubbub is history. You have to promise—on your word of honor as an ex-PhD candidate!"

Impulsively, she reached up, kissing him on the cheek. He didn't recoil.

"I promise," he replied, realizing that his grandmother knew much more about him and the pitfalls in his life than she let on.

Mission accomplished, she shooed him outside. Carefully, she replaced the padlock, snapping it shut. Together they headed toward the fully repaired Mustang. Earlier he surmised the presence of the bedraggled classic might embarrass the elegant socialite. Now he realized that Iron Pants had shifted priorities to the realm of the spirit, where material trappings did not impress.

Too soon old, but by no means, too late smart!

He left her on the veranda steps, bidding good-bye as he walked in the direction of his car. Turning around, he called, "Thanks for the goodies! And thanks to your man for fixin' the buggy!"

She responded with a final goodwill gesture. "Oh, by the way, I almost forgot to mention I've planned a Friday evening party at the Hampshire's Sam Houston Morgan dining room. Seemed timely to commemorate *Monkey II's* final curtain. Dinner will be served promptly at 8:00 p.m. I'm also inviting Judge Stone; your friend Jonathan Thomas Daniels and his grandfather Brock Daniels; and your father Michael." Feigning a touch of absent mindedness, she added with a twinkle, "Oh yes, that young woman...What's her name?...Traci something or other...?"

Intuitively, Josh suspected that the Duchess knew much more than the given name of "that young woman."

As he reached his car, she called sagely, "Unless I miss my guess, that young woman is in love with you."

Undone, Josh headed for his clunker wondering if his ears had deceived him when he heard his grandmother whisper, "I love you, Josh."

Once in the car's cockpit, he felt a wave of catharsis.

Starting to drive away, the sight of a wide swath of ugly orange stain left behind by the boiling water from the rusty radiator loomed accusingly in the rear-view mirror. Dutifully, he reversed and returned to the scene, determined to erase the unsightly patch. He scrounged a golf towel from the trunk, borrowed a nearby garden hose, and started scrubbing. Despite his best efforts, the stubborn stain persisted. Perspiring profusely, he surrendered to reality and headed home. He hoped his improvised fastidiousness had impressed the Duchess.

En route to Fort Ryan, his musings drifted to the bitter estrangement and leftover resentment still staining his soul. He wondered if, given a generous dose of gambler's luck, that aching hurt might heal and eventually succumb to some emotional elbow grease, dry up and fade away forever.

# 15

When Josh showed up thirty minutes early for the 2:30 p.m. meeting called by Jessica Saunders, he found Slade Lassiter occupying the Science Center's executive suite. Pompous and cunning, Lassiter typically emerged with high-profile assignments, despite a grade-point average that always teetered on the edge of academic oblivion. Rumors were rife that he escaped the ax thanks to a contrived selection of electives.

Lassiter qualified as a genius, of sorts, but not in scientific achievements. He had escaped extinction by mastering the art of bureaucratic *survival of the fittest*. He thus earned the sobriquet, "slippery sycophant," bestowed by less-than-admiring colleagues.

Lassiter's slick mastery of language and a willingness to work long hours had won him the position of executive editor of the Science Research Foundation's trendy quarterly, *Sci/Tek*. Consequently, Karl Striker had picked Lassiter to represent the interests of the department and to offset Josh Ryan's intolerable heresy.

Jessica Saunders had notified Josh earlier that Lassiter would join the party. From Josh Ryan's perspective, an afternoon with Slade Lassiter promised to be something less than a day at the beach.

While sitting in the waiting room, Lassiter looked up from the latest issue of *Discover* at Josh and barked, "Whatever you do, don't blow it, Ryan. Everyone knows the foundation needs funds desperately."

With the expression of a canary-swallowing cat, Lassiter added with a sneer, "It's not exactly a secret that you've been dumped as a Ph.D. candidate for recklessly disloyal conduct."

Josh ignored the biting attack. Turning his back on Lassiter, he embarked on a self-guided tour of the center. Despite an annual budget of two million dollars beyond the salaries and expenses of teaching faculty and grad assistants, the foundation consistently failed to provide earth-shattering research. Even more sobering, during its past two years, the foundation produced rivers of red ink.

Library stacks bulged impressively. Computers were updated or replaced annually. A row of state-of-the-art microscopes sparkled like props for a Hollywood movie. An MRI probed the secrets of fossilized dinosaur eggs. A sprinkling of lackadaisical students occupied scattered work stations, displaying little excitement or urgency. Josh noted no sign that the atmosphere was charged with the electricity of discovery. Judging by a quick browse, any pretense of innovative action should be considered a facade.

By contrast, a bedazzled but misinformed public caught carefully crafted glimpses of goings-on in bigger-than-life perspective. The widely circulated

*Sci/Tek,* published under the advisory eye of Gus Morley, breathlessly hyped the trail-blazing adventures of Striker and his minions in the magazine's torrent of self-serving ink. Extravagant briefings alleging cutting-edge achievements above and beyond reality were dispatched every three months to an international list of distinguished intellectuals, compliments of the foundation.

Wading through the periodical's hype, Josh spotted a subtle trend: Recent issues had been published in two, rather than four, colors and the format had been cut by eight pages.

"No wonder the good Dr. Striker pitched a fit at news of the disappearance of my project," reasoned Josh. "This joint is careening downhill. His foundation cash cow may be going dry!"

Completing his spur-of-the-moment tour, Josh returned to the site's nerve center, resigned to his fate. With another twenty minutes to kill before Ms. Saunders appeared, Josh sought to drag the clueless mind of his antagonist into the zone of confrontation. He launched the contest with an innocuous-sounding query: "How do you define *scientific method*, Slade?"

"What do you mean?"

"I'm asking *you* the meaning!"

"Surely you jest! Any fool knows the scientific method requires objective field research and laboratory analysis to confirm fact."

Smugly satisfied with his answer, the sycophant relaxed his guard.

"But what if the research uncovers evidence contrary to traditional scientific opinion?" persisted Josh. Lassiter lunged at the bait.

"That's absurd! It's axiomatic that established science is based upon prior objective research, discovery, and analysis!"

"So your scientific method always builds on a foundation requiring allegiance to prior conclusions?"

"Absolutely! No scientist worth his salt would be disloyal to or contemptuous of the traditions that launched his career."

Lassiter smirked satisfaction, oblivious to the looming verbal checkmate.

"Let's see then, at one time the majority believed the world to be flat; declared as gospel that the sun rotated around the Earth; thought Pasteur's germ theory occupied never-never land; and assumed that a swamp miasma caused malaria."

Lassiter's face flushed as Josh pushed on. "Barely two centuries ago, physicians were taught that blood should be drained from a sick person to cleanse the system of the poison causing the illness. Dedicated followers of conventional 'scientific fact' then siphoned an equal amount of 'bad' blood from the victim's other arm, to balance it out. Some say this 'cure' hastened the death of the father of our country."

Smart-aleck Dar Ryan was up to his old mind-games, but Lassiter refused to play straight man. "Ryan, you're a threat to legitimate scientific progress. You scoff at tradition and undercut long-accepted convention. You show contempt for knowledgeable professors."

The chastisement came wrapped with admonishment. "Even though your career is down the tubes, the least you can do is show some class. Take it like a man, and support the rest of us in trying to patch up a woefully under funded program."

Josh responded with sarcastic condescension, "You have my solemn word, Mr. Lassiter, that any information I offer to Jessica Saunders will be beneficial to Mid-Atlantic University."

Before Lassiter could relax, Josh loosed a disconcerting observation. "Of course, it's altogether possible that the best thing that could happen to the Science Research Foundation is to offer it a decent burial and transfer the funds to other areas of development on the MAU campus."

The entrance of a cordial Jessica Saunders assured a merciful end to the dialogue. Carrying a designer leather satchel, the assistant to the university development officer invited the students to follow her to the privacy of a mini-conference room immediately adjacent to the lab. She wasted no time getting down to business.

Seating herself at a round table with four chairs, she invited the two sparring students to follow suit. Placing an unmarked legal notepad on the table in front of her, she minced no words.

"Gentlemen, there never is enough money to fund higher education. This university has been fortunate, thanks to a century of generous contributions by founders and alumni. It's a pleasure to share with you the confidential news that a major cash injection, reportedly in excess of one million dollars, looms as a strong possibility. The potential donor has been studying the possibility of providing the fresh capital either for new campus construction or to underwrite Science Research Foundation operations.

"My assignment at this juncture is to ask for input from two representative students. Your words will be assessed and summarized by the development office for submission to the potential philanthropist as background reference.

"Dr. Striker assures me that Slade Lassiter is his personal nominee to represent the official interests of the department in this discussion."

Lassiter beamed at the endorsement.

"Dr. Striker acknowledges, that you, Mr. Ryan, enjoy the highest grade-point average of all his graduate students and in fairness, your views should be heard. He warns that you can be disagreeably unpredictable."

Josh's expression didn't change at the left-handed recognition.

As usual, Lassiter raced to board the Karl Striker bandwagon and to champion the pitch for foundation funding. Extolling the program's needs, he smoothly stroked the reputations of Drs. Striker and Morley, implying that an Albert Einstein might be jealous of their sterling reputations.

"Do you have anything to add to that?" Jessica turned to Josh.

"Not particularly. Of course I believe in scientific research...if...!"

"'If' what, Mr. Ryan?"

"If it's the real thing—open, honest, and objective—and not some spinmeister rehashing preconceived garbage. Academic freedom sounds good on paper but doesn't mean a tinker's toot unless the powers-that-be endorse the right to disagree!"

Ms. Saunders admired the feisty rebel. Lassiter wrinkled his nose as if gagging on sour wine. Rarely at a loss for words, Josh pressed on, completing part two of his "if" with a zinger.

"Nor do I believe it's honest or fair for the work of a student researcher to be usurped and marketed fraudulently as the primary product of the mind and work of a peripherally involved sponsoring professor."

"Are you suggesting that such practices exist within the MAU Science Research Foundation?"

"I'm not suggesting, I'm confirming!"

Lassiter squirmed, then retaliated. "You'll have to forgive Mr. Ryan for his bitterness at having been dumped from the university's Ph.D. program."

Josh reacted with caustic vigor. "With advocates like you, Slade, every defendant would hang."

The lady interrupted evenly, "Now, gentlemen, I expected differences of opinion. Be assured that all factors will be carefully evaluated. If you don't mind, however, I'd like to shift attention to economic reality."

She pulled a tattered folder from her satchel. Crammed with financial statistics, the bulging file overflowed, spilling its secrets in profusion. Scanning the numbers, the stern CPA demanded answers.

"Gentlemen, I've been looking at these numbers and frankly wonder if the two-million-dollar-a-year budget is being managed effectively."

Eyeing them both for a reaction, she went on. "For example, Mr. Ryan, I'm at a loss to understand how your Tennessee project could consume as much funding as you seem to have squandered." The tone signaled puzzlement, not accusation.

"I tried to be frugal," countered Josh, struggling not to sound defensive. "I purchased a junk heap of a camper to stock field research materials. Six months of the standard $2,000 per month stipend for grad student research stoked my coffee habit but not much more. And I always billed mileage at cost."

"But how about your law student friend from California?"

"Good old Jonathan Thomas Daniels? Sure, I bought him one round-trip first-class air fare to and from Tennessee and paid him a flat $500 for a week's work. For what it's worth, the foundation never reimbursed me the five bills!"

"Is that what you recall, Mr. Ryan?" The auditor hesitated, a touch dubious.

"It's not only what I recollect, it's fact! Oh...and yes...a Silicon Valley corporation underwrote the experimental television and computer technology that monitored the research—at no cost to the university or to the foundation."

He added emphatically, "Not one lousy dime!"

The disclosure left the inquisitor unimpressed. "While it's not at all clear at this point, it's only fair to advise you, Mr. Ryan, that your memory and the hard financial data available don't correlate."

"What do you mean, 'don't correlate'?" Josh stiffened.

Jessica Saunders chose her words with delicate precision. "Do I understand correctly that, you claim your total compensation from the Tennessee project, including reimbursement, was less than $25,000?"

"Appreciably less!"

"Wouldn't you then agree that your figure would differ radically from the financial statement that shows an aggregate of nearly $250,000 charged by you to your failed project?"

Josh gasped!

Sure, he gambled. Sure, he did a lot of dumb things. Sure, he had been kicked out of the Ph.D. program at Mid-Atlantic University. But never in his life had he looted a treasury. If anything, he had been as pigheadedly honest as he was dedicated to his academic pursuits.

Jessica softened her rhetoric. "Although you enjoy a campus reputation as a happy-go-lucky gambler of sorts, no one is accusing you of wagering away a quarter of a million dollars. But I am asking for your help in unscrambling these books in order to better understand the financial needs of the Science Research Foundation."

A snide Lassiter enjoyed watching his campus nemesis squirm. Gleefully he noted the beads of perspiration on Josh's forehead. As for Josh, his heart felt like a stone, and his mind was smothered in a cottony haze. Either he had been victimized by incredibly sloppy bookkeeping or else he had been targeted as the fall guy for a monumental con job.

Leaving Josh hanging, Ms. Saunders announced the interview's end, perfunctorily thanking both young men. With stiff efficiency she gathered up the papers and folded notes.

"I'm sure there are logical answers to this puzzle, Mr. Ryan. Until we find them, I'm going to need your help." Fixing unsmiling, laser-beam eyes on Josh, she left the room with the words, "I'm counting on you! Trust me!"

The smirk returned to the face of the slippery sycophant.

Josh stared at his feet, bewildered. Without doubt he would trust Jessica Saunders. He had no choice!

# 16

By curtain time Tuesday evening, Josh had recovered emotional equilibrium, vowing to take on Charles Darwin and his disciples with renewed vigor. No more Mr. Nice Guy! The time had come to attack one of Darwin's weakest links in a fragile chain of philosophical myth.

GENETICS!

Josh claimed no expertise in this twentieth-century science but understood mega-evolution's reliance on mutations as the anchoring linchpin of its theory. To him, trying to account for the emergence of *Homo sapien*s from a microbe thanks to a multitude of random genetic flukes seemed less secure than attempting to scale the face of a rocky cliff using a frayed thread as a safety rope.

With a relaxed stretch of downtime before the evening's performance, Josh took advantage of his presence at the Science Research Foundation to use its library and sketch an outline for his testimony. On a hunch, he accessed the library's computer index and Web site, determined to top off a self-taught Genetics 101 crash course with an update.

First, he typed the keywords *genetics* and *mutations* in a general search of subjects. Nothing startling popped up. But when he tried the code words, "Gregor Mendel," a curious citation leaped from the screen: *"Robert Brooke Fielding, Letter to Regina Ann Carrington, August 23, 1969."* For an instant, he forgot "Mendel" and "mutations," consumed with fascination for correspondence addressed to the Duchess by the long-deceased, one-time Mid-Atlantic University chancellor.

Prestigious mystique enshrined the memory of Chancellor Fielding! Enrolled as a sixteen-year-old freshman biology major the year MAU opened its doors in 1896, this epitome of a scholar and a gentleman built a lifetime career at the school. He earned the first-ever Ph.D. in biology granted by Mid-Atlantic in 1903. Magna cum laude credentials led to immediate faculty appointment and ultimately to biology department chairmanship at the age of thirty-five. His people skills paid off ten years later when the trustees elected him chancellor. Secure in this power position, he championed higher education astride Maryland's lush landscape until his retirement in 1950.

Since he had been an intimate confident of every Morgan generation since Great-great grandpappy Sam shook his hand the first day classes convened in 1896, a personal letter from Fielding to the Duchess raised no eyebrows. How the contents related to Gregor Mendel and genetics stumped Josh.

Fortunately, his grad student credentials gave him access to the inner sanctum of the science library's archives, where the most delicate confidential records were stashed. The assistant librarian led him to the vault, retrieved a thin manila folder protecting several yellowing, onionskin pages, and pointed to an out-of-the-way carrel reserved for serious scholars.

Josh smoothed the crinkles gently, scanning the folds for clues to its relevance. In his haste to discover the unknown, he noted absently that the letter dated August 23, 1969, had been withheld from public view, in trust, for twenty-five years in compliance with the writer's instructions. Cataloguing and release to the archives had finally occurred in July 1996. Fascinated, Josh Ryan examined the manuscript with awe!

Typed without secretarial assistance on the chancellor's old Underwood, the faded lines of the message flowed with an old man's nostalgic reminisces.

> *My dear Mrs. Carrington,*
>
> *Where does the time go? Only yesterday, Sam Houston Morgan extended a welcoming hand of greeting to a green-as-grass MAU freshman. What happy times since! I shook the hand of General Joshua Lawrence Chamberlain; received inspiring wise counsel from your father, Joshua Chamberlain Morgan; welcomed the honor of serving as your godfather; presided over your university education; and long after retirement, strolled the campus leisurely, catching many a glimpse of the vivacious beauty of your charming daughter, Leslie Anne.*

The grand old man recognized impending physical ailments, noting sagely that *Pappy Time and Mother Nature conspire in furtive collusion, picking my pockets, showing no mercy.* This led to the bottom-line reason for the letter.

> *It's no secret that despite the honor and pleasure bestowed for service as MAU chancellor, you know my first love remains biological science and genetics. Given recent trends, I cogitate as to whether I should better have devoted my career to the university's Science Division rather than its administration. Forgive me if I sound like a crotchety victim of paranoia burdened with the onset of senility, but things I can't explain cause me to wonder.*
>
> *Please understand I'm not offended that the emerging cadre of science leadership successfully opposed renaming the science library the "Robert Brooke Fielding Library." These old bones exist oblivious to symbolic ego boost. And after all, they did consent to hang an oil portrait, proving I walked the campus early this century. My concern relates to something more sinister.*

The next sentiments revealed just why the personal letter happened to be relevant to the Gregor Mendel computer reference. Excitement brewed as Josh realized that the letter could be invaluable evidence to refute Striker's charge that his own public testimony "failed to reflect the position of the university's Science Division." None other than the distinguished former chancellor blazed a trail he now followed, a trail not only abandoned but also craftily concealed by current academic authority.

> *As you may remember, dear Regina, the guiding stars in my galaxy of scientific heroes included Harvard natural history professor Louis Agassiz and the Austrian monk/botanist, Gregor Mendel. Both brilliant Darwin contemporaries, Agassiz didn't buy Darwinian explanations of origins, and Mendel's research might well have spared Darwin embracing blatant misconceptions. Needless to say, these nineteenth-century gentlemen influenced my research and dissertation topic selection: "Darwinian Theory Revisited in the Context of Mendel's Law."*

What came next galvanized Josh Ryan!

> *The premise of my dissertation disclosed in a scholarly manner my own interpretation of the Darwin thesis. The more blunt bottom line: I thought his imaginings ponderous, bordering on foolish. You can imagine, this rocked some intellectual boats but contributed to dialogue and opened new routes to research and discovery.*
>
> *The authoritative pages of the original and two copies of the dissertation occupied science library shelf space for sixty-five years—until this spring. Mysteriously, all three volumes vanished suddenly, without a trace—not even card catalogue reference remains. I'm convinced, my dear, this disappearing act is no coincidence. Based upon a series of other related events known to me, my old man's intuition warns the volumes have been removed intentionally by emerging leaders in the science division, hostile to ideas that threaten exposure of evolution's threadbare fabric.*

Overtaken by excitement, Josh rushed to make a dozen photocopies of the faded pages, without waiting to read the letter's ultimate goal. Returning to the carrel, he reviewed the eloquent summary of a deceased scientist's prescient intellectual bequest.

> *Gregor Mendel, lived contemporaneously with Darwin (1822-1884). An experimental scientist, Mendel offered research results revolutionizing knowledge of the magic mechanism of inheritance. He deserves credit for opening doors to the more recent discoveries of genes, chromosomes, and DNA—the master genetic templates that determine and distinguish life prototypes. Ignored by nineteenth-century scientists, Mendel's research blows the cover off empty rhetoric embellishing make-believe.*
>
> *Bateson, great biologist and student of Mendelian heredity, has questioned whether Darwin would have written the Origin of Species if he had been exposed to Mendel's work...Alfred Russel Wallace, Darwin's close friend and co-worker said, "I have come to a very definite conclusion...that it is really antagonistic to evolution."*[53]

Devouring the prescient pages, Josh's thoughts surged ahead, reasoning in gleeful anticipation, "Great stuff for *Monkey II!*" Articulate lines resonated as though freshly written that day.

> *Primitive evolutionists like Darwin lacked the first clue about DNA. They studied what they could see—embryos and bones. Striking similarities were erroneously interpreted as relatedness. Embryonic transformation from misunderstood simple cells to a fully developed animal encouraged conjecture that comparable transformations of any simple cell could lead eventually to most any complex creature—given spin the bottle luck and mega years of evolutionary gestation.*
>
> *Balderdash...Nothing could be further from the truth!*
>
> *This superstitious nonsense came powered with the same medieval hot air that floated spontaneous generation fallacy. Darwinists looked in the wrong direction—toward fully developed life forms and their fossils. The world might have been spared specious speculations had molecular biology been a nineteenth-century science.*
>
> *The genetic code carried by DNA and chromosomes determines prototype—not the inherited experiences of mind and muscle. Cataclysm and environmental degradation destroy species. Mutations represent a corruption or loss of genetic*

*information, weakening species. New information is never added. A diminished gene pool assures decline in diversity— never a leap linking one life form to an entirely new and different prototype.*

*Despite Darwin's fondest dream and thousands of mutations, fruit flies remain fruit flies—forever.*

Josh inhaled the heady thinking of the chancellor/scientist with growing delight, thinking, "This old guy was sharp!"

*Uninformed, if not functionally illiterate as to the fundamentals of genetic science, the master wriggler resorted to winging it, embracing hypothetical pangenesis with its gemmules, or "undeveloped atoms," as integral components of elaborate conjecture. This hoax of an idea alleged these fictitious little rascals learned modifications from the experience of somatic cells and then passed the acquired information along to germ cells for translation to successive generations, gradually achieving permanent change.[54]*

*In his own words "there can be no doubt...that use in our domestic animals has strengthened and enlarged certain parts and disuse diminished them..." So far, so good. But then Darwin tries to cross a non-existent bridge, concluding erroneously "that such modifications are inherited."[55]*

*Next he compounds the blunder by postulating that "each modified structure tends to be inherited...each modification will not readily be quite lost, but may be again, and again further altered...during its successive adaptations to changed habits and conditions of life."[56]*

*Darwin's musings confound and tantalize sophisticates and the uninitiated alike. This fairytale supposition reaches into the treetops for redemption.*

*Allegedly some giraffe ancestor grazing on vegetation gracing the African plane saw tasty tree leaves as a more delectable, reliable food source, and launched a multi-generational neck-stretching exercise. Each successive generation of giraffes-in-the-making allegedly extended taut necks a notch or two, reaching ever higher toward the tantalizing greenery beckoning from the tree tops. And like building blocks, the stretch of each generation touched an ever higher plateau—boosted to new heights by natural selection teamed with survival of the fittest.*

*And voilà! Today's towering spectacle of a long-necked animal exists in giraffe format, faux triumph for Darwin's flawed speculation.[57] Hopelessly unexplained is just why the stretching process took place only with the giraffe? Whatever spared the neck of the lowly zebra, still limited to grazing the plain?*

By now, Josh determined to present copies of the letter to each member of the *Monkey II* jury. Not only did it effectively portray his own views, but also by sharing intellectual company with a former Mid-Atlantic chancellor, he could undercut Dr. Striker's charge that his *Monkey II* testimony was disloyal to the university.

Dr. Fielding's final insights added frosting on the evidentiary cake.

*With the gemmule hoax banished to fantasy land, innovative evolutionists embrace the unassailable reality of genetic science by substituting mutations for gemmules then linking these genetic errors to natural selection.*

*Across-the-board macro change, leading from microbe-to-man, is attributed to multi-millions of mutations—another name for gene defects. Where do so-called good mutations hide? And where is the new genetic information that produces brand-new prototypes?*

*I predict no lab will ever develop mutated fruit fly genes that produce a prototype life form other than fruit flies—albeit occasionally deformed fruit flies!*

*Forgive me, my dear, for ambling on like the old man I am. But for the record, let it be said there are no brand-new genes providing new information to the genome—only genetic mistakes, errors in coding and transmission of life's prototype templates.*

*If the information isn't present in the pre-existing gene pool of the parent, physical changes induced by exercise will not pass to offspring. The child of a body-builder does not inherit impressively developed biceps if the genes weren't in the parents from the start.*

*The genetic pattern most likely to survive simply offers a genome most compatible with the available environment. This built-in flexibility includes mechanisms that activate or deactivate genes responsive to signals that someday will be understood more clearly by science. Meanwhile, this horizontal micro-change within a prototype life form never equates mega-evolutionary vertical change—not even in 500 million years!*

*It's all in the genes!*

*Variations within a prototype result from new shuffles of pre-existing ancestral genes. But no matter how many times a dealer shuffles the pack, the original 52 cards reappear, but with differing combinations dealt to each subsequent hand. Some cards may become broken, marred or discarded but a brand-new, never-before-seen card with a never-before-seen number and symbol will never evolve in the genetic deck without intelligent intervention.*

Here the chancellor introduced the reason for the letter with one last hurrah explaining cherished opinions fortified by a lifetime of scholarship.

*By now, my dear lady, I must certainly be boring you with this outpouring of ideas presented with my woefully inadequate typing skills on display for you to behold and forgive. You know my wife of fifty years passed away five years ago, and our only child died in infancy. This leaves you Regina, granddaughter of the great Sam Houston Morgan, and presently an influential university trustee, as the one person on earth I can confidently share a most treasured legacy. To you alone I've revealed my lack of trust in the present MAU science division leadership.*

*"I'm aware that you and your daughter currently are touring England in quest of your Morgan and Carrington roots. Also, I know my health is failing and we may never have the privilege of visiting again in person. Consequently, this letter will be posted registered mail—to guarantee it reaches your hands. Enclosed is the original draft of my 1903 dissertation— possibly the only remaining evidence this manuscript ever existed. I'm counting on you to keep this letter and manuscript under lock and key for at least the next 25 years, to be released thereafter at a time and in a manner that will assure archival preservation.*

*For this gracious courtesy, I remain permanently in your debt with undying love and respect.*

The hardly legible, shaky-handed, smudged signature could still be deciphered: *R. Brooke Fielding.*

Josh felt a pang of sympathy for the elderly gentleman. Campus history confirmed Dr. Fielding's death two days after the date of the letter. While the fate of the dissertation remained unknown, at least a carbon copy of the letter to the Duchess had surfaced, in the nick of time, to be introduced as evidence during Tuesday evening's *Monkey II* performance.

After stuffing the dozen Xerox copies of the letter in his briefcase for safe-keeping, Josh returned the folder protecting the frayed document to the archivist. By the time he departed the MAU Science Research Foundation Center, minutes remained until show time.

On entering the playhouse, Josh reckoned the crowd to be the largest yet for *Monkey II*. He attributed the heightened interest to the success of Pace Terhune's orchestrated hype. But in his heart he acknowledged the drawing power of a building controversy, realizing he strode through the eye of the hurricane.

Preoccupied in thought, just inside the stage door he brushed by the intoxicating presence of Traci Kilburn, who interrupted his musing with a cheery, "Yes!"

"Pardon?" he said, caught off-guard.

She queried in a saucy overtone, "You are the infamous Darwin interloper, alias Joshua Chamberlain Ryan, are you not?"

"Now you've really lost me," Josh said, totally befuddled.

"I'm simply replying to a message left on my answering machine," she said with mock hurt and puzzlement. "You *are* the Josh Ryan who out-of-the-blue invited me to be his guest at the Hampshire County Club aquatic center tomorrow afternoon, are you not? I'm trying to accept your invitation, unless perhaps you've changed your mind...or have forgotten completely."

Josh blushed crimson. Stammering, the best he could do was to mutter sheepishly, "Guilty as charged. That is, to everything but changing my mind or forgetting. Glad you can make it...I'm looking forward to it."

He pictured himself towering two feet tall!

Still teasing, she tossed a final curve. "I can understand that someone as newly famous as you must have many girlfriends and can easily forget a small insignificant commitment on what must be a very busy social calendar."

She let him off the hook with an easy laugh and a carefree toss of her stunningly coiffured hair. Josh tried to conceal his embarrassment until the thought of an all-out encounter with the girl of his dreams unraveled the remnants of his poise. He managed a weak smile and nod while wordlessly seeking refuge in the jumbled morass of notes attacking gemmules. Summoning his powers of concentration, within moments he had shifted emotional gears, comforted and invigorated by the secret of the photocopies of the afternoon's discovery.

# 17

Josh approached the stage, preoccupied with the data gleaned from Chancellor Fielding's letter. Energized at thought of the old-time scientist, Josh moved with purpose. An instant before taking the witness stand, he spotted JT and yanked him rudely off-stage for an impromptu huddle. The eyes of the witness blazed mischievously.

"Hey, Bro, it's time to play some mind games with Ms. Kilburn!"

"Come again?"

"How about a bit of innocent sandbagging? Just for fun!"

"You're some kind of crazy nut, Dar! You can't sandbag a Harvard lady with her kind of smarts! Besides, what's the point? You want to aggravate your dream girl?"

"She thinks I'm predictable! I want to throw her a soft curve!"

JT rolled his eyes. "That photographic mind of yours is out of film and out of focus! But it's your life! What's the plan?"

"Terhune says this show is for entertainment, right?"

"Yeah, I know."

"OK, let's entertain. The day of the molecular riddle has arrived! It's time to turn to the miniature. Nothing passes muster if it doesn't track with DNA."

"So big whoop. Where are the sandbags?"

With bare seconds before curtain time, Dar's mischievous eyes danced. "The DNA, man—it's all in the genes! Without mutations, mega-evolution lies dead in the water! Send me chasing after genes with your questions, but when you reach mutations, I may sound a bit hesitant."

"And Ms. Kilburn will seize the initiative and lead you to where you'd like to go with this thing—reams of evidence conveniently available! Sandbag City! Right?"

"Something like that. OK?"

JT managed a stage-whispered reply as they took their places on the set. "Whatever you say, Mr. Ryan; it's your soap opera. For my money, I think you're bound for the loony bin."

Judge Stone brought the schemers back to reality with a gaveled call to order. The words surprised both players and the audience.

"It has come to my attention that one of the jurors has ignored my instructions and spoken freely and openly as to his opinions concerning the issues in this case. While it is understood by all parties that the rules of procedure are subject to broad judicial discretion, when I exercise that discretion I require my instructions to be followed expressly."

Focusing on Slade Lassiter, the judge delivered the message in icy tones. "Mr. Lassiter, you have spoken disparagingly of the opinions of the key witness in this case. This conduct disqualifies you from further service on this jury. You are hereby dismissed forthwith."

Then addressing Terhune he added, "Bailiff, you may lead Mr. Lassiter to the exit."

Pace complied dutifully. Lassiter's jaw dropped. The ex-juror looked daggers at Josh while retreating, locked in Terhune's vice-like grip. The drama caught Josh off guard as well. He knew Lassiter served as a Striker stooge, and was unlikely to support a Ryan position, but he had never complained of the juror's patent bias to Judge Stone.

Just before he reached the door, Judge Stone added, "Just another moment, Mr. Lassiter, if you please. I assume you are aware that the university's Science Division chairman requests your appearance as a witness in these proceedings?"

Looking like a deer trapped in the path of onrushing headlights, Lassiter stuttered out a reluctant, "Yeah, sure...Yes, Sir!" Slade Lassiter usually shrank from the spotlight, wilting noticeably under its glare.

"Then I presume you will take the time to discuss the matter with Dr. Striker in order to organize a presentation palatable to his tastes?"

The newly nominated expert nodded meekly.

"Be assured that this court intends to respect the wishes of the university's science faculty and to open the record to another point of view. Accordingly, you are hereby ordered to appear as a witness in this courtroom at 8:00 p.m. tomorrow, Wednesday, September 4, 1996. You will enjoy a free and open opportunity to present your views as to evolution's validity. Your testimony will be introduced under the auspices of Ms. Kilburn. Mr. Daniels will cross-examine. Can you stipulate to this procedure?"

Stony silence betrayed Lassiter's discomfort. He eventually stammered a response supportive of the Science Division leadership. "Understand, Your Honor, I carry no personal desire to intrude in these proceedings, but for the good name of this university and its science faculty, I'll accept the assignment."

Josh concealed a smile, thinking, "Without crib sheets, Slade is doomed. Striker may regret getting what he wished for!"

Judge Stone didn't bother to look up, decreeing, "Thank you, Mr. Lassiter. We'll see you tomorrow evening." Bailiff Terhune took the cue, ceremoniously completing Slade's banishment from the evening's performance.

The six remaining jurors presented a contrasting composite. Gretchen Carmady, elderly widow of a diplomat and Striker admirer; Terrence Iwata, crack salesman of pharmaceutical products who had spent childhood time in a stark, World War II relocation camp; Marie Mackin and Ira Lonergan, MAU student activists; and Weeb Davis, convenience-store owner. Shelley Brogosi, Sherwood High coed, rounded out the half-dozen survivors.

The judge instructed JT, "You may proceed, Counsel."

Dressed to the nines, Jonathan Thomas Daniels rose in full voice. Assigned to take direct testimony, he planned powder-puff leading questions, deliberately accommodating to the comfort level of his best friend.

"Mr. Ryan, Sunday you testified to the mathematical impossibility of a life-friendly ecological niche evolving by random chance. Correct?"

"Yup!"

"And Monday you attacked the impossibility of *spontaneous generation* evolving first life?"

"Sure did!"

"Apart from those arguments, are you aware of the least shred of evidence confirming that all plant and animal life descended from a single-celled common ancestor?"

"Ridiculous!" Spiritedly, Josh continued, "No way did an orchid and a whale descend from the same great-great grand-ancestor bacterium! I can go on if you'd like."

"Please do."

Given the green light, Josh took off, running in a direction certain to arouse hostility in Karl Striker's world.

"*It's all in the genes!*" he exclaimed, exposing Darwin's primitive perceptions of life's invisible realm. He took aim at the old philosopher's *pangenesis*, poking fun at the neck-stretching antics of the giraffe wanna-be, much to the merriment of the audience. Each outburst of snickers fed the intellectual fire.

"*Pangenesis* doesn't fly, walk, or swim! Theoretical transfer of learned physical traits from the *somatic* to the *germ* cells, compliments of non-existent *gemmules,* never made it across unbridgeable genetic chasms.

"Gregor Mendel's genetic experiments showed the right stuff! That pioneer opened doors leading to the bar codes of molecules, genes, DNA, and chromosomes. Fossil bones of giant dinosaurs get attention and deserve a bow, but the ultimate key to unraveling the mysteries of origins lies in the realm of the microscopic living world—molecular biology."

"Why couldn't the first-ever gene evolve by *spontaneous generation?*" JT asked.

"DNA ribbons don't exist in a vacuum! Genes live in a membrane-encased cell, codependent on proteins and a dose of energy. The mutually interdependent relationship of these components demands simultaneous appearance in a fully functional package—no piece of the intricate puzzle present before or after the others!

"Genomes couldn't evolve—they appeared pre-packaged, *ab initio!*"

"And the role of genes in organic life formats?"

"A computer's floppy disk prints out the information stored—no more, no less! The infinitely more complex genetic code, carried by living DNA ribbons, replicates life—it's programmed to reproduce the precise format dictated by inherited genes!"

"Does that explain cloning?"

"That explains the possibility of a mass-produced, cookie-cutter result when no additional genetic information is mixed with the gene pool of one parent. Cloning reveals the limits of the genetic code. Rather than assuring the diversity offered by the gene pools of two parents, it duplicates the genetic template of a single parent—the opposite of mega-evolution."

"Since all plants and animals have genes, how come the wide diversity?"

"Each prototype genome boasts its unique number and sequence. Humans possess no less than 30,000+ genes while the *Mycoplasma genitalium* parasite has as few as 470. Ergo: the more genes, the greater the potential for complex diversity."

"And when there is a two-parent gene pool?"

"*Homo sapiens* leads the variety parade. The *'random assortment of maternal and paternal sets of chromosomes at meiosis...means that each human parent carries 8,388,608 gamete possibilities...A married couple has the possibility of producing over 70 trillion different children by this process alone (8,388,608 x 8,388,608).'*[58]

"By comparison, a puny twenty-six letters is all it takes to fashion an English vocabulary of some 200,000 words."

Arms folded, JT queried, "Haven't you just described mega-evolution?"

Josh shrugged. "Hardly! Inherited genetic information, present from the beginning in each prototype genome, assures adaptation, survival, and reproductive variety—not mega-evolution. If you buy into the microbe-to-man mentality, it's necessary to bust the DNA code."

"Doesn't Neo-Darwinism rely on mutations to do exactly that?"

"Mega-evolution lives or dies on modified genetic codes—a skimpy thread of damaged DNA ribbon!"

"A myriad of genetic mutations do exist, wouldn't you agree? Any reason mutations can't provide the steam to drive mega-evolution's engine?"

"But the question is: Do mutations add new information to genes? I'd just as soon defer arguing the point at this time, if you don't mind!"

JT read *sandbag* in the vague response and backed off.

Josh chose the moment to uncork his hot-button surprise—the obscure Fielding letter addressed to Lady Carrington. Awash in enthusiasm and short of time, he had inadvertently failed to disclose its existence to JT before curtain time.

"Your Honor, with the court's permission, I'd like to share a copy of a communication written more than twenty-five years ago but relevant to tonight's topic. Apart from the distinguished scientific credentials of its author, the letter tends to discredit the allegation that my testimony conflicts with Mid-Atlantic University's historic reputation in the scientific community."

Judge Stone looked accusingly at JT, inquiring, "Are you privy to the nature of this document, Counsel?"

Honest to the core, JT apologized, "Sorry, Your Honor, but...No." Then to cover the tracks of the overeager Josh, he added, "In fairness to the witness, I assume it is data recently acquired. In relying on the judgment of Mr. Ryan, I have no objection to the document's admission to the record."

Judge Stone tried Traci. "Do you also lack prior knowledge of this letter, Ms. Kilburn?"

"Yes, that's true, Your Honor. However, I agree with Mr. Daniels that the witness can be presumed to be acting in good faith, without ulterior motive."

Now Judge Stone trained his gaze on Josh. "Since neither counsel has seen this piece of proposed evidence, I'm asking you to share copies with both attorneys and the bench at this time."

Josh obliged; the legal fraternity deliberated. First to complete the speed read, Judge Stone addressed the student attorneys. Uncomfortable with only a photocopy of a faded onionskin carbon copy of an unverified original, the judge hesitated.

"It's up to you two. If either of you objects, I'll forbid admission."

JT jumped to his feet. "No objection here, Your Honor. The jury deserves a look!"

Traci opted to be coy. "While I have no immediate objection, I wish to reserve the right to object until after cross-examination."

Judge Stone approved, saying, "Then let the jury see copies, subject to that condition."

Josh sprang into action, passing copies to the jurors. En route back to the witness stand, he casually slipped bonus prints to the bailiff and the clerk. Confidently he touted the letter's high points.

"Dr. Fielding's message echoes reality: Mega-evolution's hope hinges on gene *mutations* as the biological bridge spanning gaping genetic chasms between disparate prototype life forms. Without billions of these theoretical sand bridges, Darwin's theory expires in primordial ooze."

Eloquent phrases tumbled like the syncopated rhythms of a cascading stream.

JT's review of the Fielding letter inspired him to ask, "Then you reject Darwin's concept of inherited change wherein a grazing prototype creature, stretching skyward to taste the foliage of a tree in order to survive, evolves into a long-necked giraffe?"

"The concept lacks credibility. Zebras also grazed but never stretched their necks giraffe-like!"

"Then how do you explain micro change diversity?"

"Professors Fielding and Striker would probably agree that variety results when genetic codes transcribe templates in at least three different ways: Random mixing of gene pools from two same-species parents during normal reproduction; hybridization, involving the controlled mixing of gene pools within the context of Mendel's law of genetic probability; and the reproductive consequences of mutated genes.

"Enormous horizontal variety: *Yes*! Vertical change without limits leading to vastly differently prototype life forms without entirely new genetic information:. *Never*!"

"What about Darwin's Galapagos finches with different-shaped beaks?"

"Darwin's finches never hatch mockingbirds, cardinals, or woodpeckers but remain true to the basic original avian prototype. The Galapagos finch gene pool carries pre-existing genetic information, including versatile beak sizes and shapes. Adaptations emerge by routinely reshuffling what already exists. Parent and descendant generations of Darwin's finches continue to be *finches*—never owls, eagles, or pigeons—much less turtles!

"Human gene pools provide a plethora of unique nose styles—all profiling *Homo sapiens*—not some entirely new species or prototype.

"No record exists of Bengal tiger genes producing a field mouse—or vice versa! English peppered moth populations shift color from predominantly light to predominantly dark, adapting to the spread of industrial smoke—in reliance on a gene pool already in place from day one. This inherent genetic adaptability never vests the lowly moth with the magic to emerge bedecked in a monarch butterfly's glory."

"Well, then," asked JT, "Do different-shaped bird beaks prove dinosaurs to be ancestors to finches? Or eagles? Or any bird? If so, how?" Exploring a popular misconception, he went on, "Can't it be argued that hybridization demonstrates mega-evolution's feasibility? Cross a grapefruit with a tangerine and you get a tangelo, right?"

"Successful hybridization, true! Mega-evolution proof? No way! The Christmas poinsettia blooms in a splashy array of sizes, colors, and exotic-shaped blooms, thanks to hybridization. But without a pre-exiting gene pool containing the desired hue, no amount of hybridization can grow a blue rose or a scarlet-red iris—unless the color-code gene is already available.

"Strawberries can be bred very small or very large, but always within the limits of biological capacity. Skillful hybridization may boost size, modify consistency and appearance, or improve flavor. With or without hybridization, gene pools for strawberries never generate apples—nor *strawpples*.

"Grapefruits and tangerines carry prototype citrus genes—that is, the same type genes with a different mix or shuffle. But don't hold your breath expecting the citrus gene pool to mix with peaches, apples, or avocados.

"The tallest corn stalk has yet to be transformed to a beanstalk for the convenience of the mythical Jack! Corn and beans can cook in the same stew, but mutations cannot link distinctly different genetic codes to evolve *coans* or *berns.*

"Horse breeders successfully produce large and small sizes, shapes, and colors to perform specialized functions. And while it's possible to cross a horse with a donkey, the stubborn mule offspring rarely can reproduce. Sterility is another collateral barrier to anything-goes variation."

Judge Stone interrupted, gently reminding the witness that his time had expired. Josh shifted into high gear. After rising to eyeball the jury, Josh moved down the short row of jurors, doing his best to mimic his own impression of Clarence Darrow, gleaned from glimpses of Spencer Tracy's *Inherit the Wind.*

"Mega-evolution is *not* fact!" he exclaimed.

Judging by the impassioned profundity of thought and scathing syllables, no spectator could have guessed that the summation came without coaching in elocution. Like a cavalryman leading the doomed charge of the Crimean War's six-hundred-man Light Brigade, Joshua Chamberlain Ryan had thrown caution to the winds. The audience caught a whiff of fire and smoke.

"The primary *fact* of evolution remains its relentless backtracking in quest of niches of refuge for discredited shibboleths. The exceptions devour Neo-Darwinian myth.

"Surmise substitutes for evidence; conjecture builds on wishful thinking; myth is extracted from speculative conjecture; assumption derives from myth; and unproved extrapolation provides the grist for unprovable theory."

The explosive power of the oratory commanded a hush.

Josh wound up with a favorite quote.

"'*Nowhere was Darwin able to point to one bona fide case of natural selection having actually generated evolutionary change in nature...Ultimately, the Darwinian theory of evolution is no more nor less than the great cosmogenic myth of the twentieth century.*'[59]

"Fervent Darwinists are hard pressed to counter that bleak assessment. Does a living, mega-evolutionary work-in-progress exist anywhere? In the words of Chancellor Fielding, *it's all in the genes!*"

Despite a performance he reckoned to be cutting edge, Josh was left with a hollow feeling rather than the expected exhilaration. It wasn't fun to step on the academic toes of Charles Darwin or Karl Striker. Brooke Fielding's letter reminded him that life was more than confrontation and debate.

But at this juncture, that meaning eluded.

Glancing at his watch, Judge Stone took control of the time, instructing, "When this court reconvenes, the balance of the evening belongs to Ms. Kilburn for cross-examination."

As the curtain descended, Josh caught a glimpse of a radiant Lady Carrington chatting amiably with friends. A couple of rows behind and to the side, he spotted a more ominous presence—an unsmiling Attorney Marcus Brogan scribbling notes.

# 18

Backstage during the break, a loquacious JT Daniels approached the giddy witness, bursting with enthusiasm. Slapping Josh on the back, he lavished praise on his friend's performance.

"Genius and brilliant, Dar, genius and brilliant! I don't know how, when, or where you confiscated that Fielding letter, but it rang some bells!"

"Sheer random luck...an atom short of *spontaneous generation*."

Arriving just in time to overhear the braggadocio, Traci chimed in, "And much to my credit, I didn't object to that outrageous stream of leading questions!"

Catching the boisterous twosome in relaxed camaraderie, the unimpressed lady waded through the deep-blue verbal haze, warning with a disarming smile, "Get ready, guys! Cross-examination could be more than perfume and roses!"

The Magruders Mill duo assumed prematurely that the not-so-subtle warning came in jest. The cast mingled jovially, consuming the fleeting moments of the intermission in animated small talk. Shortly, they returned to the stage in unison, taking their places for Day Three's closing act.

During the break, Judge Stone had advised Traci and JT that retired high school biology instructor, Ms. Lucinda Whitney, had *volunteered* to sit in with Ms. Kilburn in order "to be helpful and to submit a question or two" for Josh Ryan.

"The lady loves kids and never shirks her duty to rectify the recalcitrant! For the sake of entertainment, I'm for giving the schoolmarm her day in the sun. Agreed?"

Traci smiled her approval. JT shrugged a "whatever." By the time Judge Stone issued his call to order and requested Traci to proceed, the Harvard law student was standing patiently, ready to share the questions tendered by Ms. Whitney.

"Joining me at the counsel's table for cross-examination is a distinguished classroom biologist, prepared with a question or two for Mr. Ryan."

Stiff and professionally erect, the impeccably attired retiree didn't crack a smile as she sought to enlighten the wayward. She intended to share information spoon-fed to her in a college classroom long years before the witness was born. Rather than relay the questions through Traci as proposed, she got carried away and began to lecture the witness.

"Young man, I am here this evening because your presentation conceals, or conveniently ignores, facts persuasive of the historic evolutionary process. Understand, what I say is not intended to offend?"

"Certainly Ma'am. I understand that you share my quest for truth. Your interest flatters me. What facts do you have in mind?"

The prim white-haired cross-examiner fell for the gentle trap and slipped into the examinee mode. "Why, of course, Mr. Ryan, I'm pleased to advance your education with data apparently new to your experience. How does that sound?"

It sounded condescending, but Josh paid no mind. She meant well. Studiously polite, he urged, "Lead on, Ma'am, I'm all ears!" He added with a twinkle, "Stick it to me!"

The audience loved the sideshow. Lucinda Whitney leaned back as though evolutionary truth had evolved as self-evident. She caught herself in what passed as a tiny, triumphant smile.

"Butterflies, Mr. Ryan. Butterflies!"

"Don't worry, Ma'am, I'm not nervous, if that's what you mean." She missed the pun.

"Oh, I'm sorry, young man. I forgot your education most likely missed butterfly evolution. I'm talking about the Hawaiian *Hedylypta*. An evolved species of this beauty survives on bananas—a food product introduced to the islands in recent years."

"I'll confess, I don't know much about *Hedylypta*."

"Don't you see? An entirely new species of butterfly evolved by mutation and survived by natural selection. This is a verifiable fact that should help you understand the reality of evolution, Mr. Ryan!"

The patient instructress relaxed, convinced that her zeal had won a convert. Josh responded with the appropriate respect for the lady's age, experience, and good intentions. "You'll be happy to know that I agree with you, Ms. Whitney. The butterfly you describe proves evolution beyond a doubt—micro-evolution, that is—although the *evolution* label misleads!"

Her eyebrows arched into gray question marks. "What do you imply, young man? Don't you understand the term *evolution* as accepted by scholars since it was introduced in 1855 by Herbert Spencer?"

"I didn't mean any disrespect, Ma'am. I'm only suggesting that it's a misnomer to label inherited genetic adaptability as mega-evolution. You can't bootstrap Darwin's evolution into the real world of fact by extrapolating the hypothetical from the reality of genetic adaptability.

"The genetic information powering the adaptation resided within the prototype ancestor butterfly's DNA from the beginning. Isn't it true, Ma'am, that butterflies produce only butterflies?"

She stared blankly as Josh switched roles.

"An educated lady of your experience certainly is aware of computer technology?"

She took the bait. "Yes, of course, young man—I'm not that ancient, you know!"

The audience tittered.

"I'll bet a sophisticated educator like you even has a personal computer with your own Internet access code! Am I right?"

"Doesn't everyone? For the past five years I've been sending and receiving E-mail!"

"If a single number or letter of your access code was entered incorrectly, would the Internet pick up your signal anyway?"

"Heaven knows, it wouldn't work!" Too late, she sensed the trap.

"DNA code is infinitely more complex than a computer code. That's why butterflies breed only butterflies—it's all in the genes!"

The teacher's benign gaze dissolved. No shrinking violet, she muttered to no one in particular, "No wonder you irk Dr. Striker. You would be in big trouble in any class of mine!"

*Hedylypta* exhausted her repertoire.

Traci adroitly disarmed the exchange by picking up the cross-examination. Welcoming the reprieve, Ms. Whitney retreated and reasserted her stern, fountain-of-wisdom demeanor.

"Mr. Ryan, is it true you have a dual concentration in geology and paleontology?" Traci asked.

"True...at least prior to this week."

Despite Traci's threats during the recess, the opening was more powder puff than gunpowder. Josh could barely concentrate, distracted by her appeal and recurring thoughts of the country-club rendezvous arranged for the morrow.

Unsettling questions flowed in rapid succession.

"Given those credentials, do you fancy yourself a genetics expert?"

"Hardly, Ma'am! I apologize if I gave that impression. My knowledge relies on researching available sources, scholarly opinions, and recent discoveries. But an expert? No way!"

Then as an afterthought, he added, "At least I've learned not to fall for the gemmule hoax backing Darwin's claim *'there can be no doubt that use in our domestic animals has strengthened and enlarged certain parts and disuse diminished them; and that such modifications are inherited.* '[60]

"That's tripe! Pennsylvania's Longwood Gardens boasts a collection of bonsai trees trained to remain tiny, but years of miniaturizing never translates genetically to descendant trees. It's an intellectual con job to extrapolate using the real to assert the never-was!"

A smattering of applause signaled a scintilla of audience approval.

"Get real, Mr. Ryan. The world knows Neo-Darwinism doesn't tout *gemmules* but relies on *mutations* to crack genetic templates. Surely you don't deny that mutations impact DNA?"

She raised the mutation issue to allow him to play his game, thinking, "You're downright clever and cute, Mr. Ryan, but I read you like a book. You want to dabble with mutations? Then be my guest!"

Josh walked through the graciously opened door to a *coup de grace* answer, blissfully unaware of the gift.

"Nope! But I challenge the assumption that genetic mutations deliver the goods critical to mega-evolution's claims!"

"How so?"

Josh hammered at the inadequacy of mutations as mega-evolution's bastion. "Mutations not only crack the template but also decrease functional complexity and transmit genetic error; they are negative intrusions that degrade the genome. Mutations don't provide brand-new genetic information; they wreak havoc on what already exists.

"Mutations debase the genome! Genetic information is lost. Less is not better! The nexus of Darwin's dream dangles by a tenuous thread of suspect DNA—which goes down the tubes without substantiating evidence of good-luck mutations. Genetic deterioration doesn't account for the *progress* promised by Darwin!"

"What triggers mutations?"

"Radiation, chemical reactions, human intervention...even environmental stress!"[61]

"With what result?"

"Bleak disaster!"

"What are the odds for a happy result?"

"Big problem here! Beneficial mutations are rare!"

"What do you mean by 'rare,' Mr. Ryan?" challenged Traci.

"Even if you find a billion good ones, what are the odds that those favorable micro-changes will combine in a related sequence? How likely is it that a random chain of mutations will build an organ as complex as the human eye or ear?

"Where is the living evidence for mutations linking one life form to something entirely new and different? Darwinists have yet to identify, much less quantify mutated DNA essential to leap from one prototype life to another different prototype."

Pausing to let the jury cogitate, Josh raised the stakes.

"If mega-evolution can't jump that first hurdle to transitionals in the fossil record, and it can't be identified as a continuing living process today, where's the evidence that it qualifies as *fact*? Show me one example of a mutation process now in progress, leading to an entirely new life form!"

Traci cut into his impromptu lecture by asking, "What about the *Drosophila*'s laboratory-induced mutations? And how about those fruit fly experiments? Can't those mutations, combined with natural selection, demonstrate mega-evolutionary *fact*?"

Josh welcomed the gold-plated invitation. "Mutations produce hundreds of generations of deformed fruit flies. The experiments result in an extra leg here or a bonus pair of wings there—hardly a shining example of *progress toward perfection*. Their crippled product exhibits genomic deterioration—not mega-evolution! Years of fruit-fly experiments have proved fruitless—never have researchers succeeded in leapfrogging the insect pedigree to an unrelated prototype. Myriads of dreary defects may produce a monstrous interpretation of a fruit fly—but never a dragonfly, monarch butterfly, or honeybee!

"Where's the corroborating evidence proving that the ruby-throated hummingbird traces its ancestry to *Tyrannosaurus rex* or any dinosaur, thanks to mutations producing defective genes?

"Rarely has a threshold issue been argued so vehemently relying on evidence so flimsy!" Josh threw up his hands in mock horror.

"Apparently you reject mutations as a cornerstone for mega-evolution?" Traci challenged.

"Where's evidence to the contrary? Dr. Henry M. Morris, a distinguished scientist, lays it on the line.

"'*Mutations take place, but they are either reversible, deteriorative, or neutral. Recombinations of existing genes...are 'horizontal' changes that do not result in reproductive isolation. Natural selection takes place, but this is a conservative phenomenon, which weeds out defective mutants and keeps the population stable. Adaptations take place, but these are horizontal changes which conserve the species against extinction, but do not produce new species.*'[62]

"Forget dinosaurs and hummingbirds, for the moment," Josh continued. "What about the human genome? Every cell in the human body comes coded with as more than 30,000+ genes sequenced in helix-shaped strings of DNA. Each cell contains twenty-three chromosome pairs—a total of forty-six pairs from two parents. The twenty-third pair determines sex: Females carry two X chromosomes while males show an X and a Y."

Content with the explanation, Josh leaned back, only to hear Traci divert to an issue far beyond his experience or expertise. "What a pleasant surprise to learn, Mr. Ryan, that your sophistication encompasses the fascinating world of sex!"

The audience roared. Josh smiled weakly, unable to conceal the crimson that crept over his face. Regaining his composure, he sidestepped the diversion. Obligingly, Traci let him off the hook before JT could object to the irrelevancy.

"And you contend that the human genome does not escape mutations?"

"Absolutely! Genetic errors have been spotted—duplications, translocations, and deletions in the genetic machinery. Debilitation and susceptibility to disease can accompany these defects."

"Traceable directly to gene mutations?"

"Yep!"

"And are you prepared to share a laundry list of human defects caused by mutated genes?"

The interrogator caught the witness reaching for a pad of handwritten notes and concluded the obvious. Sheepishly, Josh allowed as how, "several references" were available, "if that's OK?"

She smiled a condescending, "Go to it, Mr. Ryan...if you promise not to bore us!"

He rattled off an unappetizing smorgasbord of birth defects. "Spinabifida, cystic fibrosis, Huntington's disease, hemachromatosis, muscular dystrophy, Down's Syndrome...Leukemia may result when a piece of chromosome translocates to another chromosome in the midst of cell division."

"Is this evidence that you yourself have observed in laboratory or field research, or is this data so much hearsay you gleaned from a journal?"

Traci had attacked the young scholar's Achilles heel—reminding the jury that Josh's strong, occasionally abrasive opinions were only that—a student's opinion. He countered by putting a twist on a retort from Will Rogers. Resignedly, he confessed, "All I know is what I read in the newspapers...and find on the Internet!"

Pretending to ignore his attempts at humor, Traci challenged, "You must have bleak reading habits, Mr. Ryan. Tell us, if you must, what your newspaper and Internet research has unveiled!"

The witness stood up, hoping to animate the scholarly collection of information. "Parkinson's Disease[63]; *'genetic flaws that make people fat'*[64]; Werner's syndrome results from a mutated site on human chromosome 8 that causes victims to *'age prematurely fast and usually die before they reach 50'*[65]; *'Best's macular dystrophy...destroys the part of the retina responsible for the sharpest vision'* has been linked to mutations *'in the gene now called bestrophin;'*[66] and, since 1990, *'discoveries of heart-handicapping mutations have been pouring out of numerous labs at an ever-increasing rate, yielding more than 100 mutations in more than a dozen genes.'*[67] *'A genetic polymorphism called 11307K in either of their two APC genes'* doubles the risk of colon cancer. [68] Spontaneous blood clots can form with the power to cause sudden death where a *'patient with the disorder has inherited at least one defective gene encoding protein C.'*[69] A mutated gene on chromosome 11 contributes to inherited hearing impairment."[70]

She challenged, "Try not to weary this patient jury with old news they've likely read. How about an update?"

Josh ticked off a dismal summary of harmful gene mutations discovered "only this year": "Progressive myoclonus epilepsy is caused by a gene mutation on chromosome 21. A defective gene on chromosome 6 is linked to hemochromatosis. A gene mutation on chromosome 5 causes Treacher Callins Syndrome, which generates deformities of the face, ears, down-slanting eyes, and deafness. Progressive blindness described as retinitis pigmentosa is a disorder linked to a gene from the X chromosome. A chromosome 9 gene mutation causes skin cancer. A mutated gene on chromosome 16 is tied to fanconi anemia, which affects children, who rarely live past their sixteenth birthday. A gene missing from chromosome 7 causes Williams Syndrome. Mutation on the X chromosome can cause baldness, and leave victims unable to sweat or even toothless when afflicted with anhidrotic ectodermal dysplasia."[71]

"Do you really understand these fancy medical terms you so glibly toss around?" Traci challenged.

"Frankly, no! But they sure don't sound like a good-luck day at the races!"

Savoring the mild titters that greeted the wit, Josh chose the moment to nonchalantly release his silver bullet. "Mutations have marked humans with 4,500 bad results—and still counting *'not one mutation that increased the efficiency of a genetically coded human protein has been found. Instead of a 'blind watchmaker,' the mutations behave like a 'blind gunman,' a destroyer who shoots his deadly 'bullets' randomly into beautifully designed models of living molecular machinery. Sometimes the 'bullets' only cause minor damage; sometime they maim and cripple; sometimes they kill.'*"[72]

"What's with this gloom and doom? Doesn't the mutated gene that causes sickle cell anemia sometimes produce a resistance to malaria?"

"Big whoop! The upside comes saddled with a major-league downside."

"Surely there is more?"

"There's hope a human gene with a defect in its center may create a partial barrier to the deadly HIV virus. Word is, Caucasians carrying two copies of a faulty CCR5 receptor gene appear to resist HIV taking root—despite exposure to the infection."[73]

"Can't you report any more 'good' mutation news for humans in 1996?"

"Only if you care to talk about sex!"

The audience sat up, energized by this bolt from the blue. The casual response caught Traci off-guard. She countered, equally matter-of-factly, "Why not?"

"Researchers *'calculated an unusually high rate of 4.2 mutations per generation, of which 1.6 diminish the fitness of the species...the species must survive in part because people who have accumulated dangerous mutations are*

*least likely to successfully have children...the human reproductive strategy helps*
*purge harmful mutations in batches...they mix their genes with another's,* and
presumably some *of the worst defects aren't passed along. That wouldn't happen*
*if humans reproduced asexually.'"*[74]

"Is that it?"

"Of course! What did you expect!"

She dodged the innuendo with saucy flair.

"I expect nothing but truth, Mr. Ryan. Does that suit your mood?"

He stuck to his guns, blazing away at mega-evolution's phantom evidence.
"Mutations deliver a double whammy: Not only are these genetic flaws
overwhelmingly bad news, but also, where's the mutational sequence promising
a new and improved species destined to replace *Homo sapiens*?

"For that matter, what keeps Darwin's continuous process from moving
laterally or even reversing direction and returning to the genetic game board's
'Go'? Can mutations remain invulnerable to genomic retreat? What's to prevent
mega-evolution from leading a rearward march to a primordial sea?"

Traci interrupted the filibuster. "You forget, Mr. Ryan, I'm asking the
questions!" Determined to wind up the mutation exposé, she offered him a last
chance to grandstand.

"What about mutated bacteria that develop an immunity to antibiotics?"

"Spare me the rehash of the tired cliché claiming that mutated bacteria prove
mega-evolution. Since the first-ever bacteria, mutated bacteria reproduce only
additional generations of bacteria—time and time again, millions of times over.

*"'There aren't any known, clear, examples of a mutation that has added*
*information.'* Instead, mutations lead *'to a loss of sensitivity to the drug...the*
*effect is heritable, and a whole strain of resistant bacteria can arise from the*
*mutation...A change in one of its proteins is then likely to degrade the*
*organism...Information cannot be built up by mutations that lose it.'*[75]

"Loss of information such as *'loss of a control gene may enhance resistance*
*to penicillin...but would be a disadvantage otherwise.'* While *'we see mutation*
*and natural selection in bacterial populations happening all the time,'* some
germs possessed genetic immunity prior to rather than as the result of the
discovery of penicillin. *'Bacteria revived from the frozen intestines of explorers*
*who died in polar expeditions carried resistance to several modern antibiotics,*
*which had not been invented when the explorer died.'*[76]

"Those pesky human enemies started as bacteria and millions of bacteria
generations later *remain* bacteria—irrespective of adaptations, mutations, or
natural selection! It's merely another example of micro-modification diversity
within pre-existing genetic limits.

"To extrapolate across-the-board mega-evolution from micro-variation constructs exaggerated hyperbole on a vacuous foundation. Zero multiplied by zero, mega-millions of times, remains forever a perpetual zip!"

"So what's your point?"

"Back to the gaming tables, Charles D! The odds defy calculation!"

"Spare me your extravagant rhetoric, Mr. Ryan. Let's jump from the microscopic to the gigantic! Doesn't the fossil record demonstrate *continuous progress* from simple to complex?"

"If you don't mind deferring the question, I plan to focus on fossils Wednesday and Thursday evenings. Would you settle for a passing observation?"

She could have pushed, but with time running, she chose to simply nod assent.

"Thank you, Ma'am! Fossil remains provide entertaining exhibits in natural history museums. These show-and-tell displays have dominated the viewing public's imagination since Darwin published his views. The enormity of a reconstructed dinosaur impresses all comers.

"But today, peeks into the past come magnified with new clarity, thanks to technologies that overpower a primitive nineteenth century that didn't see an electric light bulb until 1879. The emergence of molecular biology, DNA sequences, chromosomes, and genetic science contribute to a Darwinian Waterloo.

"Bottom line: All the dry bones in the world can't compete with microscopic genes so infinitesimal that all the genes for every organic species ever to live on Planet Earth might not fill a teaspoon.

"With or without mutations, *it's all in the genes!*"

*******

Traci had used up most of the time allotted for cross-examination. To the relief of the witness, the threats of hard-nosed interrogation began to fade. Most of her dexterous jabs landed glancing blows and floated by, soft as summer breezes. The overconfident Josh concluded prematurely, "This is like shooting fish in a rain barrel!"

The powder-puff breezes proved to be the calm before the storm. Sweet Traci unleashed the gale: "Your Honor, earlier this evening I reserved the right to examine the witness further, and if appropriate, enter an objection to admission of the Fielding letter. With the court's permission, I would like to address that subject with the witness."

Judge Stone seemed pleased to end the proceedings on a high note of legal tension. "By all means, you may proceed, Counsel. The document raises issues of interest to this court."

Traci hit the ground running. JT Daniels could do little more than sit on his hands, prepared to object in the event of unfair insinuation. Josh merely swallowed hard, wondering what kind of torture his dream girl intended.

"Tell us, Mr. Ryan, is Dr. Robert Brooke Fielding a friend, or possibly an informal acquaintance of yours?"

Thinking "She knows the guy died before I was born," he responded with courteous restraint. "Regrettably, I never enjoyed the privilege of meeting Dr. Fielding or claiming his friendship. He passed away in 1969—four years before I was born!"

"Can you vouch for the authenticity of his handwritten signature?"

"Well, uh,...I can't say that I can, having never received a letter from him addressed to me...My opinion is based upon information and belief," Josh said, trying to sound knowledgeable.

JT rested his chin on clasped hands, groaning inwardly, but impotent to impede legitimate inquiry.

"Whose belief, Mr. Ryan? Your own random chance? Blind luck?"

The audience snickered at the ploy while the witness sought refuge in silence.

"Can you testify with certainty and authority that the signature on this letter is in fact the signature of the late Dr. Fielding?"

"Well, I presume that..."

She cut him off. "I'm not asking you to presume or assume anything, Mr. Ryan. The question is forthright! Let me repeat: Can you certify, from your personal knowledge, that the written signature on this letter is in fact that of Dr. Fielding?"

For one fleeting moment he sympathized with the plight of a fox cornered by the hounds in a Hampshire Hunt.

"If you put it that way, I'm not in a position to guarantee authenticity, however..."

"No 'howevers,' Mr. Ryan. You don't claim to testify as a handwriting expert, do you?"

"I never meant to suggest..."

"Did you ask any independent handwriting expert to look at this document, and compare with other verifiable samples of Dr. Fielding's signature to authenticate the handwriting appended to this letter?"

"No, I just never thought...and besides, there wouldn't have been time."

"What do you mean 'wouldn't have been time'? Just when did the revelation of this alleged letter surface?" The biting, cross-examining voice offered no quarter.

"I found it unexpectedly, late this afternoon."

"Pray tell, where did you find this alleged bit of 'late afternoon' evidence?"

"In the vault of the MAU science library's archives."

"As I recall, you are stickler for truth?"

"Absolutely!"

"Supported by falsifiable evidence?"

"You bet!"

The semi-naïve witness squirmed in discomfort, too late sensing the trap.

"Given your commendable commitment to untampered evidence, no doubt you discovered this prized archival record in its original form? Is that right?"

"In a manner of speaking, yes. I located a well-preserved onionskin copy. I've never actually seen the original, but I am optimistic that it exists."

JT dropped his face into his hands and sighed. The lady went for the jugular. "Let me get this straight! You beat up on the memory of poor dead Darwin for unproven conjecture but feel no qualms when you rely on wishful surmise?"

While the witness groped for a defense, the lady snapped the trap shut!

"So you chose to present this court and this jury with multiple photocopies of an onionskin carbon copy of a document you allege exists based upon nothing but your blind hope and optimism? But, in fact, you never have seen the original?

"Really, Mr. Ryan, what do you take us for? You assured this jury that you required valid evidence in matters of science, did you not? And now you expect less in a court of law?"

The chastisement rankled. He responded gamely, managing only to compound the blunder.

"Like I said, it was an original, onionskin, carbon copy!"

Audible laughter unnerved the witness. Desperately searching for a substantive objection to slow the assault and buy time, JT rose to the occasion.

"Objection, Your Honor. Counsel is browbeating the witness. We are prepared to stipulate to obvious facts without this charade. If Counsel intends to object to the admission of the Fielding letter into evidence, I see no reason she can't get to the point and spare us Barnum and Bailey stunts."

Traci pretended not to enjoy the maneuver, disdainfully ruffling the "copy-of-a-copy" of the disputed and quite missing original document. The judge paused briefly, savoring the skill of the Cabot, Calvert, and Stone recruits. Then he spoke.

"You have heard the comments of Mr. Daniels, Ms. Kilburn. In view of the limited time remaining before adjournment, are you prepared to summarize your thinking as to the Fielding letter?"

The lady bought the judicial cue. "Certainly, Your Honor! Let the record show that I entertain the highest regard for the absolute integrity of this witness. His candid admissions confirm that he never intended to mislead the court and this jury. I don't accuse Mr. Ryan of forgery. But while it is altogether true that an original copy of the Fielding letter with demonstrably authentic signature may in fact exist, it has yet to be submitted to the examination of this court.

"Numerous times, I have understood Mr. Ryan to demand the use of verifiable evidence to corroborate scientific claims. In harmony with his own lofty standard, I have no choice but to object to the inclusion of the 1969 Fielding letter as valid evidence to be used in jury deliberations, unless and until the duly authenticated original is made available for consideration."

JT nodded reluctant agreement. A shell-shocked Josh scanned the faces of the legal players as though watching a tennis match with unfamiliar rules. Judge Stone required no more than split-second deliberation.

"Objection sustained. Bailiff Terhune will recover all copies of the letter presently in the possession of the jury and retain them under lock and key until further order of this court." Then he cautioned the impossible. "The jury is instructed to ignore the contents of the Fielding letter in its deliberations."

Edward Anthony Stone's gavel rapped an adjourning whack, announcing dismissal of *Monkey II* until 8 p.m., Wednesday, September 4, 1996.

The curtain dropped.

For the first time, the audience applauded with something akin to enthusiasm, inspired as much by the lively cross-examination as by the audacious pronouncements of the witness.

Josh hoped by the time Attorney Marcus Brogan appeared Thursday evening, his mounting experience as a witness would at least equip him to be a savvy moving target.

But nothing prepared Josh for the encounter awaiting him backstage. None other than the fabled Attorney Brogan himself stood in his path.

"Hello, young man," he gushed. "I've been looking forward to meeting you." Extending a bejeweled, beefy palm, he added, "Joshua Chamberlain Ryan, I just wanted to shake your hand before joining you on stage Thursday evening."

Dutifully taking the extended paw of the lawyer reputed to roar like a tiger, Josh reciprocated, "I've noticed you in the audience. And I heard you are a powerful good lawyer."

The legal tiger purred. "Don't believe everything you hear—or everything you read in the papers, young fellow."

Starting to walk away, Marcus Brogan stopped in his tracks, hesitated, then swung around in a fluid motion, leaving behind a memento and a few syrupy words.

"Oh, I almost forgot. This is for you, Mr. Ryan."

He offered an envelope with Josh's name scrawled across the front. Josh accepted innocently while a still gregarious Brogan strode to the exit, calling back, "Good night, young man...See you in court."

By the time a startled Josh pulled open the subpoena and complaint designating him as a "defendant" in civil litigation, JT had appeared at his elbow. "Better let me have a look, Dar."

Befuddled, Josh surrendered the complaint meekly.

"This is not make-believe, Josh. Seems as though an organization described as *Mid-Atlantic University Science Research Foundation* wields a legal ax aimed at your neck. Marcus Brogan is plaintiff's attorney. And you, my friend, are being sued for a cool million-and-a-half bucks due to alleged *gross negligence resulting in serious financial loss to the plaintiff.* It looks as though your buddy, Dr. Karl Striker, intends to hide behind the cover of MAU's Science Research Foundation to hit you up for claimed *mismanagement of a costly investment in a Tennessee scientific research project.*"

"So, do you recommend that I run for the Canadian border tonight?"

JT ignored the gallows humor. "This is serious, Josh. Not *Monkey II* stuff. With your permission, I'd like to show this to the big guy and let your future rest in the capable hands of Cabot, Calvert, and Stone."

Josh protested lamely, "But it will take a year of my hard-earned allowance to buy an hour of that firm's time."

JT brushed aside his concern with an inconsequential line. "With a nod to Scarlet O'Hara, 'Let's worry about that tomorrow.'"

Accepting his friend's wordless head bob as assent, JT took off in the direction of Judge Stone, assuring over his shoulder, "Don't worry—we have thirty days to answer."

By the time Josh reached Fort Ryan and again entered the comfort of the mill house, he encountered a new kind of emotion—stress-induced exhaustion. Within five minutes, he collapsed into the comfort of a familiar bed, but it took nearly an hour to drift into uncertain sleep.

On the dot of midnight, the telephone's piercing jangle jarred him awake. When he answered, the same eerie silence that had interrupted his rest the previous night rolled in like black fog. In disgust, he yanked the plug, disconnecting the electronic intruder. Still, the adrenaline infusion kept sleep at bay.

Who was this stalking nemesis? And why?

After rehashing plausible explanations, he ruled out the possibility that some disgruntled fanatic sought retribution for his controversial *Monkey II* performance.

Nothing made sense! But one fact rang clear as a bell: He had one or more unidentified enemies! Slade Lassiter qualified as a suspect, but his candidacy seemed too obvious. Besides, sycophant Slade lacked the courage to confront on his own initiative. Nor would he risk exposure to satisfy a personal grievance!

And for sure, anonymous harassment wasn't Dr. Striker's style. The professor preferred eyeball-to-eyeball showdowns.

Eventually, Josh drew comfort from the inspiration of the late Chancellor Robert Brooke Fielding. Despite the cross-examination shellacking at the hands of his irresistible love interest, he believed fervently in the authenticity of the letter. He determined to seek the original as well as the missing doctoral dissertation. Perhaps the Duchess could help.

While pondering strategy and with the still-indecipherable mystery blanking out his mind's personal computer screen, sleep mercifully enveloped him.

Paraphrasing Scarlet, "Tomorrow would be another day."

# IV
# Day Four: Wednesday

## *The Blind Staggers*

**AUTHOR AT WORK**

Copyright 1998 Warren LeRoi Johns

**"WHAT'S WITH THIS #@!%# GIBBERISH?
YOUR FANCY DANCE ON MY COMPUTER
WORKED FOR MY FIRST BOOK!"**

# 19

Josh Ryan awoke refreshed at sunrise, undisturbed except by exotic dreams of Traci Kilburn. Never before had he felt so exhilarated. An epiphany dawned with the day: He had slipped over the brink! His heart surged at the prospect of their promised encounter at the Hampshire Country Club. But, knowing he had a strenuous day ahead of him, Josh ordered his three billion brain cells to "get a grip." Years of mental discipline empowered him to banish the titillating image to a memory bank where it would draw compound interest. He rolled out of bed, stretched, and headed for the comfort of a wake-up shower, basking in the glory of the fragrant autumn morning.

While under the shower, Josh remembered an August 1996, news story claiming that scientists had discovered bacteria-like shapes imbedded in a chunk of meteorite in Antarctica. Enthusiasts gave the rock a Martian pedigree and speculated that the tiny fossil shapes might even represent the fingerprints of extra-terrestrial life—overlooking the likelihood of contamination after the years of exposure to Earth's life-forms.

Josh's mind whirled at the possible scenario of life beyond planet Earth. He remembered that within the year, *Pathfinder* would bounce down on the bleak, rock-strewn landscape of the red planet so *Sojourner* could explore and *Rover* could roam.

Could there be a cosmic war pitting humans against beings in the universe with superior intelligence? Could an evil adversary undermine the forces of good by sinister manipulation of genetic codes? Could a mosquito represent a mutated life form of a once-beneficial insect? Overwhelmed by his ignorance of the big picture, Josh felt as if he had hit an intellectual wall.

Returning to the here and now, he left the shower and dressed. Drifting off into exotic imaginings beyond the reach of current scientific knowledge, he nearly forgot that *Monkey II* exposure had snared him in a sticky web of civil litigation.

Suddenly, an incessant pounding exploded at the front entry. Startled, Josh swung the door open to see JT Daniels.

"Where you been, man? I've been trying to phone you all morning. I figured perhaps you're smarter than I gave you credit for, so you'd seen the light, and had headed for some Caribbean hangout."

Not in the mood to joke, Josh invited his guest inside, and grumbled, "You must be trackin' London time! It's barely 7 a.m. here in Maryland's backwoods. I may not be savvy enough to have fled to the Caribbean...But now that you mention it..."

Once they had gotten comfortable, JT's face sobered. "My instructions are to take you into protective custody immediately and deliver you post-haste to the D.C. offices of Cabot, Calvert, and Stone."

Josh waited for details.

"While I know your waking—and probably your sleeping—moments are consumed with recounting the intellectual qualities of my colleague, Traci Kilburn, the fact is, you are being sued for a cool half-million bucks—plus another million punitives for good measure. You should be honored to know that the law firm of Cabot, Calvert, and Stone is rising to your defense to spare you capital punishment—even as we speak."

"But I can't afford $500 an hour!"

"It shouldn't cost you much more than a filing fee, provided you're willing."

"Willing for what?"

JT unleashed one of his patented grins denoting macho confidence. "Willing to accept the likes of your trusted friend Jonathan Thomas Daniels as the lawyer doing all the scutwork while Cabot, Calvert, and Stone looks over my shoulder to underwrite the quality of my skills! Lest you forget, I'm still a law student and can't practice law nor bill you a single buck for services rendered."

"Thanks, I think."

They both chuckled.

Josh did indeed need a lawyer, and no law firm stood more superbly qualified to represent his interest than Cabot, Calvert, and Stone. Under the auspices of that prestigious legal team, JT promised to be an innovative point man.

"What does Judge Stone think about this?"

"It was his idea after I gave him the complaint last night. He ordered me to deliver your tail to his District of Columbia office by 8 a.m. this morning. And if we don't get cranking, he's gonna' find us both in contempt."

Josh grabbed a jacket and obediently followed JT out the door to a waiting limousine. Once wrapped in the elegant comfort of the Cadillac, pointed in the direction of the United States Capitol, Josh adjusted to his first-time journey in a chauffeured vehicle, momentarily forgetting that he had skipped breakfast. Gliding southward on Route 97 in the direction of historic Washington, the fledgling client sensed that his fate rested in capable hands.

Promptly at five minutes before eight, the limo deposited the duo at the grand entry of a granite-sheathed office tower within sight of the White House. Burnished bronze artwork accented the lobby's white marble floors and wainscot. As they strode past the sleepy concierge and a uniformed guard to a bank of elevators, JT whipped out a key to the penthouse level.

Once in the penthouse, Josh looked around in amazement. Every window offered picturesque vistas of the nerve center of the free world. Senior partners occupied the preferred corner offices, which were embellished with the exquisite trappings of power. Judge Stone, the surviving name partner, sat ensconced in the southeastern corner, overlooking the White House to the south and the domed Capitol on the gently rising hill to the east.

JT steered his client through a double-doored entry into a conference room designed to accommodate twelve people. Immediately adjacent to Judge Stone's office, this room gained its inspiration from a cinerama-sized overview of the seat of the United States Government. Its walls were covered with elaborately sculpted black walnut paneling, complementing the matching conference table and shelves that displayed obsolete legal texts reminiscent of nineteenth-century Washington. Lining the solid walnut table were executive chairs of glove-soft, royal-blue leather, color-coordinated with the sensuously soft carpet.

Mounted on opposite walls, life-size oil portraits of the two deceased founding partners, Cornelius Cabot and Richard Calvert, gazed unblinking across the conference table, epitomizing commitment to equal justice under law.

Josh and JT entered promptly at eight as muted chimes, struck by a grandfather's clock officiously announced the time. Judge Stone presided from a chair stationed at the head of the conference table, reputedly occupied at one time by a long-since departed Supreme Court associate justice. Josh was surprised to see Jessica Saunders seated immediately to the right of Judge Stone.

The quartet complete, the judge wasted no time getting down to business, offering a friendly welcome with the instruction, "Right on the dot, gentlemen. Let's go to work."

Formalities dispatched, the young men took their places by the file folders highlighted with their names.

With a twinkle in his voice, Judge Stone looked intently at Josh, observing, "I assume, Mr. Ryan, that you are not inclined to write a $500,000 check to Dr. Striker's foundation to settle this case?"

Josh gulped. "I could write one, but the big bounce would guarantee another legal battle!"

Getting no more than a raised eyebrow from the judge, the sober defendant inquired with trepidation, "Is that what Striker really wants?"

"Not really! In fact, he is asking for an additional one million dollars for alleged punitive damages. Unless you have more resources than most of us did at your age, I'm assuming you are reluctant to offer any form of financial settlement?"

Reality struck like lightening! Josh began to recognize the stark implications. In a feeble attempt at humor, he reacted humbly. "Unless there are some checks that haven't cleared, at last count I could put together something like $459.23— which leaves me three or four zeros short."

"Then in light of glum monetary reality, I take it that the proposition for your legal representation outlined to you by Mr. Daniels is acceptable?"

"Where do I sign?"

"The first page in that folder in front of you authorizes this firm to defend your interests, if you so choose."

Sure enough, the first page summarized a boiler-plate retainer agreement that made polite reference to Josh's fragile financial condition; the fact that law student Jonathan Thomas Daniels would be expected to build a defense; and the welcome assurance that Cabot, Calvert, and Stone would undertake the representation at no fee other than actual costs as incurred, to be advanced by the client.

Pro bono at its finest!

Josh took out a pen, ready to sign, until interrupted by Judge Stone.

"To avoid even the appearance of a conflict of interest, it's imperative that I call your attention to the paragraph disclosing that not only am I a Mid-Atlantic University trustee, but also that this firm provides general counsel legal services for that institution. Although the plaintiff uses the name of the university in the complaint, Striker's foundation functions as an academic enterprise, legally and financially independent of the Mid-Atlantic University corporation.

"Cabot, Calvert, and Stone has never performed legal services for Karl Striker personally nor for his science foundation. Still, Dr. Striker serves as an influential member of the university faculty. It is my duty to provide you with this information before you retain this firm to represent your interests, however remote the potential for conflict of interest may be."

Josh looked at JT, who mouthed the assurance, "No problem."

Josh signed the retainer with a flourish.

Listening to the judge's instructions, Josh figured out that Jessica Saunders was present because of her detailed knowledge about the financial details of the Science Research Foundation. A fifth person joined the consultation when Judge Stone invited a court reporter employed by the firm to sit in and create a detailed record. For the next three hours the reluctant defendant found himself immersed in detail.

"Rather than go through the counts point by point, a simple summary will do at this time," Judge Stone told him.

Nothing about legal jargon seemed "simple" to Josh. He listened attentively, trying to conceal his concern.

"Dr. Striker accuses you of absconding with what may be a very valuable fossil trophy for personal profit and enhancement of your own professional reputation. You are also blamed for defrauding the foundation to the tune of a half-million dollars through dishonest billings and fraudulent reimbursements. One allegation accuses you of reckless misconduct and defaming the reputation of the foundation and its principal officers, thereby impacting negatively on its ability to solicit supporting contributions from the scientific community and the public. There is more, but that's the gist.

"Do you deny those allegations?" The jurist's steel-gray eyes bored into Josh's soul.

"False on all counts."

"Go on."

"First, as to the missing fossil, it doesn't make sense that I would deliberately sabotage the project that anchored my dissertation research and undercut my Ph.D. candidacy. Where would I sell a hot fossil, and what dollar value could it have compared to a career in geology?"

"What about the money?"

"The original budget was for $25,000 to $50,000 max. I received a basic compensation of $2,000 a month from June 1 through August 31, 1995 and from January 1 through March 31, 1996, totaling $12,000 gross. I also received reimbursement for the cost of the decrepit camper used as field headquarters and some travel expenses to and from Tennessee.

"I doubt if the grand total for everything reaches $25,000."

"Have you received any funds for costs incurred or services rendered after April 1, 1996?"

"No."

"With your permission, Judge Stone, I have a question for Mr. Ryan," interjected JT.

"Be my guest, Mr. Daniels."

"Judge Stone knows you invited me to join the Tennessee fossil dig and that you paid me a flat $500 plus round-trip airfare from Los Angeles. Weren't you ever reimbursed?"

"I knew you wouldn't keep a penny if you knew it came out of my own pocket, so I never bothered to tell you. Get real, Bro, roughing it in a rocky gully was not exactly your choice of a vacation. You showed up only as a courtesy—a treat for me, more like a treatment for you!"

Shaking his head in amazement, JT remonstrated gently, "You are a hopeless incorrigible," quickly adding, "Consider those bucks you squandered as your prepaid fee for any quasi-legal services I render in this case."

Judge Stone reassumed the questioner's role. "Mr. Ryan, your estimation of the compensation given you as totaling less than $25,000 appears plausible, in view of the present paltry state of your personal bank account. Attached to the plaintiff's complaint is a receipt for a laundry list of reimbursements to you approaching $500,000, purportedly signed by Joshua C. Ryan. Why don't you take a look."

Josh fumbled through the pages in quest of he knew not what. Sure enough, he found a puzzling scramble of funded reimbursements endorsed with the big-as-you-please signature: *Joshua C. Ryan.*

"Take your time. Read carefully."

"I don't need time."

Josh's speed-reading capability, paired with a photographic mind had aided his ascent to academic excellence. Nonetheless, Judge Stone took special pains to caution the naïve client. "This inventory is a persuasive piece of evidence critical to the plaintiff's argument. Your position could be in serious jeopardy if you signed, ratifying the reimbursement."

"That's just the point—I never signed...Nor did I receive 500,000 big ones!"

Still smarting from the previous evening's shambles of a cross-examination, he exclaimed, "I may not recognize the sig of Dr. Brooke Fielding, but sure as shootin' I can vouch for my own!"

"Have you ever seen this document?"

"I recognize the cover page with the signature line, but I've never seen the attached itemized record of payments, nor have I ever signed anything with or without the attachment."

"Are you positive?"

"Absolutely. Admittedly, it resembles my signature, but never did I ever sign that phony baloney!"

"But you said you recognized the cover page? Explain, Mr. Ryan."

"I saw that page...or one just like it, last April. Immediately after I returned from Tennessee without the fossil, I briefed the department on the details of the theft. Although Dr. Striker made himself unavailable, Dr. Morley was gracious enough to suggest reimbursement for my unpaid costs and offered a form, like that one, to sign as a pre-condition to reimbursement.

"Frankly, I felt embarrassed to go back requesting further payment for a failed mission and consequently never signed."

"Do you still have the form?"

"Probably. But, to find it, I'd have to do some heavy-duty looking."

"I'm counting on you to find it right away. If you succeed, give the original to Mr. Daniels immediately." With a wink, Judge Stone added pointedly, "After last evening's cross-examination, Mr. Ryan, I'm sure you understand the significance of an original!"

The friendly witticism encouraged the defendant, who replied with jocular bravado, "I'll do my best to find it, Judge Stone. But you know the reputation of mad scientists for their allegiance to chaos."

"This guy is as organized as a CPA," assured JT with a glance toward Jessica Saunders. She smiled, having said nothing up to this point.

"What about that third allegation, Mr. Ryan? The charge that you impugned the reputation of the Science Research Foundation and its officers?"

"If asking questions in class and offering personal opinions as to truth in *Monkey II* qualifies, I'll have to plead guilty. Dr. Striker commands a public platform through his rigid editorial control over *Sci/Tek*. His quarterly journal tolerates no dissent. Hopefully he isn't so insecure as to believe the opinions of a single no-name student endanger his professional reputation."

"You can be sure, young man, that he does not control the judicial system of the United States in any way, shape, or form."

"I didn't mean to imply..."

"I know you didn't. I just didn't want you to lose any sleep worrying over the boundaries of his domain. Even the influence of a Dr. Karl Striker has limits."

The conference room clock clicked just before chiming eleven times as Judge Stone completed the preliminary interview. Jessica Saunders sat stoically throughout, not saying a word. The court reporter concentrated drafting a stenographic record and never looked up. Most of the time, JT sat hunched over, furiously jotting notes. The show belonged to Judge Stone. In conclusion, the silver-haired jurist admonished his new client, "Mr. Ryan, please rest assured that this law firm will represent your interests with all our resources, irrespective of fees."

Josh whispered "Thank you" in appreciation.

"What's more, you should be advised that your continued participation as a witness in what Mr. Terhune described as *Monkey II* is not without serious legal risk. Attorney Marcus Brogan has been present in the playhouse for two nights running; he will likely be there tonight; and he will cross-examine you abrasively Thursday evening.

"My counsel to you is simple: Pull the plug now! Let that innovative rascal Terhune scout around for a replacement.

"I make that recommendation despite my personal interest and pleasure in watching you perform. You have brains and guts, my boy! I'm not saying that it will be fatal to your legal position to continue participation in *Monkey II*, but as your attorney of record, I have to warn you that further testimony may expose you to precipitate risk."

Josh Ryan had a gunfighter mentality and rarely retreated. "I respect your counsel, Judge Stone. No doubt you're correct. But seeing as how I'm already in some difficulty, I won't back off.

"I'm reminded of the man who slipped from a ten-floor building, and once airborne shouted, 'Five floors down, and all's well so far!'

"In the words of my Granddaddy Ryan, 'I might as well be hanged for a goat as a lamb!' Lamb or goat, I think I'll see this thing through, wherever it goes!"

"See what I mean, Your Honor?" JT had correctly predicted his friend's response.

"So be it," pronounced Judge Stone. "We've got work to do, Mr. Daniels."

Judge Stone slipped through the side door to his office while JT ushered Josh to the elevator. Brimming with confidence, he disclosed the depth of his personal commitment by promising smashing success.

"Judge Stone wasn't kidding. Your tail is on the line. I have to go to work for truth, justice, and Joshua Chamberlain Ryan." Then as an afterthought, he added "And, oh yes, you have wheels. Your limo waits at the street entrance, even as we speak. All you have to do is climb into the back seat and pronounce the magic, 'Home, James'—or wherever you choose."

With that, JT disappeared in the direction of the legal inner sanctum.

Three hours in the presence of the great jurist, Edward Anthony Stone, had brought a measure of peace. Both the jurist and JT understood: No way would Josh Ryan shrink from confrontation.

Multiple motivations fortified his feisty mood. For one thing, Josh sensed that his dad, old salt Michael Ryan, would be disappointed to arrive home Thursday morning to discover that his son had abandoned a principled commitment in order to save his own skin. He had agreed to perform. He couldn't betray that commitment.

And far above and beyond *Monkey II* drama, preoccupation with Traci Kilburn absorbed his every imagination.

Suddenly jolted back to reality, Josh noted that it was high noon when the limo turned up Fort Ryan's winding drive. A gnawing hunger reminded him that he had gone the entire morning without a bite of breakfast. He decided on a leisurely lunch at the university's Student Commons to satiate his ravenous appetite.

Aware that the limousine jaunt had been prepaid by a generous Cabot, Calvert, and Stone, he nevertheless peeled off $10 from his meager stash of bills and handed it to the appreciative chauffeur. Then with nerves atwitter at thought of an imminent rendezvous with Traci, he alighted from the limo and made a beeline to his Mustang.

# 20

Approaching the immaculately manicured MAU campus, Josh realized how rarely he paused to admire its symmetry and beauty. An expansive entryway beckoned "Welcome," inviting vehicles into the gently curving double ribbon of asphalt, appropriately christened Fielding Drive.

Spilling across the ridge of low undulating hills, the grounds glistened, masterminded by the innovative skills of nineteenth-century landscape architect, Frederick Law Olmstead. Every white-columned brick structure occupied a meticulously sculpted space, fitting the big picture as neatly as a minute piece of a Currier and Ives jigsaw puzzle.

Chamberlain Playhouse, the campus showpiece, dominated the campus entrance. Just beyond and inside a huge double-gated archway, the administrative nerve center sprawled impressively. A cluster of student residence halls dotted the far end of the campus; classrooms generating the busiest pedestrian traffic graced the master plan's core; the library, gymnasium, and Commons straddled a central garden mall.

At the heart of the Commons, a student affairs and publications suite buzzed with muted commotion. The student lounge, headquarters to most campus happenings, spread its inviting wings. A buffet-style dining area featured fast foods for anyone in a rush to socialize or to study. The Commons offered a center for romance, politics, and philosophical discussion, although, on rare occasions, serious academic pursuits prevailed.

Heading toward the Commons, it dawned on Josh that he could count on the fingers of one hand his previous visits. Never aspiring to student leadership, or even serious socializing, his past two years had been dedicated to a single obsession—a geology/paleontology Ph.D. Today's visit offered an opportunity to snack, check out the day's news, and then head off to the Hampshire Country Club for the much-anticipated encounter with Traci. By the time he devoured two cheeseburgers, a double order of fries, and an oversized chocolate shake, hunger pangs had been quenched—albeit at the price of an unhealthy dose of cholesterol and calories.

With another hour's downtime, Josh picked up Wednesday's *News Press* and scanned the front page, with its news of the United States once again targeting Iraq with Cruise missiles.

The first page of the Entertainment section revealed Pace Terhune's flamboyant handiwork. An ominous headline shouted out: *Darwin to Be Cross-Examined.* The copy parroted the *Monkey II* promo, boasting a Thursday evening extravaganza featuring an appearance by nationally acclaimed trial attorney,

Marcus Brogan. Poetic wordsmith and spin doctor Terhune had delivered the goods with finely honed professionalism.

As Josh admired the handiwork, a student walked by, flashing a dazzling smile and offering a mock bow. Not recognizing the face, Josh instinctively waved and returned to his reading. But he hesitated, recalling that since entering the Commons, he had received other unexpected salutations from complete strangers.

Before he could regain his concentration, a youngish coed with provocatively short skirt, casual air, and silky voice sauntered to the couch where he reclined, commenting, "I'm not much into science, but you've become very popular around these parts for giving fits to Old Man Striker. You probably know that most undergrads despise his guts. Keep on keeping on, Sir Darwin!" With that bubbly commentary, she flitted away.

"No, I didn't really know," thought Josh. He had never campaigned to be the people's hero for verbally bashing unpopular faculty members. But the truth dawned. Unintentionally, the erstwhile Darwin had achieved overnight campus celebrity status. He felt a tinge of self-consciousness, particularly when a troop of four or five students meandered past, offering thumbs-up approval. A chorus of modulated whispers echoed in unison, "Go get 'em, Darwin!"

Josh gamely reciprocated the smiles, waved them on, and tried to finish browsing the *News Press*. Before he could open the comics, another uninvited guest appeared. Popping up like a genie out of its bottle, the irrepressible Pace Terhune made himself at home on an adjoining chair.

"You're just the man I've been looking for. I hope you can break away from your fan club long enough to give me a few moments."

"Sure." A less-than-enthusiastic Josh discarded the *News Press* on an adjacent coffee table.

"Great news, my friend! *Monkey II* is gaining serious attention with the national media. With Brogan showing tomorrow night, several local media names, as well as one or two TV stations, have asked for a press conference late Thursday afternoon. Attorney Brogan promised to be available, and I think Dr. Striker will show. You need to be there, too—in order to tell your side of the story."

Josh suspected correctly that Terhune had set up the event himself. He declined the invitation with flair.

"I suppose I should say, 'Thanks a lot, Pace.' With everybody including Attila the Hun present, it should be a real fun party. By the way, what *is* my 'side of the story'?"

The affable Terhune laughed off the implicit turn-down. "Oh, I can see how you might feel a bit reluctant, given the delicate relationship between you and Dr.

Striker. Still, a press conference gives you a splendid opportunity to set the record straight."

Josh felt steamed. Instinctively, he knew the canny publicist relished the combat of ideas. The bloodshed was incidental. Pace marched behind the banner he had stitched from his own, hand-woven cloth—Entertainment. Josh paid dutiful homage.

"A press conference may do wonders to promote *Monkey II* and the university's centennial. But with all due respect, Mr. Promoter, I'm at a loss to understand what's in it for someone who waves good-bye to a doctorate and has the dubious honor of being slapped with a one-and-a-half-million-dollar lawsuit."

Terhune arched his eyebrows in genuine surprise. "I didn't realize it had gone that far. I'm sorry, Josh; that's news to me."

"Afraid so, Pace, it's gone that far...with no end in sight. And but for loyalty to good ol' Mid-Atlantic University and a trace of intrigue with a lady lawyer, this witness would have abandoned ship the day before its launching. My shipboard specialty seems to be walking the plank.

"And you want me to risk a bashing at a feel-good press conference?"

"Hey, fella, I'm not going to twist your arm. There's much I don't know. And I don't blame you for not wanting to stick your neck out."

"Get ready for some bubbly, feel-good blather," thought Josh.

"But you should know—for what it's worth—that you've earned my respect. Not only for that awesome intellect of yours, but also for your integrity and a lion's share of courage under fire."

Josh bought the line, satisfied that Terhune spoke from the heart. The words did have a calming effect. For some intangible reason, he trusted Terhune and felt it opportune to unburden about a pet peeve.

"Look, Pace, you've been there, done that, in real-world trench warfare. You recognize the power of words and the ability to edit and spin to reflect the writer's bias. Already this week I've been blind-sided by a local print journalist who probably can't distinguish a fossil from a turkey drumstick. He never called for an interview. Never in his life did he sweat out a field research project. He's just a flak for an establishment tradition, dishing out slick propaganda dressed in full-color graphics to better snow a gullible public."

"But, Josh, you can't believe that objective journalists don't exist?"

"Of course there are some—top-notch people like Tim Russert and Cokie Roberts—rising to the top like cream. And great scientists are out there, too— thousands of brainy men and women who objectively explore the cutting-edge of truth. But quacks and self-serving phonies ride those coattails. They crave attention and are willing to sell their souls to advance their own agendas."

"Are you saying Charles Robert Darwin was a phony?" Terhune held few scientific preconceptions.

"I'm saying only that he relied on conjecture, rather than verifiable scientific evidence, to advance unproven, sometimes erroneous pet biases. My beef rests more with subsequent supplicants who enjoy access to superior technology but still kiss Darwin's boots. It amazes me that his ardent defenders never bat an eye when back-peddling from flawed dogma as they say, unblushingly, 'Sure, key points of Darwinian theory may be shambles, but nevertheless, mega-evolution is fact!'"

Josh welcomed the captive audience, directing a five-minute monologue at the attentive but emotionally detached listener. With fevered eloquence, he argued that a century of discovery had punctured Darwin's hot-air balloon fantasy, "leaving evolutionary surmise blowing willy-nilly in the breeze."

Trying to humor his witness protégé, Terhune inquired whether "the fact that components of the theory are flawed means that its house of cards collapses when it asserts that *evolution is fact*?"

Josh retorted, "Not if the faithful seek refuge in verbal gymnastics by pinning new labels on repudiated ideas."

"You've lost me!"

"Last evening I harped on Darwin's flawed faith in non-existent *gemmules* as agents to modify germ cells whereby '*some intelligent actions...performed during many generations, become converted into instincts and are inherited.*'[77] Today's evolutionists ignore that junk science, instead embracing *mutations* as theoretical agents of macro change. Is there empirical evidence proving evolution is fact? Hardly!"

Less than fascinated, Terhune interrupted, adroitly escaping additional unsolicited sound bites by steering the conversation toward academic ethics.

"Just because a Darwinist is biased doesn't imply fraud. Be fair now, Josh!"

"You're right, Pace. But fraud sometimes sneaks in, clouding the debate."

"As in deliberate, intentional deceit?"

"Big time! The *Scopes* trial released the written testimony of a scientist referencing the marvels of the 1912 *Piltdown Man* discovery. Conspirators foisted the fraudulent fossil on the British Museum, presumably to bolster mega-evolutionary theory. Not only was the fake *Piltdown Man* touted to be 1925 state-of-the-art science, but also droves of paleoanthropologists bought the con job hook, line, and sinker. Ironically, neither Darrow nor Bryan lived to see the fraud uncovered in 1953.

"That's why I don't have much faith in unproven opinions. It's tempting to endorse and promote, without authenticating research; to misinterpret or suppress valid evidence; and to skewer a voice daring to dissent. My point is this: Once misinterpreted data makes it into the conventional record, it can enjoy a long shelf life—until some gutsy scientist rides to the rescue."

"I take it that's a 'no' to my invitation to parade your insight before tomorrow afternoon's media barrage?"

Recognizing the student's blunt turn-down, Terhune nevertheless inquired again rhetorically, signaling readiness to return to his office. Josh's response left no room for ambiguity.

"In the inimitable words of Dr. Striker, 'Nein.'"

As if on cue, a solicitous Dr. Augustus Morley approached from nowhere and placed a firm hand on Terhune's shoulder, inviting him to hang around for a few more words. "I've been looking for you, Josh, and I'm pleased that Mr. Terhune is here, as well, to hear what I have to say."

The effusive Gus Morley often took time to befriend a student, particularly one in need of scholastic or emotional encouragement. Josh sputtered, trying not to sound unappreciative or cynical.

"You must be here to share the electrifying news that my doctorate program is back on track? Or perhaps that Karl Striker has volunteered to serve in the Peace Corps?"

Morley shook his balding head, "I wish I could report some encouraging news, Josh, but you must be patient. I have scheduled an appointment with Dr. Striker this coming Sunday and plan to plead your case. Hopefully, Karl can be persuaded that MAU can't afford to lose a prize graduate student."

"Well, I guess good intentions are worth something." said Josh, struggling to project a good-guy image. Then looking at Terhune, he added, "You needn't worry, Dr. Morley, I'm not going to show up to embarrass Dr. Striker during his fifteen-minutes-in-the-sun press conference."

"That's sensitive and perceptive, Mr. Ryan. I'm sure Mr. Terhune would prefer having you there, but my counsel is to be exceptionally careful to avoid abrasive confrontations until this whole tempest-in-a-teapot blows over."

"Is that all you wanted to tell me?"

"Well, yes and no. Primarily I wanted to share at least a hint of a positive word—light at the end of the tunnel, so to speak. It took some doing, but I'm encouraged to believe a bury-the-hatchet meeting can be arranged between you and Dr. Striker early next week."

"In my head, no doubt?"

"Be serious, young man...and keep your fingers crossed. I wanted to assure you, in confidence—and I'm counting on your confidence as well, Mr. Terhune—that the litigation naming you as a defendant was initiated without prior knowledge or input on my part."

Josh fired back, "Are you telling me that Dr. Striker alone is acting as plaintiff using the guise of the Science Research Foundation?"

"You know it wouldn't be ethical for me to comment on that. I may have said too much already, but I felt compelled to let you know I'm trying desperately to stay outside the litigation loop. Remember, I care what happens to you and your future...you can count on me!"

Without further comment, Dr. Morley waddled away toward the expansive suite of student offices in the Commons.

Rising to leave, Terhune observed, "It appears you have one friend."

"Maybe so, but I'm still on the hook in a million-dollar plus lawsuit, with or without the friendship of Dr. Morley. It remains to be seen whether he has any more clout with Striker than I do. Personally, I doubt it. I would guess that gutless Gus quakes in his boots at the thought of Karl Striker...At least I'm not burdened by that handicap," Josh added wryly.

Looking at his watch, he welcomed the chance to exit and seek the enchanting presence of Traci. Shaking hands with Pace Terhune, he wished him well in the centennial promotion, adding, "Here's to a merciful end to *Monkey II.*"

Heading for the parking lot, Josh walked the gauntlet of friendly faces. Most smiled, waved from afar, and unanimously mouthed what had become the student underground battle cry for the week, *"Go get 'em, Darwin!"*

# 21

With twenty minutes to kill before his rendezvous with Traci, Josh dropped by the Hampshire Club Pro Shop to exchange pleasantries with golf mentor Lyle Grant. When Lyle looked up from the day's tee-off-times spreadsheet, he showed no surprise.

"Been expectin' you, Darwin. There's a scintillating young woman lounging pool-side who claims to know you." Then with mischief in his eyes, he quipped, "Sure beats messin' with fossils. For what it's worth, I sure enough admire your taste."

Never a poker face, Josh turned crimson. He felt emotionally naked as a jaybird in the presence of this worldly man, twice his age, who seemed able to read his mind.

"Uh...I wasn't expecting her this soon. Where did you say she was?"

"Lounging on the deck of the pool, you poor naïve fool! What happened to that photographic mind of yours? Maybe you could use advice from a seasoned veteran: 'Never underestimate the power of a woman!'" Lyle chortled at Josh's discomfort.

"Thanks, old wise one," was the only rejoinder the young blade could muster as he stumbled off in the direction of destiny. Before he could get away, Lyle changed the subject.

"Oh, by the way, Josh, it's lucky for you the judge kicked me off that jury—my vote was leaning toward the real Darwin!"

Adroitly, the Darwin imposter reacted in kind. "Didn't want to be confused with the facts, eh? Even an old duffer like you shouldn't be that closed-minded!"

As the two friends enjoyed the banter, Lyle ordered, "Go on, now. Get out of here; your destiny lurks over yonder!"

The pool looked deserted, most of the kids having returned to school desks. But sure enough, just as Lyle predicted, the most fabulous apparition was stretched seductively on a lounge chair, contentedly soaking up whatever ultraviolet rays poked through the overcast sky. Josh's heart leaped at the sight of Traci, attired in a skimpy, coal-black bikini. Her burnished olive skin was framed by a cascading mane of chestnut-gold hair.

Spying his approach, she unwound in a languid, flowing motion, tossed back her hair, and fixed sparkling hazel eyes on her host. "Do you keep all your dates waiting like this?" she teased in pretended remonstrance.

"No, no...of course not...I mean...I thought I was arriving early, but it seems you beat me." He stammered like a five-year-old. She laughed easily.

"I know, Darwin. I didn't mean to sneak up on you. Since I had a few extra minutes, it seemed like prime time to work on my tan."

"As if she needed the work," flooded his thoughts.

"I'm glad you told them you were my guest and they welcomed you aboard."

"Oh, I didn't tell them I was your guest. The Kilburns have belonged to Hampshire for as long as I can remember!"

Feeling deflated, Josh accepted the news, thinking, "So much with trying to impress this lady with my access to the good life." But then, why had he never seen her here before?

"I know what you're thinking." Even she could read his guileless mind.

"My mom's the active member. She's interested mostly in the social goings on. My dad's too busy to show up except for an occasional dance. And, oh yes, since I rarely play golf, the chances of our paths crossing here would have been as remote as those fancy mathematical odds you toss around so glibly."

Lowering long lashes, she tested his memory. "You wouldn't have noticed a skinny high school kid anyway. And my past six years have been spent on the banks of Boston's Charles River gleaning Harvard-brand wisdom."

"Whatever...I'm glad you're here." By now, Josh had regained a semblance of control of basic communication skills. "So don't run away. I'm off to the lockers and will return with all the charm I can muster."

"I can hardly wait," she exaggerated, adding, "but I won't hold my breath...if you don't mind."

He answered with a foolish grin, backing away in the direction of the men's dressing suite. Reappearing minutes later, he strutted some stuff of his own, diving headlong into the far end of the pool while leaving hardly a ripple. Long slicing strokes stirred a caldron of bubbles reminiscent of the splash, flash, and dash of long-forgotten swim meets. Halfway across, he dipped like a dolphin, finishing the trip underwater, then shot up the pool's opposite side like a rocket exploding from the sea. He emerged in a shower of water at the feet of a cheering Traci.

Josh never hesitated to display his aquatic skills. Rigid rehabilitation from his teenage bicycle injury had required relentless practice laps. The sustained effort had produced impressive results. Except for the visible limp, his left leg had healed with little sign of atrophy or cosmetic scarring.

He dragged a recliner over next to Traci's. She rolled onto her stomach with effortless fluidity, and turned her head toward him, leaving one eye closed to the sun and the other squinting in his direction.

Scooting alongside, Josh attempted small talk. "Ever since we met in April, I've been curious about you. This week it dawned on me that although we've spent a lot of time together, *Monkey II* has taken precedence. Today I wanted to see you minus fossil clutter."

Amused by his candor, she tantalized. "Oh, I'm just a simple country girl, really. I promise to hang on every word you speak, Sir Darwin."

"Yeah, and Einstein was a dunce!" he countered with some blatant palaver of his own. "After all these months of seeing you on a more-or-less regular basis, I feel privileged and impressed—given the torrid social schedule that must engulf your Boston life."

"Now don't tell me that would bother a hunk like you?" Eyes fluttering in feigned innocence, she reached out, gently caressing his fingers.

Blushing at the innuendo, he retreated to stuttering mode. "Uh...well, no, at least it shouldn't...I mean I have no right..."

"That's rational thought, Mr. Scientist. Still, I don't mind telling you that my priority commitment at the moment is to the legal profession...in case you're wondering."

Josh knew that the legal profession would be a formidable foe: a jealous mistress or master, as the case might be! For Traci, it would never be an either/or. Romance would require coexistence with the profession—or not at all.

"What about you, Mr. Ryan? A five-foot, eleven-inch brawny scientist who can swim like a fish must attract more than his share of admirers."

He tried to make his response sound casual. "Perhaps on occasion. But never anything approaching a knockout punch variety. At least..."

The halting confession grabbed the attention of the somnolent guest. Traci propped her head at a crazy angle. The other eyelid popped open for good measure. "At least...?"

Josh struggled to fill in the blanks, as best he could, without making a blathering idiot of himself in one "swell foop."

"At least...until I get to know you better!"

"Well, what would you like to know about me?"

Posturing confidence, he explored, with gentle but awkward inquiry, the unpublished life history of Traci Kilburn. He beamed approvingly at everything he heard, while admitting to himself that his attraction to her loomed so powerful that he would have treasured any tidbit.

Her physician father had spent his entire career at Johns Hopkins Medical Center. Her mother had graduated from the University of Maryland with a degree in English. She had a ten-year-old sister named Kristi. The family lived in Columbia, Maryland. And as long as she could remember, she had cherished the dream of becoming a world-class lawyer, determining never to consider marriage until she completed her legal training.

After ten minutes of biographical detail, Josh broke the spell, issuing a challenge. "Come on," he urged, taking her hand and guiding her to the edge of the pool. "Let's sprint freestyle to the other end."

Without a hint of a "ready, set, go," Traci hit the water in a racing dive, confirming his suspicions that she was no stranger to swimming competition.

Instantly, he gave chase. Though out of practice, he harnessed the same old moves that in teenage years had earned him more than one gold medal.

To his chagrin, she matched him stroke for stroke. As they hit the wall in a virtual tie, he imagined victory by a split second but banished the inclination to exult. Gripped by a strange new power, he joined her in body-shaking laughter and spontaneous hugs. Sensing the wet warmth of her body so close to his, he encountered something more than the adrenaline rush of the race. Fervently he hoped that she shared a faint whiff of the lingering ecstasy.

Climbing out of the pool, gasping for breath, he took her hand, gave her a boost, and then clung protectively as they strolled back to the lounge chairs. This time they propped up the chair backs, shoving them tight together. Josh dashed off and returned with a couple of Pepsis.

"Thank you, Josh. You read my mind."

"Don't I wish."

"Men are so blind!"

Missing the oblique overture of encouragement, the romantically inept suitor laughed it off, admitting, "Deaf, dumb, and stupid as well!"

She admired his self-deprecating wit and healthy self-image. Trying to be gallant, he added, "I've been fascinated by the story of your life. It's only fair that I tell you anything you might care to know about me—providing, of course, you might have some remote interest."

A leading question no less. Her bombshell response astounded him. "I already know more about you than you might think."

"What do you mean? You couldn't have learned much from watching me being wheeled into emergency surgery at Johns Hopkins."

"True enough! But remember, I told you my mother has been socially active here at the Hampshire. She's been a close friend of your grandmother, Regina Ann, since before I was born. In fact, Mom has basked under the watchful eye of Lady Carrington for longer than I can recall."

"Amazing!"

"Not really so amazing when you realize that long before my mom and dad were married, your mother, Leslie Anne, and my mom were best friends. Mom frequented the club regularly as a Lady Carrington guest, sort of a social protégé. I suspect your grandmother went out of her way because it spun fuzzy reminders of the good times with Leslie Anne."

Josh Ryan listened spellbound, awed by the coincidence.

"I know you're probably shocked, but Mom and I have enjoyed the hospitality of Morgan Manor more times than I can remember."

"Overnight visits?"

"Sure. Entire weekends. I already know a lot about your mom and your dad—and even about my handpicked, swim-like-a-fish witness!"

She giggled at the disclosure. Josh responded with restraint. "Until this week, I've only seen the Duchess from afar. She couldn't have known much about me, and I doubt she would have sung my praises!"

"Quite the contrary. Your grandmother knows you inside out. She always sings your praises." Still, the estrangement rankled.

"I really wouldn't know. She shut me out of her life from day one. I don't mean to sound bitter, but it would have been a bonus to have a caring grandmother—since I didn't have the privilege of meeting my mother."

"But she did care, Josh. The 'problem,' as she called it, started with a serious misunderstanding with your dad. She admits now that it's her own fault that she made a horrible misjudgment, unfair to your mother, father, and you."

"So?"

"So, incredibly proud woman that she is, she's battled admitting the error of her ways. Now, she's desperately afraid her behavior may have alienated both you and your father for keeps."

"That's not true. My dad never badmouths the Duchess. As for me, well, it's hard to love somebody you've never known."

With empathy, she assured him, "I understand."

"Do you know what created the 'problem,' as you describe it?"

"Only your grandmother can answer that one. There is much even my mother doesn't know. Mrs. Carrington may have held the crazy notion that Michael Ryan reminded her of Lord Carrington—a gold-digger, grasping for the Morgan family fortune. I suspect she worried he would eventually leave your mom in the lurch when the fantasy wore thin."

"No way! Dad is a Ryan, loyal to the death. In fact, he's the greatest person I've met in my life—present company excepted, of course."

"Lady Carrington believes that now. Perhaps the day will come when your grandmother will open up. I know she is trying desperately to reach out to heal wounds."

"You mean like the dinner she's scheduled this Friday evening? She's invited both me and Dad."

"I may have mentioned earlier that I received an invite, too. In fact, she's invited me for a long weekend, starting tonight."

"She must like you big time!"

"At least she likes me enough to tell me spicy tales about her only grandson."

"Spicy?"

"Even you can't be that dense." She laughed out loud. "It doesn't take a rocket scientist to surmise that a powerful woman like the Duchess is not above setting the stage for social accommodations that meet her approval, to say nothing of all-systems-go endorsement."

Slowly the light dawned. His strenuous efforts at pursuing romance with this dream woman were only one scene in a complicated script. Not only did the Duchess approve of this charmer, but she had apparently been planting the seeds of the Josh/Traci *tête-à-tête* for some time.

But he didn't mind. He appreciated all the help he could get. In these few, short moments he had gained a lifetime's briefing about the Duchess. He welcomed the implied reality: No way would Traci have joined him for the afternoon if she didn't harbor a hint of mutual attraction!

Time evaporated under magic's spell. Josh longed to stay forever. But he had to break away to tend to Turbo and gear up for the evening's *Monkey II* stint.

"I hate to abandon you for a hungry horse, but duty calls. Unless there is objection from Counsel, my motion to the court calls for a recess until the *Monkey II* verdict is in and a return to this discussion Friday afternoon."

Josh's heart leaped at the immediacy of her rejoinder.

"Given the stipulation of both parties, I expect the court to grant an order to that effect."

He had wished for a dramatic occasion to surprise Traci with a keepsake gift, but the present would have to do. Reaching into the folds of a plush Hampshire Club beach towel, he removed a slightly soiled jewelry box. Traci's eyes widened.

"I want to give you something to guarantee that you will remember me."

Noting her wonderment, and pleased with the subtle power inherent in the gesture, he savored the ritual. "My first choice was a piece of a dinosaur fossil. Then it dawned that a necklace strung with chunks of ancient bones wouldn't do you justice."

She broke into a grin, secure in the knowledge he meant the comment as a joke.

Opening the jewel box slowly, he removed the prize. Mounted on an intricate gold chain was a carefully preserved, childhood treasure. Barely an inch in diameter, the championship medal awarded its holder *First Place, 1987 Junior Golf Tournament, Hampshire Club.* The obverse carried the engraved name of the winner: *Joshua Chamberlain Ryan.* Following the starry-eyed athlete's instructions, a jeweler had replaced the medallion's tiny golden golf ball with a sparkling diamond. Diamond chips circled the perimeter.

Displaying the trophy proudly, Josh explained in bashful understatement, "I tried to come up with something that was part of me. I hope you approve."

Holding both ends of the necklace to either side of her neck, he whispered, "May I?"

Traci beamed unspoken affirmation. Josh wasn't sure, but he thought he spotted the beginning of a tear. As he leaned forward, he caught the delicious aroma of her fresh-scented hair, and romantic fantasies began running through his head again. Having practiced until his fingertips ached, he managed to snap the necklace clasp shut with minimal fumbling. As he pulled away, their eyes locked. Salty tears had replaced the moisture in her eyes.

Without a word, she slid caressing fingers behind his neck, grasped his head, and cupped his face in her hands. Slowly, imperceptibly, with mascara running down her cheeks, she moved moist lips next to his. Her hazel eyes searched his face as she answered his gesture with deft, feminine charm. "Thank you, Joshua Chamberlain Ryan."

A torrent of never-before-experienced emotion engulfed his being. Inevitably he succumbed to the whirlpool tugging at his heart. Alternating waves of hot and cold played tag up and down his spine.

Strange paradox, to drown in a living cauldron of burning fire!

With that, their lips touched and arms entwined in a passionate embrace. Temporarily forgetting that they presented a spectacle in a semi-public environment, the twosome looked at each other in wonderment, then pushed and pulled each other back into the pool, electing to swim off the burst of heat with another cool lap.

And not a moment too soon! Unobserved at the far end of the pool, a lone eavesdropper giggled in wide-eyed glee—absorbing chapter and verse for some future show-and-tell. Traci had neglected to alert Josh to her afternoon's role as sitter for a tag-along sister, Kristi, who had just been treated to a glimpse of education's fourth "R."

By the time they exited poolside, a discombobulated Josh recognized the depth of his own encounter with pure, primal love. Mind and heart merged to spin and whirl in unison. He felt like throwing caution to the winds, scrambling to the top of the clubhouse roof and shouting with reckless abandon to the world, "I love Traci Kilburn...and Traci Kilburn loves me."

Time prevented *Monkey II's* star witness from following his crazy impulse and risking further damage to his crippled leg, to say nothing of his finely honed reputation for academic sanity. Nothing remained but a shock-traumatized memory to navigate the drive home to Fort Ryan!

# 22

Turbo rollicked in crazy, crisscross patterns at the Mustang's approach, as though intuitively sharing the joy overflowing the heart of his owner. Chuckling at the exhibition, Josh detoured to the fence to pat the nose of his equine chum. A brisk inspection of the barn and pasture confirmed a more-than-ample supply of hay, grain, and water. To the tune of exuberant whinnies and snorts, Josh returned to the cottage while the frisky Turbo galloped in erratic frenzy, rejoicing at the return of his human keeper.

Still savoring delicious fantasies of Traci Kilburn, Josh harnessed his ego with a word of caution: "Don't blow it now, old boy!" His wishful thinking came crashing back to earth when he remembered that he had forgotten to share the distracting news of the impending litigation with Traci. He dismissed the thought, reasoning that in light of her affiliation with Judge Stone's law firm, she had probably heard the news already.

Josh resolutely suppressed self-pity at his diminished career prospects. Even if Traci loved him as much as he desperately hoped, the most heartfelt commitment might unravel at sight of his meager $200 a month overseer income. In humbling contrast, starting salaries for newly recruited attorneys of Traci's skill at Cabot, Calvert, and Stone were reputed to exceed eighty-grand per annum.

Jolted back to reality, Josh abandoned his disconcerting reveries and ambled toward the living room to sort through his notes for the evening's hot-seat. Passing the clutter spilling over the edges of a battered desk, he recalled the judge's admonition to recover the original draft of the unsigned reimbursement request from the previous April. Rummaging through the haphazard stack of accumulated paper was a daunting prospect. Then he spotted his battered briefcase leaning against the desk.

Repository of all hastily categorized important stuff, the scuffed carry-all bulged at the seams with long-forgotten articles, letters, and a scattering of heavily marked theme-paper drafts. Frustrated at his own incompetence in the basic rudiments of data organization, he fumbled through the mass, finding nothing. Almost as a last resort, he reached into an outside pocket and unearthed a single, wrinkled scrap of paper—the missing form!

Gratified at his good fortune, he scanned it to confirm that the signature line remained blank. Quickly he folded the critical original and deposited it in an envelope for hand-delivery to JT.

Feeling vindicated, Josh sighed with relief.

Looking across the breakfast table, he spotted the family heirlooms delivered by the Duchess to his custody the day before. Dutifully putting the diary aside until after Monkey II's final curtain, as his grandmother had requested, he decided to look for permanent display sites for the rest of the treasured items.

He set the diary on a sagging bookshelf beside some high-priced leather-bound volumes. Although this newest treasure had no discernable market value, he wouldn't trade it for all the gold in California. It represented the only copy of his mother's handwriting that he possessed. And it chronicled the last tumultuous years in the life of Leslie Anne Carrington-Ryan.

The oil portrait of his mother depicted a wiry young woman fashionably attired in the gear of the Hampshire Hunt, her Arabian mount captured in mid-stride, jumping a fence. The picture fit snugly above the crudely crafted, rarely used fireplace where it enhanced the décor, looking as though it had hung there forever.

The Ryan heir stood back, admiring the finely chiseled features of the woman he forever craved to know and to love. His appetite for information had been whetted when a door to his mother's life had been pried open innocently earlier this day, by Traci Kilburn. He guessed correctly that Traci's mom could offer tempting additional tidbits, far beyond Michael Ryan's knowledge.

Next he turned his attention to the hand-hewn candlesticks. The masterpiece pair, created by an anonymous artisan, would command a pretty penny at public auction. But like the diary and the painting, they would never be sold at any price during Josh's lifetime. Rather than placing them in the window sill fronting Georgia Avenue, Josh chose to bracket the fireplace mantle with the artwork. To clear a place of honor, he removed a jumble of fossils, demoting all but two or three of the more illustrious specimens to a less-prominent site in the mill house. Nothing unsightly remained to detract from the belated memorial.

Lacking fancy new candles to complete the scene, he scrounged around kitchen drawers, turning up a couple of faded remnants. By shaving and trimming the oversized bases, he managed to fit the stubby fragments for temporary service. In the process, he noticed handwriting etched on the underside of each candlestick base. In tiny, almost indistinguishable print, one carried a complete genealogy of the Morgan/Carrington clan commencing with Sam Houston Morgan. To his amazement, his own name and birth date appeared as the last entry, immediately after *Leslie Anne Carrington-Ryan*.

The second candlestick base presented another surprise. Penned in delicate longhand and dated Tuesday, September 3, 1996, the inscription read: *To my grandson, Joshua Chamberlain Ryan, with all my love. The Duchess.*

Satisfied with his efforts at interior design, Josh set to work on Wednesday evening's stage performance. Gazing out the mill-house window, he caught a glimpse of a cardinal's scarlet plumage, bird feeder bound. Mesmerized by its lavish beauty, he marveled at Fort Ryan's daily parade of birds: black-capped goldfinches clad in brilliant hues of yellow; iridescent indigo buntings flaunting shimmering shades of purple and deep blue; and the aerobatic stunts of tiny, ruby-throated hummingbirds.

Josh felt delight at the sight, despite Charles Darwin's insistence that it would be fatal to his theory if *"organic beings have been created beautiful for the delight of man."*[78]

Nothing grated Josh's sensibilities more than the outrageous assertion that birds were descended from dinosaurs. He mused, "Just think of the hummer hovering in the face of an alleged great-grand pappy dino, commenting, 'What big teeth you have, Gramps.' Grumpy grandfather dino stares back astonished, wondering rhetorically, 'What absurd quirk of genetic fate produced this runt-of-the-litter offspring—with darting wings, no less!'"

Josh laughed out loud at the outlandish caricature, which reminded him that most birds are unable to sustain for long the enormous energy required to hover, but *"hummingbirds...have an unusual shoulder design that allows them to generate lift on both down-beat and up-beat."*[79]

"Random luck of a bazillion mutations from dinosaur DNA? No way! Talk about generation gap!" The young scientist heehawed at the absurdity.

Like the prior events of the day, time swept by like a tornado. Inspired by the clock, the witness shifted his trial prep into high gear, brushed his hair, grabbed a jacket, and headed out the door. He had missed another meal, but shrugged, rationalizing that hungry minds were sharper.

The derelict Mustang exterior belied the finely tuned quality of the rebuilt engine. Next to Turbo, this mechanical horse was Josh's most treasured possession. As always, the engine kicked over at the twist of the key.

But today, something seemed seriously amiss. When he shifted into reverse, his classic car crawled sluggishly down the drive, then ground to a halt. Josh sputtered with frustration, "Talk about lousy luck!"

Slipping from behind the wheel, he saw to his chagrin that the front left tire was flat as a pancake, wheel rim resting on gravel. Hustling to the trunk for a jack and spare, he confronted an equally squashed and forlorn left rear tire. Swallowing panic, he circled around to the right side to confirm his worst fear—tires three and four matched the first two. All four showed signs of having been punctured.

Josh remembered a local tale featuring a fastidious, Marauders Mill Quaker facing a similar dilemma before the era of roadside service. Church-bound in Sunday-go-to-meetin' clothes, a sequence of three flat tires en route left the old-timer no choice but resorting to fix-it-yourself. Exasperated by the husband's take-it-in-stride demeanor, the prim wife finally nagged: "Can't thee at least think, damn?"

With no time to enjoy the gallows humor or to assess the implications of sabotage, Josh turned to the house, intending to call a taxi. Remembering that it normally took a minimum of twenty minutes for the cab to show, he abandoned that idea. Next he thought of his dad's car locked in the manor house garage on the hill, before recalling that he had relinquished the ignition keys the day he took title to the Mustang.

Trying to keep calm, he spied the slim lines of a mountain bike propped against the wall of the barn. While he had ridden these country roads recreationally many times, he knew the circuitous route to Mid-Atlantic University meandered up and down steep hills. There just wasn't enough time.

He kicked himself for his shortsighted reliance on split-second timing. Now, feeling isolated and very much alone, he felt cold fury at the realization that someone had tried to keep him from making tonight's appearance. More than upsetting the momentum of the stage drama, the substitute Darwin would appear irresponsible. Worse yet, the public might construe his absence as an act of cowardice—and no Ryan ever chickened out!

Josh reacted indignantly. Propelled by the realization that desperate circumstances demand desperate measures, Josh burst out laughing. Gazing across open pasture, an answer stared him in the face.

"Here I come, Turbo, ready or not!"

Many a moon had elapsed since Josh had last ridden Turbo in the Hampshire Hunts. While every inch of the trails used in the old days had been indelibly tattooed in memory, he lacked the remotest idea whether the fields remained open to uncharted gallops.

Now was not the time to quibble. With precious seconds ticking away, Josh streaked toward the meadow, coattails flying. He knew that saddling Turbo was out of the question, even if the mount proved amenable. The best he could do would be to grab a bridle as he raced through the tack room and hope the horse would not reject the overture.

Josh dared not consider how long it had been since a rider attempted to take on the fickle Turbo bareback. His only chance of reaching the campus by curtain time was to travel as the crow flies, albeit closer to the ground. In his growing years, Josh had heard Michael Ryan borrow a family phrase that depicted Sam Morgan's dash into history as a *ride to glory*. The wise father added a philosophical touch.

"Remember, Son, sooner or later every one of us is offered a chance to take that ride." If the mule-headed Turbo only shared the proper mood, Josh would willingly relinquish the glory.

A hopeful Josh welcomed the sight of Turbo trotting to the fence. Casually vaulting the oak slats, Josh slipped the bridle on the now-docile mustang. With no time for ceremony or the trumpet's call to derby starting gates, he slid onto Turbo's back and grabbed the reins. Without so much as a word or a heel in the flank, a super-charged Turbo bolted cross-country in the general direction of Mid-Atlantic University.

Thus launched, Josh borrowed a line from radio's Lone Ranger, and shouted encouragement. "Hi O' Turbo...awaaaay!"

The unpredictable mount pranced exuberantly at the prospect of this one-man/one-horse private hunt. At first, Josh worried about the impact of age on the smooth-muscled Arabian. Within minutes, his concern shifted to personal survival.

Glory or no, it proved a ride to remember: mad dashes through dense thickets; headlong leaps over tumble-down fences; and wild plunges into shallow stream beds, spraying sheets of water in all directions. It tested the rider's strength to cling precariously to Turbo's bony back while succumbing to the animal's rambunctious zest!

Josh needn't have worried about the aging steed's stamina—Arabians were bred for endurance. Tearing at breakneck speed past the bewildered stares of grazing cattle while cutting like a scythe through the jagged edges of wheat fields, the bold rider's confidence grew with each stride. The narrow time frame could be conquered—if he could just hang on to the finish line!

Occasionally Josh caught sight of waving bystanders, intrigued by the spectacle. Forgoing the courtesy of returning the salutations, he clung to the horse's mane for dear life. To Josh's mind, the ride had become a desperation dash, trying to hang to the sleek spine of the horse while ducking protruding tree branches and searching for familiar landmarks. If the destination proved to be somewhere short of eternity and midway between downtown Baltimore and the District of Columbia, the horseman would claim a win, place, and show—with a clean sweep.

"So far, so good," thought Josh, calculating 50/50 odds for a timely arrival—if Turbo's boundless energy didn't dissipate or some new subdivision had not devoured the familiar paths of the hunts.

For one anxious moment the wildly flying pair had to reroute, gamely bypassing a field carpeted with groundhog condominiums. Any one of the network sieve of gaping holes could have brought down both horse and rider. At another point they had to jump a broad expanse of wetland without a clue as to the opposite bank's landing site. Josh viewed it as a blue-ribbon, championship challenge—which Turbo performed admirably.

Josh and Turbo sustained multiple scratches and gouges during their desperate ride, and both were drenched with sweat. Rain had begun to fall, adding to their discomfort. But nothing would deter them from reaching their goal. The intrepid duo kept on until the goal loomed in sight. Chamberlain Playhouse emerged, bathed in a golden light affirming its landmark status.

Josh tasted victory!

Out of consideration for Turbo, he attempted to slow the blazing pace, but the Arabian refused to be denied. His clattering hoofs accepted but two speeds: floorboard and stop! And it was floorboard from the start until he turned into the university's iron-gated entry. As if on cue in a play with a script of his own creation, Turbo finished the run by strutting up the hill in haughty triumph. Here he halted, standing shoulder-to-shoulder with the bronze replica of General Chamberlain's rearing equine, basking in the forever salute of the general's sword.

Josh relaxed, more proud than relieved. Sliding deftly to the ground, he took the reins and twisted them in a single loop through the stirrup of the general's bronze mount. Then he embraced Turbo.

Knowing that even a great Arabian needed meticulous grooming following such a furious chase, Josh faced a new conundrum. He needn't have worried. A smiling Lady Carrington suddenly appeared at his side, dressed dazzlingly for her grand entrance to the evening's performance. She signaled her chauffeur to stand by as she beguiled her grandson with lavish praise.

"Land's sakes, boy—you ride like a Morgan!"

And then, caught up in the spirit of the moment, she added with intuitive perception, "I'm sure there's a reason for your unconventional arrival on the scene beyond some impetuous craving for publicity."

"You got that, Ma'am. I've added new meaning to the old saw, *skin of your teeth.*"

Looking at the gallant Turbo, the Duchess' nurturing instincts took over. "Looks like this dear creature is in serious need of a rubdown. You know, of course, he wears the champion heart of his sire, the horse your mother rode for the oil portrait?"

"Yup, I know, on both counts. Turbo does have a champion's heart, and the poor guy does deserve some pampered grooming. I'm just debating whether to put the show on hold while I take care of him."

"Nonsense, Josh. Don't worry that handsome Morgan head of yours. It's not a problem."

The Duchess whipped out a cell phone, called a foreman at the Morgan Manor stable, and ordered her driver to "pick up the groom with all necessary tools of his trade to take care of this magnificent blue-ribbon Arabian."

Turning back to Josh, she said, "We'll take him to the Morgan stable. This baby deserves quality R&R at the home place. You can pick him up later in the weekend. Meanwhile, you go in there and get the show on the road. I'll keep him company until my man shows. Then I plan to strut my grand entrance."

As an afterthought, she barked out a last directive. "Give 'em hell, young man—Morgan style!"

He chuckled at her brazen sassiness. "Thanks, Duchess. I've got five minutes to comb my hair, find a clean shirt, and rush to the witness stand to smirk for the cameras.

"One thing for sure, somebody is going to be surprised and disappointed to see me safely on stage, exposing missing fossil links. I'll fill you in later!"

He took off running.

The curtain rose precisely at 8 p.m. with Josh at ease in the hot-seat—despite the fact that his desperate search for a fresh shirt had failed abysmally. Vainly he hoped that no one would notice his bedraggled state.

# 23

Josh wished fervently he could escape the searching finger of the center stage spotlight. "Perhaps my disheveled appearance may pass as part of the act," he thought.

He needn't have worried. An oblivious audience saw the witness settled serenely, poised to unleash another verbal barrage aimed at *fact-free science*. Still, the performers noticed something askew. Judge Stone suspected that Josh's rain-soaked attire could have some connection to the drama involving academic dispute and civil litigation. Before inviting testimony, he informed the audience that he found himself the ringmaster of a circus being played out in at least two venues.

*Monkey II* had been designed to produce entertaining confrontation. But the million-dollar-plus lawsuit targeting the student witness was not some promotional hype. Noting the capacity crowd, no doubt buoyed by newspaper reports and campus television coverage, his judicial instincts compelled Edward Anthony Stone to comment on information leaking into public domain.

"Before I convene this evening's performance, you should know that a potential conflict of interest has arisen relating to my role in this pretend courtroom. In real life I would recuse myself from judicial responsibility.

"Earlier today, my law firm undertook to represent Mr. Ryan in the defense of a claim for damages filed against him in civil litigation. It remains to be seen whether testimony given here earlier this week inspired the litigation. The plaintiff in the lawsuit happens to be an entity exploiting the university's name in its title while claiming to function independently. As a trustee of Mid-Atlantic University it would, of course, have been impossible for my firm to represent a defendant being sued by the university.

"I believe the litigation against Mr. Ryan is without merit and in no way represents the university or its best interests. The plaintiff's counsel has stipulated to Cabot, Calvert, and Stone's representation of the defendant. Regardless, I've advised Mr. Ryan that it would be to his advantage to discontinue testifying in this staged show trial.

"As you can see, he has ignored the advice of counsel. His presence here this evening strikes me as an act smacking of either pig-headed presumption or of courageous commitment."

Applause erupted. Friendly shouts of "Go, Darwin" erupted from a surging band of student supporters, punctuated by polite claps from the more reserved.

Judge Stone ended with a judgment call, not without risk to himself. "Since the attorney for the plaintiff in the pending civil litigation has been invited to cross-examine our witness on this stage Thursday evening, the conflict issue cuts two ways. The plaintiff encouraged the courtesy of that appearance, fully aware of my judicial role here. I have no intention of abandoning my client to a risky fish hunt in a barrel.

"Consequently, I intend to remain as the presiding judge in this staged drama until its conclusion, at noon on Friday."

More applause, spiced this time by boisterous shouts from students, *"Go get 'em, Darwin!"*

Seated next to JT at the counsel's table, the soggy Josh learned belatedly that his tardy arrival would neither derail nor delay the evening's show. Preoccupied by his pell-mell dash to beat the clock, he had forgotten that Striker's peon would be appearing to rebut prior testimony. At first kicking himself for the mental lapse, he finally sat back, grateful for the reprieve.

Dropping the oak gavel, Judge Stone ordered Traci to proceed. "You may call Mr. Lassiter to the witness stand, Ms. Kilburn. After he is duly sworn in, you are entitled to employ the same open-ended direct examination of the witness that has been indulged by this court previously between Mr. Daniels and Mr. Ryan. This includes leading the witness, expression of personal opinion, and even a touch of history—providing you avoid the totally outrageous! Is that understood?"

"Yes, of course, Your Honor," replied Traci. JT signaled that he understood and agreed.

"You may take the stand and be sworn in, Mr. Lassiter."

Slade Lassiter swaggered to the witness box, in stark contrast to the hesitancy he had displayed when removed from jury duty. The previously shy student seemed eager to perform. His overnight transformation baffled onlookers. On closer observation, he seemed unnaturally calm, even spacy. A suspicious JT Daniels leaned closer to Josh, saying, "Hey, pal, this chameleon needs a breathalyzer test. Unless I miss my guess, ol' Slade's been sniffing forbidden smoke or guzzling heavy-duty firewater, trying to fortify himself to unload his sleaze."

Josh responded with a playful fist to his friend's shoulder, remarking, "Well then, *Judge Daniels*, you should sentence Mr. Slippery to thirty days in the tank to cool his heels for contempt!"

It was all the irrepressible duo could do to suppress muffled snickers, anathema to the decorum demanded by the court setting. Returning to reality, they shared the delicious humor in labored silence. Josh's follow-up wisecrack nearly destroyed their composure.

"With Slade's intellect, he'd best try to keep those sparse faculties unimpaired!"

Lassiter took the oath in modulated tones, doing his best to appear the fountain-of-wisdom academic, determined to refute error and champion truth.

184

Traci went to work, enjoying the chance to play lawyer under the gaze of a renowned jurist but with rules of evidence deliberately lax. "Is it true, Mr. Lassiter," she asked, "that your appearance as a witness this evening is at the behest of MAU's Science Division chairman, Dr. Karl Striker?"

"Yes, absolutely true. And I consider it an honor to be present in that capacity!"

"Would you characterize your capacity as official representation of the university's Science Division?"

"Yes, of course. I'm here to testify officially!"

"Would you say, then, that Mr. Ryan's testimony is something less than official?"

"Not only are Mr. Ryan's careless remarks unofficial, but also his presence here violates the express wishes of Science Division leadership. He no longer qualifies even as a credentialed Ph.D. candidate at Mid-Atlantic University. His conjectures about whether or not evolution is a fact are based entirely on his own views. He speaks exclusively for himself...just like the rest of the unscientific crackpots in the world. Joshua Ryan is fully aware that he is *persona non grata* as far as Science Division leadership is concerned."

Soft mumblings echoed from the audience.

"The rumor persists that Mr. Ryan's grade transcript reflects nothing but a straight 'A' average," Traci challenged. "If that is true, how do you reconcile that record with the attitude of officialdom?"

"Understandably, I'm not privy to details of Mr. Ryan's records and relationships. Perhaps this coincidence reflects the objective scholarship and magnanimous fairness of MAU's presiding science leadership in view of this gentleman's persistently obstinate and offensive behavior."

"Given your duly declared *official* status, to which I've heard no objection from Mr. Daniels, would it be a fair assessment that you speak for Dr. Striker regarding the core issue: *whether or not evolution is a fact*?"

"Yes! Absolutely fair...and correct! It's obvious as the nose on your face: *Evolution is fact*! Any fool with a first-grade education knows as much."

"Do you judge Mr. Ryan to be a fool?"

Lassiter squirmed and looked reprovingly at Traci as if to say, "You're supposed to be on my side." Chastened, he backed off from the hyperbole.

"Maybe irrational or unrealistic would be more accurate."

"Mr. Ryan claims that there is little empirical evidence supporting the claim that *evolution is fact*. What evidence can you introduce demonstrating conclusively to the contrary—that *evolution is fact*?"

"Perhaps we could short-circuit the process and go straight to the bottom line without a boring rehash of the junk the jury has been subjected to for the past three days?"

"It's your testimony, Mr. Lassiter. Feel free to introduce your 'bottom line.' Remember, we are seeking falsifiable evidence."

"Although it pains me to do this, I have no choice but to present solid proof that the so-called expert in this case speaks with the credentials of a first-class fabricating fraud!"

Jurors looked startled at the icy blast, while the bold accusation drew an audible gasp from the audience.

A white-faced Josh Ryan seethed. JT's powerful restraining hand on his shoulder was all that kept Josh from erupting in self-righteous anger.

"Don't play the fool and pull a Striker tantrum!" JT warned. "Sit on it, pal! Let your lawyer do your lawyering. Consider the source, man. You'll have your day!"

Only partially appeased, Josh bought the prudent counsel and sat silently, glaring at the witness.

Traci hesitated, having no idea where the venomous attack might lead, then asked, "Could you be more specific, Mr. Lassiter?"

"I certainly can, and intend to do so. I carry in my possession a duly executed, authentic document—an original, I should add—that impeaches the integrity of so-called Mr. Clean. Anything Mr. Ryan may have testified to in this hearing should be seen as so much hyped prevarication!"

"Is Dr. Striker aware that his official representative possesses such dynamite documentation?" Traci asked.

"Aware of it? His notarized signature guarantees it!" Lassiter answered haughtily.

"Please describe this document to the court, if you would."

Lassiter smugly reached inside the breast pocket of his jacket and whipped out a fresh, unsealed envelope. Opening the flap, he withdrew a crisp sheet of parchment bearing the embossed seal of Mid-Atlantic University, and continued triumphantly, "This letter, dated today, is addressed to me, and was hand-delivered by the big man himself."

"Big man?"

"You know, Dr. Karl Striker—a *real* scientist of international stature!"

"And what did the letter reveal that might be of interest to this court?"

"Why don't I read it for the benefit of the jury? It's short."

"Go ahead, if it's OK with Judge Stone and Mr. Daniels."

Traci looked inquiringly at one, then the other.

"I can hardly wait," shot back JT.

"You may proceed," assured the judge.

Clearing his throat self-consciously, Slade Lassiter spit out each phrase with sarcastic innuendo:

*Dear Mr. Lassiter:*

*Effective as of 2:33 PM today, Wednesday, September 4, 1996, I completed an inspection of the Science Library's archives and its computer index reference. I found no evidence of the existence of or reference to any alleged August 1969 letter written by the distinguished, late chancellor, Robert Brooke Fielding, to Lady Regina Ann Carrington, respected trustee of this university.*

*Sincerely,*
*Dr. Karl Striker*
*Chairman, Science Division*

Stony silence greeted the bombshell, followed by noisy commotion. Lassiter beamed at the rolling wave of consternation. His verbal home run had cast serious doubt on the three previous days' testimony. Lassiter's evidence pointed an accusing finger at Josh Ryan, erstwhile hero of "whole truth, and nothing but."

White fury engulfed the star witness. JT used his considerable muscle to restrain Josh from unraveling in anger, whispering, "Don't be an idiot, Josh. If you lose it now and explode, think of the smirks on the faces of Lassiter and Striker. That's what they want you to do, man—self-destruct and erode your credibility. There's something rotten in Denmark, and it ain't stale cheese. We may not find out what it is tonight, but remember, that's why you pay us lawyers ridiculously high fees. Trust me!"

Though roiling inside, Josh managed a semblance of outward composure as Judge Stone restored order with a few, sharp raps and warned: "This court tolerates nothing less than absolute order!"

Unsmiling, he looked to JT, inviting comment. "Perhaps, Mr. Daniels, you would care to review the Striker letter before addressing the witness."

Taking the elegant piece of official stationery brandished in Slade's pudgy hand, Traci delivered it ceremoniously to JT, who scanned it quickly while moving to the witness stand to confront Lassiter face-to-face.

First, he addressed the bench, "Thank you, Your Honor, for giving me the opportunity to examine the witness. Before I do, however, am I correct in assuming that this court lacks subpoena power and that Dr. Striker has rejected all invitations to appear here voluntarily and testify as to his beliefs?"

"Correct on both counts, Counsel."

"In that case, my questions to Mr. Lassiter will be brief."

JT swung around to confront Slade. "Do you, Sir, claim to be a handwriting expert?"

"No, of course not. But I personally witnessed Dr. Striker's execution of this letter and therefore can vouch for the authenticity of his signature."

Sensibly, JT abandoned this line of questioning, choosing to explore elsewhere for an armor chink.

"Did you accompany Dr. Striker when he researched the Science Library's archives and computer reference index?"

"No, I did not."

"Did you ever check the index yourself?"

"No, I did not."

"Could you explain just how Dr. Striker happened to check the index?"

"Certainly. I urged him to do so, since his authoritative opinion carries probative value!"

"When did you make that request?"

"I don't remember for sure, probably shortly before noon today."

"And although you had never checked the index prior to that request, were you confident about what the doctor might or might not find?"

"While I couldn't know for sure, Dr. Striker's findings—or lack thereof—were no big surprise, given the reputation of the star witness for gambling and his contempt for scientific tradition."

"Is it then your assertion that Mr. Ryan fabricated evidence for this hearing in the form of a make-believe letter from the late Dr. Fielding to Lady Carrington?"

"Sadly, there seems to be no other conclusion," Lassiter said in mournful tones, sorrow worthy of a soap opera.

"I notice this letter affirms Dr. Striker found no evidence of the letter in question as of 2:33 p.m. today. Is that correct?"

"Dr. Striker is a stickler for precise detail!"

"And Mr. Ryan testified previously he discovered the existence of the letter not quite 24 hours previously—is that not correct?"

"Of course it's correct—unless you prefer to swallow the hype of a proven perjurer! For sure, it implies that one or the other is lying. My money is on the integrity of Dr. Striker. He is no prevaricator! Obviously, Mr. Ryan's position appears quite untenable."

Satisfied with the logic, Lassiter eyed each juror, seeking hints of approval.

"You've overlooked one other possible conclusion, Sir. Is it not conceivable that both gentlemen could be telling the truth?"

"Impossible!" Lassiter said with indignation.

"Yesterday Mr. Ryan found the letter. Today Dr. Striker states he could not find the letter. Could it be that Mr. Ryan researched more skillfully than Dr. Striker?"

"That's outrageous. Dr. Karl Striker possesses unmatched scholarly credentials!"

"Seems to me, Mr. Lassiter, with two scholars skilled in research, quest for an archival artifact should not result in two conflicting results, unless..." JT stopped in mid-sentence, moving in nose-to-nose with the witness.

"Unless, Sir, sometime between late yesterday afternoon and 2:33 p.m. today, the Fielding letter, along with its index reference, were deliberately removed from the Science Library!"

"Preposterous! That's absolutely absurd, Mr. Daniels!" the witness sputtered, nostrils flaring. Judging by the stirrings and hummings coming from the jury, JT had planted the seed that might produce a alternate conclusion.

Turning abruptly to the judge, JT said, "No further questions of this witness, Your Honor. And I have no objections if Ms. Kilburn chooses to submit the Striker letter to the jury as evidence—bearing in mind some pertinent, unanswered questions: Which document is forged? The Fielding letter? The Striker memo? Or both? Also, who could be the perjurer? Dr. Karl Striker? Joshua C. Ryan? How about our own self-appointed investigator, Mr. Slade Lassiter? Or could all of the above be guilty of obstruction of justice?"

A frowning Josh wrinkled his nose with distaste at having his name associated with Striker's and Lassiter's, but held his peace.

A poker-faced Judge Stone peered over his spectacles as he turned to Traci. "Very well, Counsel. What is your pleasure, Ms. Kilburn?"

"Understanding that I had no prior knowledge of the Striker memo until its presentation here, and given the gracious response of Mr. Daniels, I move the admission of this letter into the record as evidence for jury consideration."

"So ordered. That will be all, Mr. Lassiter. You're excused."

Looking at his watch, Judge Stone decided the time had come to take a brief recess and gaveled for the curtain.

The inquisitive buzzing of voices echoed in the acoustically sensitive auditorium. As of this instant, *Monkey II* had become a hot ticket.

# 24

"Did you hear that poster boy for fraud perjure himself with a straight face? How can you let him get away with that stuff, JT? And what's with this lumping me in with that rogues' gallery of perjury suspects?"

JT had never seen his usually restrained friend so livid. To avoid a backstage scene, he guided the incensed witness out the exit to the soothing embrace of the evening air.

"Get a grip, guy! A public explosion in the presence of a thousand witnesses would fulfill the fondest dreams of Slade and Striker. I promise to mount a counterattack by Friday at the latest—but I need time to find out what's been going on! Meanwhile, you, my friend, need to get your act together. Forget the insults, and focus on the chinks in evolution's armor. Keep hitting on the target. They wouldn't risk public prevarication had they not been reeling from the intellectual nerves you've struck."

The ad-libbed lecture continued uninterrupted until the distraught witness reclaimed his composure. After a few moments, Josh said grimly, "Let's go sic 'em, JT."

Relieved, JT Daniels steered his distraught pal back to the probe of the spotlight. Disheveled and damp from his cross-country dash, Josh looked the part of a victim. The skin-drenching ride and verbal shellacking had taken their toll.

Backstage, Joshua's sorry appearance evoked queries from the cast, who gathered around him as though drawn by a magnet. All faces shared a common concern: How come the soggy look cloaking the normally chic student scientist?

Josh shrugged off their inquiries, trying to appear flip and jovial. "I suppose you've heard of the guy who was too stupid to come in out of the rain? Truthfully, I did the best I could, but the horseback ride lasted longer than I expected!"

Alarmed, Traci said, "I don't understand. Are you OK?"

"Well, it's a long story...and there's no time now. There's a lot I don't understand either. But for sure—someone out there wanted me AWOL from tonight's show!"

Judge Stone stepped in to sternly remind the young Darwin, "Remember, young man, I warned you. More than one agenda is in play here. Marcus Brogan can perform in a courtroom with the cunning of a tiger, but stupid he is not. Nor is his client, Karl Striker."

After nodding polite assent, Josh rushed to the dressing room to make a last, vain effort to brush his unruly hair, congenitally resistant to the clean-cut look. Tugging and smoothing at his soggy, rain-wrinkled clothes proved equally fruitless. Surrendering to the inevitable, he hastened to center stage, looking decidedly worse for the wear and barely outpacing the rising curtain.

With his clear view of the audience, Josh noticed Dr. Karl Striker posted at the elbow of Attorney Marcus Brogan. Internally he raged at the sight, thinking, "Striker, you gutless wonder! You make your boot-licking sycophants do your dirty work."

Front and center, a regally splendorous Duchess presided over the Row G seats allocated to her guests. Unfazed by Slade Lassiter's brazen assault on the credibility of her grandson, she basked in the thought of their earlier encounter at the base of the Chamberlain statute and the opportunity to lavish royal treatment on Turbo.

She caught Josh's gaze, giving him a dazzling smile and a spirited thumbs up. The Duchess' body might be approaching the four-score-year anniversary, but her mind and spirit were those of a twenty-year-old.

Once Josh settled in, JT responded on cue to Judge Stone's order to proceed. "You've made a big deal of the microscopic to trash Darwin's ideas. It's time to look at the macro picture! Tell us, Mr. Ryan, what's the deal on fossils?"

Josh reached to the floor of the witness box and lifted out a fossilized dinosaur bone fragment he had discovered while tramping through Colorado's high country. With the court's permission, he passed the prize to the jury for hands-on inspection.

"Darwin imagined a process starting with a single cell featuring gradual incremental change, over millions of years. *'Natural selection acts only by taking advantage of slight successive variations; she can never take a great and sudden leap, but must advance by short and sure, though slow steps.'* [80]

"Fossil discoveries since *Origin* suggest the precise opposite. Complex life had an across-the-board beginning: abrupt, simultaneous, worldwide. Multi-celled animal embryos no bigger than a grain of sand have been discovered in China, preserved in calcium phosphate and dated at the edge of Cambrian/Precambrian—marking the prolific explosion of organic life during which *'virtually all the major animal body plans seen on Earth today blossomed in a sudden riotous evolutionary springtime.'* [81]

"An estimated 7,640 [82] complex animal species saturate the fossil record of the Cambrian without a trace of prior ancestors. This explosion of complex life forms covers the waterfront as to invertebrate animal phyla, both living and extinct. I'm not aware of any new invertebrate phyla appearing since that original beginning.

"The gradualism theory, which calls for continuous, incremental change, remains only a vague enigma! Darwin put his whole ideology on the line, admitting, *'If it could be demonstrated that any complex organ existed, which could not possibly have been formed by numerous, successive, slight modifications, my theory would absolutely break down.* [83]

"The Cambrian Explosion exposes the theory to *breakdown*! Unlike conclusions produced by the Darwinian analysis, Cambrian life appears complete and complex—a *sudden leap*, simultaneous and worldwide! Arthropods, mollusks, and echinoderms—there they are, without a fossil clue as to prior *numerous, successive, slight modifications*."

Jurors exchanged knowing nods while JT suggested an alternative explanation. "Could Cambrian species have been intermediate varieties, transitional from their fossil ancestors and leading to radically different future prototypes?"

"What ancestors? That's just the point! The Precambrian Era preceding the Cambrian Period is a paleontological desert, virtually barren of identifiable fossil ancestors. We're talking blank screen here! No *intermediates;* no *transitionals*!"

"Not even something to argue about?" JT stood back, allowing the witness to vent.

"Sure, some animal-like Ediacaran fauna and bits and pieces of stuff like bacteria—maybe OK for somebody who wants to honor a molecule as the first ancestor in a family's genealogy!"

"But no sign of transitionals?"

"None to my knowledge!" Josh asserted.

"Suppose we forget the missing intermediates preceding the Cambrian. What about fossil intermediates since then?"

"The *numberless intermediate varieties* Darwin declared *must assuredly have existed*, still evade discovery in fossil format! In a game of paleontological hide-and-seek, even a century-and-a-half of serious seeking has not been enough to discover the hiding place of creatures that never existed!"

Jonathan kept prodding the witness. "Still, it's widely believed that mutations presage change! What's wrong with mutations as agents for mega-evolution, with or without intermediates?"

"Evolution latches onto mutations as the mechanism for change, a magic elixir that lubricates natural selection. But mutations make pitifully poor agents for *progress toward perfection.* If gene defects guard evolution's crossroads, all transitional trains derail and Neo-Darwinism goes extinct!"

"Please explain, Mr. Ryan. The jury awaits enlightenment."

"Mutations deal mega-evolutionary theory a double whammy: Evidence for beneficial mutations is scarce, if not non-existent; and mutations are flaws that don't add any new genetic information."

"Go on..."

"'*Major functional disorders in humans, animals and plants are caused by the loss or displacement of a single DNA molecule, or even a single nucleotide within that molecule.'*[84] *'There is no evidence for beneficial spontaneous genetic mutations.'*[85]

"And the second whammy?"

"Mutations degrade the genome by displacement, damage, or loss of information. They provide no new genes to fashion an entirely different prototype life. It's like shuffling a deck of cards and dealing new hands using the same old stack. Some may be lost or turn up bent and torn, but nothing new is added to the original deck. You just get a variety of hands with assorted combinations."

"Then you do admit change?"

"Of course!"

"Do my ears deceive? Does this mean you and Charles D agree for once?"

Josh guffawed, shaking his head in protest. "Diversity each time the original genetic card deck is shuffled, sure! An existing genome's gene pool makes possible different-shaped beaks among Galapagos finches. A genome degraded by mutation may induce disastrous change. But you can't extrapolate to build a bridge for mega-evolution in either case because no entirely new genetic info is produced.

"Micro-evolution is actually a deceptive misnomer. Inherited potential for genetic adaptability by shuffling and dealing the pre-existing genetic card deck doesn't equate to evolution of any kind. As for extrapolation from natural diversity to mega-evolution, not a chance!

"*'No one has ever bred a new species artificially—and both plant and animal breeders have been trying for hundreds of years, as have the scientists.'* The *Drosophila* with its four pairs of chromosomes *'can breed a new generation in less than a month. Using the same stock with an average of 36 bristles...over thirty generations, the experimenters were able to reduce the average carried by the offspring to 25 bristles.'* Efforts to increase the bristle count also hit the outer limits to change: *'Over twenty generations the average rose from 36 to 56.'*[86] And there was an undesirable by-product—the experiments typically led to diminished genetic fitness and sterility."

Josh paused to dramatize the punch line. "*'Genetic homeostasis will prevent morphological change beyond a certain point...there is no evidence for gradual change leading to macroevolution in the fossil record...no transitional species showing evolution in progress has ever been found.*[87]

"Darwin's cumulative slow steps lack inevitability. Without a master plan, what's to prevent that micro step reverting to the original format, slipping laterally, meandering aimlessly or eventually hitting the barrier to change—as with the *Drosophila*? Mega-evolution flounders, trapped in a catch-22. The theory requires millions of years but the necessary intermediate steps, with incomplete change, render any halfway critter vulnerable to extinction. And what if the postulated millions of years are not available?

"*The Earth may have been made from materials that are 4,500 million years old and yet still have been formed relatively recently.*'[88] The application of Willard Libby's '*own data*' to the '*radiocarbon technique*' suggests that '*the age of the atmosphere is around 10,000 years.*'"[89]

The audience and jury buzzed.

"Could this skimpy scheme offer anything more than mega-myth malarkey?" Josh queried rhetorically.

"You're sure it's malarkey?" asked JT.

"Beyond the built-in limits to genetic codes jumping prototypes, the interdependence of all plant and animal life demands a synchronized ecosystem! Even the honeybee can't sweeten the Darwin recipe."

"What about the bee, Mr. Ryan?"

"The bee defies evolution! Where is its intermediate ancestor? How did the honey-making process evolve? What inspired the engineered design for a comb capsulated for honey storage and a hierarchy of a special-diet queen, workers, and drones? What about flight? Stings? Swarming?

"What if bees evolved before clover or clover before bees? Each depends on the other for survival."

"Mr. Ryan, apart from your knowledge of the birds and the bees...," JT said with a grin, "can you explain the core problem with halfway transitionals?"

"Sure! How could the eye concurrently evolve eyelids, tear ducts, a retina, or bony sockets over millions of years without the pathetic transitional creature suffering disaster? Imagine a mouse-like critter trying to evolve wings like a bat over millions of years. When stuck halfway through the process, lacking working wings to fly or front legs to walk, the creature would perish! Picture a critter intermediate between dinosaur and bird—until all systems are go, crash-landing wipe-outs would destroy the monstrosity!"

"Without a complete, functional design in place, the system fails. A part-way process would destroy a species in transit. A multi-million year transition time offers a self-destructing environment!"

Traci jumped up. "Objection, Your Honor. The witness insists on rattling off personal opinion!"

The jurist reacted nonchalantly, without bothering to look up. "Overruled!"

Traci had anticipated the judge's ruling, wanting only to jiggle Josh's concentration.

JT spurred the witness to offer more opinions. "So if Cambrian phyla appeared abruptly, without a certifiable trace of fossil ancestry, what about Post-Cambrian living *transitional*s after the period?"

194

"Can't find them in today's plants or animals," responded Josh. "At the molecular level, there's not a valid trace of those elusive evolutionary *intermediates* between fish, amphibians, reptiles, and mammals. An absolute blank! Nothing!

"'*Instead of revealing a multitude of transition forms through which the evolution of the cell might have occurred, molecular biology has served only to emphasize the enormity of the gap. We now know not only of the existence of a break between the living and the non-living world, but also that it represents the most dramatic and fundamental of all the discontinuities in nature.*'"[90]

"Perhaps the intermediates Darwin predicted are still hiding in the strata somewhere?" JT challenged.

"Millions of fossils have been found—in fact, more than 250,000 species, including some still living and others long extinct. But pitifully few even arguably intermediate. Darwin boasted that '*we may safely infer that not one living species will transmit its unaltered likeness to a distant futurity.*'[91] Instead, stasis prevails: Clams remain clams; bacteria continue as bacteria; and coelacanths still roam the seas as fish.

"'*The general picture is that no fossil connecting links occur between most groups of animals...The vertebrate record, in many cases, does not contain convincing series of evolutionary links between orders and classes.*'[92] Fragmentary specimens such as *Archaeopteryx*, mammal-like reptiles, amphibians, and whales with small hind limbs prove nothing without the supportive evidence of molecular biology and comparative analysis of DNA sequences. These meager fossil remnants most likely represent extinct species, not missing links!"

"And is this a big secret?" JT asked.

"Some evolutionists do admit the discontinuity. '*The curious thing is that there is a consistency about the fossil gaps: the fossils go missing in all the important places.*'"[93]

"Then what about intermediates in the plant kingdom?"

"What intermediates? '*In plants, the lack of connecting links is perhaps even more striking.*'[94]

"Evolution expires as an unscientific bad guess, a dead-end. Today's orthodoxy can be tomorrow's heresy! Charles D argued for *continuity*. The still-missing fossil links shout *discontinuity*."

Josh was soaring high in the intellectual stratosphere, content to spin his yarn till midnight. JT was reminded of a Bishop Daniels' story about a long-winded preacher. After an unusually tedious sermon, the cleric shook the hand of a parishioner, asking what he thought of the message. The guileless reply struck home.

"The hay seemed OK, but did you have to dump the whole load?"

195

Before the jury could show the faintest trace of a yawn, JT halted the testimony—as abruptly as the Cambrian Explosion.

"No further questions, Your Honor!"

Judge and jury were caught off guard, but Judge Stone recovered quickly.

"Very well, then. Ms. Kilburn, would you care to cross-examine? I recall your indicating earlier you might have a question or two for Mr. Ryan."

Secretly, Josh wished Traci's wind-up questioning could somehow vouch for his integrity and rehabilitate his reputation after the slings and arrows of the Lassiter testimony. Lost in thought, he missed the coming curves.

"You've been cruisin' in cyberspace, Mr. Ryan. Why don't you skim the treetops of knowledge and speak in terms familiar to us humans?"

The rhetorical question drew snickers as Traci zeroed in. "Don't Darwinists argue that mammals evolved from reptiles because reptiles precede mammals in most geologic strata and because some intervening fossils show increasingly mammal-like traits?"

"Yes, but the argument raises a raft of collateral questions."

"For example, Mr. Ryan?"

"Is it conclusive that mammals never co-existed with reptiles from day one? Is it logical to speculate that any intervening fossil in the geologic column could have bridged the gap from cold-blooded reptiles to warm-blooded mammals? Does a similar skeletal structure prove ancestry rather than revealing just another extinct species? Would reptilian ancestry of mammals be proven feasible if measured within the molecular domain of DNA sequences?"

"Can you answer any of your own questions with evidence?"

"'*At a molecular level...there is no trace of the evolutionary transition from fish to amphibian to reptile to mammal. So amphibia, always traditionally considered intermediate between fish and the other terrestrial vertebrates, are in molecular terms as far from fish as any group of reptiles or mammals.*'[95]

"Conventional dating of the geologic column does not allow enough time for the theoretical gene mutations to show up in order to evolve a warm-blooded mammal from a cold-blooded reptile. An ancestral relationship leaping from reptiles to mammals has yet to be proved. Last I looked, reptiles are still coexisting with mammals!"

"But what about homology?" Traci asked. "Doesn't resemblance in the mechanical structure of fins, legs, arms, or wings of unrelated prototypes suggest continuity and common ancestry?"

"The short answer is that neither homology nor embryology equate to genealogy!"

"And the long answer?"

"Structural and mechanical resemblance indicates efficient design! Neither common genetic ancestry nor molecular composition can be assumed from morphological similarities. Reliance on similar design or chemical content to prove evolution substitutes verbal gymnastics for verifiable evidence. Where are the homologous genes?

"The chemical composition of a redwood and a rabbit both contain carbon. Does that chemical commonality suggest a genetic relationship?"

"You sound quite confident, Mr. Ryan."

"What are the alternatives for efficient motion? Should a rabbit hop with a pogo stick? Could a bird fly better with a propeller?"

Josh cited the wheel as an inorganic example of design commonality that exploits the mechanical efficiency of a circle—obviously vastly better suited to the task than other geometric shapes such as a triangle, rectangle, or cube.

"Design efficiency doesn't prove commonality of molecular composition or genetic relationship. It does suggest intelligent design. Ever since Darwin, when people look at a femur of an ape and a human, they say, 'These leg bones sure do resemble each other. Look at their similar shape! Eureka! Common ancestry hikes on these leg bones! Case closed!' Nothing could be further from the truth. Fossil bone resemblance proves zip!

"'*If the Darwinian interpretation of homology is correct, then you would expect to find at the microscopic level the same homologies that are found at the macroscopic level.*' '*In fact, that is not what has been found.*'[96]

"Old-time Darwinists focused on the visible while the invisible held the key. Conjectures collapse with the molecular world of DNA and the genetic composition of cells. No mater how badly scientists want to claim that a monkey is somebody's remote uncle, molecular reality assures that never the twain shall meet—past, present, or anytime in the future.

"Of course *Homo sapiens* don't own the patent on the leg bone design, but bone design offers anemic support of mega-evolution's claims about common ancestry."

Forlorn figure that he was, sitting soaked and disheveled, Traci had second thoughts at heaping any further embarrassment on the witness she had recruited. The hazel eyes, which had smoldered in passionate emotion earlier in the day, now danced mischievously. Throwing caution to the winds, she couldn't resist playing lawyer, hoping he would understand and forgive.

"Is it not true, Mr. Ryan, that you have repeatedly assured this jury of your veracity?"

"Absolutely!"

"Promising to tell the truth...?"

"Yes, Ma'am!"

"...the whole truth...and nothing but the truth?"

"Of course!"

"I was certain that was the case. Then, perhaps, I assume you might be willing to answer a question of a personal nature—under oath?"

"I'll do my best."

She thought, "Poor guy, lamb to the slaughter." Her sympathy surfaced too late to spare the unsuspecting witness.

"You accepted sponsorship money, did you not, from MAU's Science Research Foundation on the basis that you honestly believed you might have discovered a fossil *Archaeopteryx.* Is that not true?"

"Well, sure...That's standard procedure."

"Presumably you never saw the critter creep, crawl, flutter, or fly?"

"Obviously not!"

"Then when you accepted backing to chase *Archaeopteryx*, were you duplicating Darwin's tendency to conjecture by relying on self-serving phrases like...*as far as I can make out...in my best judgment...*"

The audience roared at the bold humor while the stammering witness struggled to salvage a modicum of composure. Flabbergasted and feeling betrayed, Josh interpreted the query as a "Have you stopped beating your wife?" booby-trap, cynically calculated to divert attention from his carefully crafted testimony.

He needn't have worried. No answer had been expected and none was allowed. JT bounced to his feet, objecting strenuously. "Mr. Ryan's personal life history lacks relevance to the primary issue before this court. And I can assure Ms. Kilburn that before the end of this trial, Josh Ryan's veracity will be established beyond dispute!"

"Sustained!"

The ruling reverberated while the judge choked back a chuckle at the lady's brassy maneuver. Rather than admonishing her to cease and desist, Judge Stone announced adjournment.

# 25

As the curtain descended, the Wurlitzer blasted out a medley of football fight songs with beats familiar to USC and Notre Dame alums. Maestro Redondo Calizar again performed at the console. A rousing rendition of "Hail to the Redskins" shook the rafters and brought Washingtonians to their feet in a spontaneous sing-a-long. The appreciative audience sat glued in place for another ten minutes, roaring approval until the artist took to the keyboard for a final encore, once again pounding out the stirring strains of "The Battle Hymn of the Republic."

Moments earlier, a crescendo of loud, sustained applause had greeted the day's adjournment of *Monkey II.*

Josh Ryan hardly noticed. Two hours of verbal jousting while attired in damp clothing had cooled his ardor considerably. Bidding a quick farewell to the rest of the cast, he charged for the exit and his Mustang, forgetting for a moment that he lacked wheels.

Just outside the stage door, a feminine voice inquired, "Need a taxi, Mr. Ryan?" Gratefully, he accepted Jessica Saunders' offer.

"Thanks a lot. I almost forgot the wheels fell off." Then he added, "That is, if it's OK with you. Other than being slightly musty and smelling more like a horse than a human, I'm good to go."

"Don't worry, my three kids clutter up this buggy with all manner of creatures—dogs, cats, turtles, iguanas, you name it...although a horse would be a first."

Exhausted and appreciative, Josh sank into the van's passenger seat. Compared to the early evening's rambunctious, cross-country dash, the leisurely trip home felt like a stroll in the park. Josh reflected that in one day, he had been chauffeured in a limousine; piloted his own vintage Mustang; ridden the faithful Turbo bareback in a once-in-a-lifetime dash across trackless meadows; and here at 10:17 p.m. he found himself escorted by a petite CPA.

As they took off, bound for Fort Ryan, she broke the silence. "There is something you should know."

"There's a lot I should know," he muttered.

"At the break, Judge Stone asked me to share a scoop with you."

"Here's to the wisdom of Judge Stone," he volunteered, raising his hand in mock toast.

"Seriously, Josh. Mind your manners, now." Only about fifteen years his senior, she didn't hesitate to play the role of disciplining matriarch, mirroring her real-life, child-rearing routines.

"Perhaps this morning you wondered why the assistant development officer of Mid-Atlantic University joined the exploratory conference with your legal team. You have a right to know the answer. For one thing, I'm a private investigator for Cabot, Calvert, and Stone."

He didn't stir at the news.

She continued without pause. "For another, I served as special agent for the Federal Bureau of Investigation for more than 12 years. I still enjoy close connections with the Bureau. *Very close!*"

At this, he peered at her curiously.

"I'm not free to share much more with you at this time except to confirm that I'm working your side of the fence—all the way. Anything you hear from me as well as any information you care to share demands absolute confidentiality. Understood?"

A now very attentive Josh replied in the affirmative.

"It shouldn't surprise you that your personal record has been scrutinized."

In fact, it surprised him very much. "What's there to know?"

"I'm aware of the fact that you've received five speeding tickets in three different jurisdictions. I was pleased to learn you paid your fines honorably, on time, and without protest."

The old memory chafed. "One Delaware speed trap deserved a protest."

"Be that as it may, you acted with integrity. You also pay your taxes, pitifully small as they are."

He laughed at that one, confirming, "That's a fact, especially the *pitifully small* part."

"You have a reputation for gambling on the horses. And you enjoy the respect of the pros at the Atlantic City and Vegas 21 tables. Apparently you've put your mathematical genius to productive use. But you've always faced your losses honorably and even paid taxes on your occasional winnings. In short, without rehashing a lot of intricate detail, you appear to shape up pretty much the way most of your neighbors see you—a bright young guy with a high degree of discipline and old-fashioned integrity."

"Does this mean that Striker the tyrant now cowers in terror, begging to withdraw his litigation against me?"

Her sobriety disappeared in a grin at the feisty wit of her passenger. "Don't you wish. All it means is I'm inclined to believe whatever you choose to tell me."

"What do you want to know?"

She didn't beat around the bush. "Josh, do you do drugs? I know you don't smoke and you drink only a touch of wine on social occasions. But I have to know...Do you do illegal drugs?"

The question struck like a bomb. "Are you telling me Striker's complaint threatens criminal repercussions?"

"That's not a big concern, Josh. I just need you to answer the question. I will accept your answer. Nothing you say will be used against you."

"Well then, in the words of our Rhodes-scholar U.S. president, I'll admit to taking a breath of a joint at a party back in high school days, but *I didn't inhale*. Well, maybe a sniff or two, but never tried it again. I knew better, but sometimes when you try to be cool you come off super-dumb.

"Except for that stupid moment, drugs have never been, are not now, and never will be a part of the life of Joshua Chamberlain Ryan. If my dad had even gotten wind of that high school foul-up, I might have been drummed out of the Ryan tribe."

He meant it as a joke, but her reaction came as a warning shot across the bow. She cherished a rich pride for her native American roots. "When you speak of tribes, Josh, remember the Cherokee nation. Once a Cherokee, always a Cherokee. There's no such thing as being drummed out of the tribe."

"I'm sorry...I didn't mean..." He fumbled for the right words, but she understood.

"You're a fine young man. I believe everything you've shared. But this is important...A trace of cocaine showed up in the camper you lost in Tennessee."

"You're kidding!" Josh stiffened at the news.

"That's the other thing Judge Stone wanted you to know. Earlier today we received a fax from a William Bradford Magruder, deputy sheriff of DeKalb County, Tennessee. We're not talking about an unsophisticated good ol' boy in middle Tennessee law enforcement, Josh. These guys are polite and smooth as molasses, but they play tough and smart. Seems as though Sheriff Magruder took your theft report seriously, did some investigation, and spoke to neighbors who were witnesses to the workings of Camp Ryan. Fortunately, one of the neighbors jotted down the license number of an out-of-state pickup seen toting the camper northbound. Since the truck had Maryland license tags, Magruder sent the report to Maryland law enforcement, which completed its investigation last week."

"They found my camper?"

"Sure did. In the woods. A charred heap, burned to a crisp."

"Do they know who stole it?"

"It's been narrowed down to a handful of suspects."

"And my fossil, Archy?"

"No sign of him. But some search warrants have been issued, so it's too early for you to toss in the towel."

Feeling depressed at the loss of the raw material for his dissertation, Josh said nothing.

"But they did find traces of cocaine. That's the reason for my rude questioning."

"You didn't think..."

"No, indeed. You're clean as a whistle in my book. I'm ready to place my bets. The Indians should whip the cowboys on this one."

Pausing, she added, "By the way, that Magruder guy claims to be a shirttail relative of yours. And you know what it means to be kin south of the Mason Dixon! Loyalty to kinfolks ranks as an eleventh commandment—even if you do come from 'off.'"

"What's this *from off* lingo?"

"If you don't speak Southern, Josh, you're missing a rich mix of language enhancers. *Off* means anywhere in the world other than *here*—your home town. Even if you're born *here,* leave, and eventually return—your forever fate is to be branded *from off*—an outsider for sure. Being *kin* at least partially erases the stigma!"

He brightened. "Looks like this lucky Yankee gambler hit the jackpot by sharing a genealogical heritage with Sheriff Magruder!"

"You don't know the half of it, Josh! Tennessee is the volunteer state, and Deputy Sheriff Magruder takes that volunteer stuff seriously. He even promised to travel here to testify on your behalf, at his own expense, if necessary!"

"I'm impressed—and appreciative," Josh said, trying to take it all in.

The minivan's headlights illuminated the spacious Fort Ryan entry. Wheeling into the cottage yard, Jessica stopped just behind the disabled Mustang. Josh popped open the passenger door, stepped out, then stuck his head back inside.

"Many thanks, Ms. Saunders. It all seems downright perplexing. Puts my mind in a whirl to where I feel like a guy with *the blind staggers.*"

Her easy laughter reassured. "Sheriff Magruder would be relieved to know his Yankee kin speaks bits and pieces of Southern."

Jessica Saunders inspired trust, experienced as she was in the ways of the world. He responded in kind. "I plan to sleep like a baby tonight. I figure that with the Cherokee nation, the Federal Bureau of Investigation, and the South's finest constabulary working my side of the street, the odds promise as close to a sure thing as I'll ever see."

He shut the door to the van with authority and turned toward the cottage porch. Jessica waved, backed out to the main drive, and melted into the night.

Before hitting the sack, Josh noticed he had neglected to reconnect the phone. As he plugged it in, it jangled loudly. Picking up the receiver, he heard nothing but sinister silence. Disgusted, he slammed it down, and again yanked the plug from the wall.

A potpourri of conundrums lurked. But tonight, sleep swept his troubles aside, leaving no time to fret or to suffer the *blind staggers.*

*Warren LeRoi Johns*

# V
# Day Five: Thursday

# *"Darwin" Cross-examined*

# 26

Thoughts of Traci's impertinent cross-examination kept running through Josh's head. How could the girl of his dreams threaten him with public humiliation, even if merely to poke fun? With the morning's sun barely peeking through the trees around the cottage, Josh reactivated the phone, and called yesterday's football hero.

JT responded in a voice raspy with sleep.

"Sorry to impose farmer's hours on your schedule, Bro, but that's the price you rich lawyers pay for serving emotionally distraught clients! Right?" Josh joked.

"Whatever you say, old chap. What can I do for you?" JT sounded more weary than amused, trying to clear his scratchy vocal cords.

"Glad you asked, Counselor! It's this thing about that lady's questioning last evening. You know, the noble chance to prove my own credibility...or lack thereof...saddling me with those Darwin-type equivocations?"

Awakened by the comic relief, JT chortled. "Traci was just funning you—playing with your mind. She wanted to spice up the action. Could be the lady sees you as a first-class hunk."

As JT's snickers escalated, Josh regretted initiating the call. Still, he plunged on, flailing for answers. Hopelessly agitated, he unloaded on JT, willing to settle for any shred of comfort.

"It may be a big joke to you, Bro, but it's your client, Dar, who's hanging out to dry. If I had answered, 'Yes, of course this fossil is *Archaeopteryx*,' the jury would think, 'Uh huh, we hear you...But you can't expect us to accept unproven evidence!' If I said, 'I don't know,' they would have branded me the fool described by Lassiter! So when I say, '*I think*,' the lady lawyer implies I'm just another *master wriggler* in the Charles Darwin tradition."

The predicament drew belly laughs from the amused law student.

Josh summarized his frustration with a plea. "Good grief, man, I'm trusting you as you insisted, but would you mind letting me in on the grand strategy—if you have one?"

At last swallowing his mirth, JT edged away from dark comedy, offering his emotionally besieged friend assurance that deliverance awaited.

"It's already taken care of, Josh. Relax. Enjoy the day!"

"So, you plan to re-establish my credibility tonight?"

"Not tonight, Josh," he said. "But expect a big surprise, mañana. Like I said, 'Trust me!' And we won't have to be sidetracked with that bogus lose/lose detour that sandbags you as a Darwin-speak clone. Your integrity can shine like a beacon...made in the shade...Well, you get the idea!

207

"Now, if you don't mind, *nod* craves a touch more shut-eye, even though *winken* and *blinken* have succumbed to dawn's early light!"

"Thanks, JT. But don't pull anything drastic. For what it's worth, I'm nuts about that lady—so don't scare her away. Sorry for unloading my anxiety attack on you."

JT soothed Josh's overwrought sensitivities, assuring, "The lady lawyer's reputation is safe with me. Gotta go. See ya this evening."

The agitated client enjoyed genuine peace for the first time since Slade Lassiter took his free shot. Josh remembered that his dad likely had returned to the Fort Ryan manor house by now.

As his son mused, Michael Ryan sifted through the daunting stack of mail that had accumulated during a four-week absence. An experienced world traveler, he prioritized time and motion. Shuffling with the speed and precision of a Vegas dealer, he separated first-class from monotonous reams of catalogs, magazines, and promos.

One hand-addressed letter riveted his attention.

The envelope carried a "Morgan Manor" return address embossed in elegant script. Never before had Michael Ryan received a written communication from Iron Pants Carrington!

Curiosity outpaced surprise! The envelope seal popped with one slice of the opener. Unfolding the single-page note, he confirmed that the hand-written message was indeed addressed to Michael Joseph Ryan and duly executed with Regina Ann Carrington's characteristic flourish. It began with a folksy salutation:

> *Dear Michael,*
>
> *In recent days it has been my privilege to become informally acquainted with your son, Joshua Chamberlain Ryan. No doubt you have heard that he is the featured star witness in Mid-Atlantic University's stage production, The People v. Charles Robert Darwin.*
>
> *It is my pleasure to report that he has acquitted himself admirably in the presentation of firmly held convictions while pressured by intense public scrutiny.*
>
> *It occurred to me that you would enjoy seeing Joshua in action. Since the production's growing acclaim puts a premium on choice seats, I have taken the liberty of enclosing complimentary tickets for the Thursday and Friday performances, September 5 and 6. Hopefully your frenzied itinerary will allow you to share in the proud sentiments offered by these moments.*

> *To crown the memory of this extraordinary week-long centennial event, I have scheduled a private dinner party for a select few at the Hampshire Club this Friday at 8:00 p.m., immediately following the annual MAU awards reception and presentation..*
>
> *You are cordially invited as my personal guest.*
>
> *However difficult our past relationship may have been, I cherish the thought that you will find it in your heart to indulge the grandmother of Josh Ryan—at least on this auspicious occasion.*
>
> *Affectionately,*
> *Regina Ann Carrington*

Bewildered, Michael read the note again. And then again.

Temporarily forgetting the huge stack of unopened mail, he walked out onto the front porch of the Fort Ryan manor house, still clutching the note. A torrent of thoughts and emotions surged. The *whys*, *hows*, and *whats* eluded. For a five-minute stretch, Michael inhaled the rich autumn air, blended with the heady aromas of pasture lands and woods.

Michael's lexicon excluded *vindictive* or *bitter*. His son had been taught to travel the same track. No biting words derogatory of Grandmother Carrington had ever been spoken. Now was not the time to abandon that tradition. Although the communication from the Duchess arrived a quarter of a century tardy, he reasoned that it was better late than never.

Since he was in town for at least the next week, Michael mentally accepted the invitation. Calming exhilaration replaced debilitating jet lag. Savoring the emotion, Michael's glance swept the pasture, looking vainly for a glimpse of a romping Turbo. Turning to re-enter the house, he noticed the Olney Garage's red tow truck parked below, stinkering with the Mustang.

The sight reminded him to call Josh to announce his safe arrival, and to make sure he was still planned to come to their 9:00 a.m. father-son homecoming ritual. When the phone rang only once before pick-up, Michael guessed correctly that Josh had risen early.

"The coffee perks as we speak! If you inhale deeply, you might catch a whiff of a new Sumatra blend. Time to follow your nose up the trail!"

"Keep the java hot! I'll be there just as soon as I give a quick call to a special girl I want you to meet one of these days. Welcome home, Dad!"

The casual nuance caught the father's attention.

"That's a switch! You've got my attention! I'll look forward to checking out this charmer."

"Incidentally, where's Turbo? And what's wrong with the Mustang?"

"Don't worry about Turbo; he's in good hands. As a matter of fact, he's feeling his oats in a first-class stable for a day or two. As for the Mustang, it will be as good as new by noon. Nothing to worry about. It's a long story."

Still intrigued by the letter from the Duchess, Michael told Josh of his surprise invitation, adding, "You obviously know more about this than I."

"You've got that right," Josh said. "She's even had me over to her place. Maybe she feels sorry for me. This week I've become a *persona non grata,* a virtual pariah, and all for speaking my mind in a fake trial. I'll tell you about it after I climb Mount Ryan."

Hanging up, Michael smiled. He could think of nothing in his son he would care to change—even if he could. He knew Leslie Anne would also have approved. Temporarily bypassing the stack of unopened mail, he settled back in his favorite leather recliner to scan Thursday's *News Press.*

The U.S. presidential race dragged on monotonously; saber-rattling over Iraqi skies continued; the Baltimore Orioles had whipped the California Angels in Anaheim, 4 to 2; and stock market indicators hadn't changed appreciably since the previous June. But before Michael could page over to his favorite comics, a front-page headline in the Montgomery County section caught his eye: *Ride to Glory.* The subhead added: *Scholar-Horseman Defies Odds, Beats Curtain Time.*

Splashy headlines and story highlights spilled out in a three-column spread. Four punctured tires on the Mustang; no emergency transportation available in time for the Wednesday *Monkey II* performance; reckless bareback ride astride stout-hearted Arabian horse; and triumphant, nick-of-time testimony by unflappable student witness, undeterred by bedraggled, rain-soaked appearance. And finally, a picture of the stalwart Turbo reined loosely to the stirrup of General Joshua Lawrence Chamberlain's imposing bronze charger.

By the time the *News Press* had been put aside and the door opened to a beaming Josh, Michael Ryan had caught a glimmer of the unfolding campus drama. Hearty hugs followed glad handshakes, shoulder slapping, and laughter. The reunion heralded a great morning!

Pausing for an instant to assess his offspring, Michael congratulated himself inwardly. In Michael's hopelessly biased view, Josh represented the ultimate that natural selection could have garnered from parental gene pools. The finely chiseled features resembled Leslie Anne more than himself. Josh had also inherited his mother's raven black hair framing iridescent, blue-gray eyes.

By contrast, Michael's eyes blazed the emerald-green of his Celtic heritage. Though he was blessed with a full head of sandy blonde hair, gray now liberally salted its fringes and tips. Despite his stocky build, the senior Ryan enjoyed remarkable success in evading the middle-aged ravages of thinning scalp and bulging waistline.

The son mirrored the father mostly in speech and mannerisms. But beyond resemblance or blood ties, the two shared a close bond.

After pouring cups of steaming coffee and getting comfortable, Josh quipped, "Now about that surprise you mentioned on the phone! Don't tell me you're throwing your hat in the ring as a write-in candidate challenging Clinton and Dole for the Presidency?"

Josh's teasing tone drew a grin from his dad. Michael deflected the question. "From the sound of things, you may have a surprise or two for me. Let's start with your social life."

Straining to appear cool, the normally articulate Josh stumbled in quest of a few select phrases to do the subject justice. His scarlet face didn't help.

Michael Ryan chuckled at the signs. "Looks to me, Son, as if this story of yours has more going for it than any surprise of mine!"

Josh's trusting heart opened wide. Eagerly he shared the highlights of his infatuation with Traci Kilburn—and his hopes for a steamy romance, if he had his way. His previous evening's cross-examination emerged as little more than a footnote.

Wishing fervently that the young woman was everything Josh believed her to be and that she shared his romantic dreams, Michael tried not to pry or to spout benign counsel.

"I'm looking forward to future episodes of your soap-opera. If you're that smitten, she must be a terrific gal. Not that you need my approval," he added with a smile. "I've always had faith in your judgment...just as long as you don't let your hormones run away with your head."

The classic parental caution induced mutual grins as Josh blushed a deeper shade of scarlet. Michael decided to back away from the romance's racier innuendoes. Changing the subject, he asked, "Why don't you brief me about the status of your academic standing with Dr. Striker and the curious phenomena hounding *Monkey II's* star witness."

Retrieving the Montgomery County section of the *News Press*, he volunteered a tongue-in-cheek analysis. "Looks like you've achieved instant fame as a red-necked cowboy."

Josh summarized the week's misadventures: Dr. Striker had neither forgiven him for the loss of the Tennessee project nor restored his academic eligibility; and *Monkey II* had exacerbated the always shaky teacher-student relationship.

His appearance as *Monkey II* witness had prompted on-campus rivalries and unexpected media attention.

Dr. Striker's Science Research Foundation had filed a civil suit alleging negligence and claiming damages exceeding a million big ones.

And finally, Dr. Striker's handpicked attorney would appear onstage in *Monkey II* this very evening to cross-examine and to "set the scientific record straight."

Michael listened attentively as Josh recounted the harassing night phone calls, the punctured tires, and the desperate resort to horseback transportation to meet Wednesday evening's curtain. By the time Josh reported Jessica Saunders' association with Judge Stone in defense of the lawsuit against him, along with the jolting news about traces of cocaine found in the burned-out camper, Michael's expression of amused fascination had been replaced by anger and dismay.

"You're describing something more than a teacher-student dispute, Josh. I taught you to look for the good in people, but this is the other side of that coin. Genuinely rotten people walk this earth. Some of the worst put on phony masks of good.

"There's a lot going on beyond our knowledge. Take comfort that you have Judge Stone and Jessica Saunders working your corner. I remember her in an advanced computer class—she was one of the all-time brightest programmer/analysts I ever trained. Still, you need to watch your step."

"Don't worry...I didn't exactly volunteer for the honor of this attention. Even if the light at the end of the tunnel reveals that my career has been deep-sixed, leaving me stuck with a million-dollar judgment, it may be worth it—if that's what it takes for me to cross paths with Traci Kilburn."

Michael smiled at his son's upbeat fatalism. He recognized both danger and the compelling power of young love—having been there, done that. He tried to be encouraging, "Good luck, Son...Shadows sometimes follow sunshine. I'll be in town for the rest of the month, so we can sort this out later. Meanwhile, tell me what I should know about your grandmother."

Josh recounted the startling events of the week and the unexpected outreach of Regina Ann Carrington. Michael listened, intrigued. "You think I should accept the invitation to be her guest at the remaining *Monkey II* presentations?"

"I'd really like for you to be there."

"And Friday evening's dinner?"

"I think you should come, Dad!"

"Then I'll be there. And I promise to be as well behaved as any old sea salt."

With the coffee pot empty and his dad now up to speed on the week's events, Josh's curiosity resurfaced. "After spilling my guts for the past hour, it seems to me I've earned the right to see the surprise you dangled earlier."

"You'll find it worth waiting for, my boy."

Pointing to several cartons stamped with the logo of a Silicon Valley computer firm, Michael smugly downplayed the mystery. "Don't get your hopes up—it's not paper money. There's really nothing much in those boxes—just computer disks, photos, and stacks of paper printouts."

Michael grinned and added, "On second thought, given your current plight, this stuff could be worth more than bucks to you." Motioning Josh to have a look, Michael unlatched a leather briefcase resting atop the cartons.

"Help yourself, Son...It's all yours."

Josh stared open-mouthed. The briefcase was bulging at the seams with color photographs. Picking up the top two or three, he scrutinized each, hardly daring to believe his eyes. The digital photographs shimmered in laser-jet color. His fingertips caressed the graphic depictions of the missing fossil Archy as well as other scenes from his ill-fated Tennessee expedition.

"It's all there, Son! That experimental TV mini-cam mounted on your helmet as a courtesy to me and the sponsoring Silicon Valley computer company recorded everything you said and did throughout the dig. The two hundred or so pictures in this briefcase give a bird's-eye chronicle of your expedition. The other boxes contain unedited transcriptions of your conversations...It's all there in that pack of CDs."

"Awesome! What can I say?!"

Taking the pictures from Josh, Michael flipped through and pulled out one. "Look at that, Josh...even a record of your first formal, or should I say informal, introduction to Traci when she stumbled into your arms in that Center Hill Lake cove."

"Oh, boy," thought Josh, "Dad's known about Traci from ground zero."

Father and son reviewed the evidence as they admired the latest marvel of the computer age—a detailed history validating the hard work at the research site—all in CD and printout format—despite the theft of the camper and its cargo.

Josh scanned the mass of data in awe. It was all there—dramatic discoveries, dusty drudgery, and long-forgotten small talk. Despite the daunting task of wading through all the trivia to correlate scientific reality, Josh rejoiced at the prospect of a resurrection of his doctoral degree candidacy.

"Seems to me you possess stuff Dr. Striker would very much like to get his hands on," Michael cautioned. "Make haste slowly, Son. Check with Judge Stone before you do anything rash!"

Overwhelmed by his good fortune, Josh said, "I'd like to spring the hottest pictures as surprise evidence this evening. It may be a hard sell to Judge Stone, but it sure would make me feel good. And if it's OK by you, I'll leave the boxes here while I spend this afternoon scanning the CDs."

Michael raised an eyebrow, but offered no objection.

"Before you get bogged down wading through fossil history, how about joining me for some head-clearing exercise on the Hampshire Club's driving range? Jet lag leaves me groggy."

The pair relished the time-honored drill. Since they had called ahead, a golf cart stashed with both sets of clubs awaited curbside at the Hampshire. In moments, they had reached the range. They spotted Lyle Grant instructing a young woman in the rudiments of the sport. Before they were close enough to wave, Josh recognized the pupil.

"That's *her,* Dad," Josh whispered, nudging his father. "I hadn't planned it this way, but it looks as though this is as good a time as any to introduce you." Handshakes and salutations followed in quick order. If Traci Kilburn felt embarrassed at being caught in the beginner's role, it didn't show. And when Josh introduced her to his father, she came across as a study in aplomb, revealing no residual regrets for the past evening's aggressive cross-examination of the witness.

"It's a pleasure, Mr. Ryan. I've heard many good things about you from your newly famous son."

Michael liked her dignity, poise, and spontaneous wit. No doubt about it, something more profound than hormones had attracted his son's attention.

Within the hour, the Ryans had launched a shower of golf balls and had bid adieu to Lyle Grant and his pupil. On the way home, Michael resorted to dry humor. "I thought it extraordinary that a fledgling golfer, like this young lady, managed to have won a Hampshire Country Club junior championship some years ago—at least, judging by the gold medal hanging on her necklace." He added wryly, "I didn't realize the club awarded first-place medals decorated with diamonds disguised as golf balls."

Josh mustered a grin as the two drove home in silent contemplation. Excitement at the prospect of researching the computer data for the Tennessee project competed with Traci for Josh's thoughts.

Michael felt gratification at his success in parenting, along with a pang at the recognition that life between father and son was approaching a new threshold. He sensed that Leslie Anne would have applauded his achievements.

Approaching the winding, tree-studded driveway leading to the Fort Ryan manor house, Michael looked over at Josh and said, after a pause, "There's something else I've been saving to give you for a long, long time. Today that time has come."

Josh reacted in silence, without a clue.

Once inside the house, Michael disappeared. When he returned, his hands cupped something lovingly. Still, without speaking, he led Josh to the dining room and pointed to a chair. Once they were seated, he opened his palm to reveal a cherished heirloom. Pushing it gently toward Josh, Michael smiled and said reassuringly, "Go ahead and open it, Son. It's your mother's gift to you. She would want you to have it sooner or later. I think today's the day!"

Josh reached out tentatively to accept the small jewelry box covered in faded blue velvet. Carefully yanking the wrinkled bow and removing the silver-white ribbon, he pried the lid open. Inside he found a sculpted golden wedding band with a matching engagement ring. The diamond, though modest in size, sparkled dazzlingly in the sun's rays. While not ostentatious, the set represented an extravagance that strained the purse strings of a young ex-Navy man.

The diamond flashed and winked enticingly. Josh pointedly avoided his father's eyes, dreading the sight of tears—even when shed in joy. It was enough to feel the love that surpassed time and words.

A hushed, "Thanks, Dad," sufficed.

# 27

As the 4 p.m. news conference deadline approached, the atmosphere engulfing Mid-Atlantic University's community throbbed with excitement. Headlines promising *Darwin to Be Cross-Examined* whetted the appetites of journalists weaned on disputes and confrontations. Prominent radio and television reps joined print media types; agents of national networks signed on. The anticipated appearance of pugnacious trial attorney Marcus Brogan, who boasted skills matching Clarence Darrow, had drawn them like a magnet. Pace Terhune exploited this unlikely analogy. Swallowing the bait, the media showed in force, prospecting for yet another *trial of the century*.

The turnout astonished even ringmaster Terhune. The commodious Chamberlain Playhouse foyer proved too small. An overflow of curious spectators spilled into the courtyard, aggravating the confusion. To the annoyance of Striker loyalists, Terhune had ordered the foyer decorated with a fresh batch of life-size posters satirizing Darwin's sacrosanct ideas. One scene, titled *Fact-Free Science,* depicted a Darwin caricature smugly announcing to an incredulous judge, *Who needs evidence? I admit I'm right!* The milling crowd gawked at the outrageous spectacle.

Terhune had arranged a platform at the far wall of the foyer with four executive chairs spaced strategically behind an elaborate conference table loaded with microphones. Spotlighted behind the chair reserved for the canny barrister Brogan towered a blown-up photograph of the real Charles Robert Darwin, implicitly blessing the proceedings.

Gus Morley, mild-mannered friend of students, experienced cold feet. He dreaded the least hint of confrontation. Being exposed to bright lights and cameras gave him stage fright. He would have preferred to keep a low profile, but appeared reluctantly at Striker's command. Hoping for refuge in numbers, Morley had ordered Slade Lassiter to accompany him.

Buoyed by his performance as a *Monkey II* witness, Lassiter obliged. The wily opportunist reasoned that prancing to his mentor's tune and impressing the media would make his already bright prospects for reward by the Striker/Morley team a virtual slam dunk.

Five minutes before four o'clock, the scene crackled with anticipation. Morley and Lassiter sat on the platform wedged like decorative bookends on either side of the central seats of honor. All appeared good-to-go but for a disconcerting *faux pas*—the chairs reserved for Brogan and Striker remained vacant.

Filling the vacuum, to the delight of the media, was a boisterous crowd of several hundred chanting, sign-carrying students who jammed the playhouse courtyard.

Morley cringed. Overnight, the nettlesome Josh Ryan had risen from obscure grad student to campus folk hero. Defiant behavior on the *Monkey II* stage contributed partially to his elevated status. But universal distaste for the dictates of Dr. Karl Striker had fueled enthusiastic endorsement.

Orchestrated by campus activists Marie Mackin and Ira Lonergan, the spontaneous party swept across the campus in a boisterous wave. Derogatory placards taunted and pilloried the Science Division chief. Hastily composed with marker pens and desktop publishing programs, placards in all colors and sizes featured the catch phrase, *Strike Striker*. Enthusiastic students, shouted, "Go, Dar," sending chills of discomfort through the ranks of the academic establishment. A pep band encouraged the students to strut an improvised version of *River Dance*. Joy and jubilation marked the scene—a far cry from the sober setting planned for the news conference.

Morley hated the circus atmosphere and shrank from its limelight. The ferment compromised his attempt to fly Striker's banner with decorous flair. Outgunned at the academic ramparts, the consummate bureaucrat fumed and fidgeted at the no-show by the big boss and the battle-scarred lawyer.

The media exulted in the spontaneous clamor. And as long as the lawn party played out peacefully, Terhune could smile innocently at the impulsive student demonstration. Morley stewed, beads of perspiration dripping from his forehead. Agitated media types, accustomed to tight deadlines, stirred impatiently.

Undismayed, Terhune soft-pedaled, stalling for time. "Moments ago I received a cell phone call from your host, Dr. Karl Striker. Earlier today, he joined Attorney Marcus Brogan for a northern Virginia luncheon appearance. He asked that I apologize for the delay...it seems they are trapped in a Beltway traffic tie-up."

Out of the corner of his eye, Terhune caught a glimpse of the anguished face of Gus Morley. His oil-on-troubled-waters routine was drawing only daggers. Morley was a stranger to impromptu ad-libs and ill-prepared for an unscripted performance.

"Anticipating the imminent arrival of Attorney Brogan, I suggest we commence the press conference at once! We'll proceed with an introductory statement from Dr. Striker that I'm sure his associate, Dr. Augustus Morley, will be happy to present."

Terhune passed a typewritten page to the apoplectic Morley, unceremoniously drafted to perform against his will. Bristling at Terhune's presumption, the professor had little choice but to move to the mikes, projecting charm but inwardly cursing Striker's absence. His voice a reedy whine, the

217

corpulent Morley avoided public speaking. Clearing his throat, he grimly launched into an expurgated version of Striker's message of welcome for the missing guest-of-honor.

> *"In recent days, this campus has been the scene of a sanctioned centennial event, proposed to be a scientific discussion designed to enlighten and entertain. Instead, the so-called case of People v. Charles Robert Darwin has evolved into an embarrassing public spectacle."*

The unappeased listeners didn't blink at the pun, so Morley plodded on.

> *"Regrettably, the farce offers nothing more than unfettered bashing of scientific views articulated brilliantly by the late Charles Darwin."*

Realizing that the press was getting antsy, Morley loosened his collar and condensed the statement, shifting gears to begin a rapid-fire delivery.

> *"Monkey II defames Mid-Atlantic University's scientific community. Mr. Terhune has graciously agreed to open this evening's performance to serious inquiry by featuring nationally acclaimed trial counsel Marcus Brogan. Hopefully his presence and participation will restore some balance to what has been a one-sided insult to scientific traditions and to this university's lofty academic commitments."*

Almost before Morley finished reciting the tired clichés, the media went to work ferreting out a story. A blizzard of questions greeted his recitation of Striker's statement.

A young man in the front row shouted, "Is it true that Darwin failed to explain how organic life generated spontaneously from inorganic matter?"

"Well, yes, that may be...but..."

"Do you personally support Darwin's claim that male *Homo sapiens* are born more intelligent than females?" interrupted a chic woman, exquisitely clad and coiffured.

Stung by the ferocity of the assault, Morley floundered, unwilling to acknowledge that he had never bothered to read Charles Darwin's writings cover-to-cover. He was spared further fumbling by a tug at his elbow.

Emboldened by the plaudits for his *Monkey II* performance, Slade Lassiter jumped to rescue the Science Division's drooping battle flag. Gratefully, Morley retreated while Lassiter tried to sound eloquent.

"Ah...yes, Darwin may have expressed those views...politically correct, perhaps...for Victorian England...However..."

Even less versed in Darwinisms than his mentor, Lassiter tried to finesse the answer. His lame retort drew few smiles from protagonists hungry for headlines.

The lady's verbal fingernails dug deeper. "The question, Sir, was whether or not you personally agree with Charles Darwin on that issue. If you agree, can you show scientific proof? Or if you disagree, is it conceivable that you share some of Josh Ryan's distaste for Charles Darwin's ideas?"

"Well...uh...I suppose there's...a...uh...difference...of opinion...among reputable scientists...Regardless, mega-evolutionary theory has moved beyond Darwin..."

Unwilling to be painted as an ally of his nemesis Dar Ryan, the voice of the beleaguered spokesman trailed into silence.

Unmoved, the sharp-tongued lady in the front row demanded clarification. Lassiter turned away to field another question.

A hoarse voice erupted from a weather-worn face in the back row. "Do you support the Darwinian theory in its entirety?"

"Of course. Absolutely! Evolution is fact!"

Prematurely, Lassiter reasoned, "So far, so good." Rejoicing at being given an academic blank check, he promptly proceeded to overdraw the account.

The next question, from a travel-hardened veteran of diverse geographical beats, offered no solace. "Then you agree with Darwin's racist views?"

Lassiter opened his mouth to reply but drew a blank. He knew nothing about social Darwinism. As he flailed to regain his equilibrium, a torrent of other questions flooded in. His poise and polish ebbed, like a snowman exposed to spring sunshine.

Despite a delivery that came across as abrasive and blunt, the articulate Karl Striker could handle questions with the skill of a politician clinging to office. But Lassiter lacked such skills. For one brief sliver of time he basked in the spotlight—as a convenient stage prop. Too late, reality dawned.

Appearing on stage as a *Monkey II* witness and being scrutinized by student lawyers shrank in comparison to the unshackled assault by sophisticated news reporters! Lassiter's ruthless interrogators zeroed in, leaving him intellectually naked as a jaybird. Pace Terhune intervened in time to rescue the university's reputation, but too late to shield the Morley/Lassiter duo.

Terhune addressed the media in soothing tones, summarily axing the charade. "Moments ago I received an update from Dr. Striker. Regrettably, he and Attorney Brogan have only now crept to the I-270 turnoff and can't reach the

MAU campus for at least another hour." Over audible groans, he added, "Remember, dinner at the commons for all accredited media reps, compliments of the university. And tonight's *Monkey II* performance guarantees newsworthy fireworks!"

TV camera lights blinked good-bye; a train of communications equipment snaked away toward the courtyard as media masters exited en masse. The sham event collapsed on its sword, without fanfare or a tune from the fat lady.

Spotting reporters filtering through the crowd, Terhune sensed that they had discovered a story more energizing than a Brogan interview. A living, breathing controversy at the Chamberlain Playhouse promised good publicity for *Monkey II*. Both Friday's newspaper headlines and Thursday newscasts would feature the rollicking student rally.

Sullen and despondent, Gus Morley sat immobile, staring blankly at his feet as though they were glued to the floor. An apprehensive Lassiter hovered nearby, hoping for some sign of approbation from the man whose boots he had licked for the past two years.

After the crowd dissolved, Morley hesitantly roused from his trance. Expressionless, he shuffled through the playhouse exit. No words were spoken: no collegial commiseration was offered. Lassiter imagined that the professor must be fashioning an epitaph to his academic career.

Unknown to all, something more sinister than a student's incompetence was gnawing at Gus Morley's gut.

# 28

Unaware of the furor unfolding on the university campus, Josh took advantage of the downtime to try to sort through the enormous quantity of data delivered to his laptop that morning. Awed by his immense good fortune, the young scholar focused on reconstructing the Tennessee project. Fort Ryan's silent comfort inspired thought-provoking perspective.

Twenty-five acres of carefully manicured greenscape crowned by a meticulously designed manor house on the hill had been the Ryan home place since before Josh's birth. A string of white pines edged the curves of the drive that wound down the hill to the estate's Georgia Avenue entrance. Waves of burnished bronze promised a banner year for wheat. Although the property was less than opulent, the site had a restrained elegance befitting Michael Ryan's success.

Michael's computer skills had long since translated into an annual six-figure income. But it was his investment savvy that had made him wealthy. Every time Michael spotted a fledgling, cash-hungry computer enterprise blessed with visionary management and cutting-edge patents, he invested at pre-public-offering prices. As share values doubled and tripled, he benefited from price-boosting splits and mergers. The one-time sailor had become a Wall Street wizard.

Josh had no idea of his father's net worth, and didn't much care. He flourished under the wings of a parent guided by a simple credo: "Give the stability of love while they are little, and the wings to fly when they are grown!"

Josh thrived, sprouting wings of his own, which he had tested successfully in the turbulent currents of *Monkey II*. Now, computer technology would keep him soaring aloft as he rebuilt his Camp Ryan chronicles.

A pack of CD-ROMs spilled from a manila envelope. A caption identified the data: *Audio/Video Record, Ryan Research, Center Hill Lake, Tennessee: June 1995—April 1996*. Reams of printouts revealed that even the most inane comment had been captured by the snooping eye and ear. His heart pounded at the prospect of academic redemption.

Random sorting confirmed that his initial encounter with Traci Kilburn had been captured in digitized words and pictures. Presumably Michael had reviewed the whole nine yards, rendering Josh's account a trifle redundant! He mused aloud, "You are something else, Dad."

It was impossible to tell from the pictures whether Archy passed muster as an *Archaeopteryx* fossil rather than some lesser species. Josh didn't know and, at the moment, didn't care. Above all, the data confirmed that Archy was a genuine fossil find and not the contrived fraud Striker implied in the civil litigation. The exquisitely enhanced sequence of video frames convinced Josh that he had more than enough evidence to win this treasure a respectable taxonomic niche.

Ryan luck had triumphed—the critical key to a doctoral dissertation had been preserved and was accessible at the touch of a button. Temporarily leaving the stack of typewritten pages in the security of his dad's library, he picked up the CD-ROMs and full-color prints, carting them off to the mill house and the efficiency of his desktop computer. He remembered to take along the blue velvet ring box entrusted to his care.

Entering the cottage, Josh breathed a sigh of exhilaration. Before settling in for a dose of heavy-duty academic analysis, he placed the ring box out of sight on the bookshelf, behind his mother's diary. Content that the day's acquisitions were secure, Josh began to explore the mysteries now converging on his PC monitor, thanks to twentieth-century technology. Musing about the ancient landscape littered with extinctions, Josh reviewed the fate of Archy.

"Wrong again, Charles Robert. Extinctions result from cataclysm or human encroachment, not because descendants exterminate parents!"

A knock at the door interrupted his soliloquy. The immaculately dressed woman waiting in the doorway identified herself as Susan Sinclair, the host of a daily radio show syndicated nationally on more than 300 stations. She explained that she had attended the news conference at the university but said she departed "vexed" by the Science Division representatives' inept evasion of the question: "Do you agree with Darwin's assessment alleging women to be intellectually inferior to men?"

With tape recorder in hand, she asked pointedly, "Frankly, Mr. Ryan, I hate to leave town without something more tangible for my show. Would you mind taking a few moments to do a brief interview?"

The genial Josh couldn't resist the bait. While he had wisely sidestepped Striker's attempted media blitz, he sensed correctly that the distinguished lady guest had no agenda. She sought background to a story certain to capture the fancy of listeners. Josh realized that the burgeoning campus dispute had attracted national attention.

Before he could invite her to take a seat, she flicked the tape recorder's "on" button and zeroed in. He admired her professional zest.

"Do you agree with Darwin's writings declaring that males are superior to females?"

"Not in a million years! My dad is a great guy—and from all I can tell, my mom must have been every bit his equal. And don't forget my Grandma Carrington! She doesn't take a back seat to any man on the planet!"

"Then I can assume that you don't believe that the survival of females is threatened by the *superior* male gender exterminating the so-called *weaker sex*?"

Josh shook his head emphatically as they both laughed easily. She added with a knowing wink, "Judging by the way I saw you looking at Traci Kilburn during last evening's *Monkey II* performance, it's safe to say your actions match your words."

Again they chuckled as Josh readily shared his dream that relationships between the genders should involve love, not war. Disarmed by the good will, he rambled on.

"For my money, an ideal, well-matched couple is possible where the man and woman live and love as equal partners."

Ms. Sinclair smiled approvingly at the insight. Next she tried another trail, blazed by the original Darwin's conjectures.

"It's my understanding, Mr. Ryan, that you reject Darwin's explanation that humans can trace their ancestry to some ancient, unknown fish. Correct?"

"I don't cotton to cannibalism or eating descendants of alleged ancestors...but I sure enough enjoy a salmon steak broiled Cajun style!"

She tried not to laugh out loud before asking, "Then do you disagree with Mr. Darwin's idea that species extinctions result when superior descendants win the war for survival against inferior parents?"

"I reject fish in the Ryan ancestry as unfounded, untrue, and unproven...Although I confess that some Ryan shirttail relative of mine is reputed to be a shark, posing as a lawyer!"

Her laughter punctuated his reply "As to extinctions, species disappear regularly at the hands of humans. The fossil record suggests species wipeouts were caused by cataclysm! Dinosaurs did not die off because they surrendered to multiple generations of allegedly improved offspring. Their extinction arrived like a bolt from the blue!"

"You seem quite certain. If Darwin could join in this conversation, what would you say to change his mind?"

"He might not understand my American accent!"

"Why don't you give it a try? Perhaps I can translate."

Josh's natural candor kicked in. "I'd say, 'Now look here, Charles, old man, fossils would be bloody rare without cataclysm. Biological decay normally follows death. Bone fragments are preserved from past life because of explosive hydraulic action. You need water; a violent shifting land mass that inundates, engulfing the specimen; and pressure enough to bury it in the firm grip of sediment without crushing it. Dinos in the Gobi prove it!'"

"What's this about dinosaur fossils in Mongolia's dry Gobi? I heard that a huge meteorite strike did in the big guys!"

"I can't prove that meteor strike erased dinosaurs but *'Dinosaur bones...had to fall into water and be buried to be preserved, and most dinosaurs spent most of their time on dry land.* [97]

"With or without the meteor, sudden hydraulic action fossilized the Gobi dinos. At the time it was *a wetter and greener Gobi*...and home to '*hundreds of dinosaurs and mammals...Avalanches of water-soaked sand buried the animals alive, creating one of the world's riches fossil sites. The fossils appear remarkably complete in that...all the bones are connected to form whole skeletons...suggesting...death was sudden, and the ill-fated creatures were quickly buried before scavenging animals could make off with the meaty bits.*'[98]

"And then I would go to my bottom line: Species extermination arrives as a by-product of violent inorganic force rather than the hostile action of a descendant competing against a parent."

"You have strong opinions, Mr. Ryan. Too bad both you and Mr. Darwin couldn't have been contemporary adversaries pitted as *Scopes* trial witnesses!"

Within thirty minutes, the inquisitive interviewer had marshaled a plethora of quotes for her Friday radio report. As she waved good-bye, Josh realized it was time to head back to the MAU campus to face Marcus Brogan. Stepping out the door, he caught a last glimpse of the oil portrait showing his mom vaulting a fence astride her pet Arabian. He spoke to the picture.

"Don't worry, Mom! Darwin wandered lost in elitist woods, without a clue as to gender equality! I'm sure he never met a lady like you!"

# 29

The grandfather clock chimed seven times as Josh left the mill house. Painstakingly, he had selected the most representative eleven-by-fourteen prints of the fossil Archy as *Monkey II* evidence. Boasting four fully restored tires, the Mustang jalopy sat in the driveway, ready to roll. Josh jumped in gratefully, intending to go by the Morgan Manor Stables to check on Turbo.

True to her word, the Duchess had seen to it that the frisky Arabian received the kingly treatment expected by His Majesty. Turbo, the ultimate show-off, galloped about with spirited abandon, cutting carefree swaths across the white-fenced pasture. Clearly, he felt that the manicured meadows befitted his royal sensibilities.

The happy steed alternately charged headlong, as though bound for some invisible victory circle, then halted abruptly to nibble at clusters of grass. In the midst of a run-for-the-roses spurt, he recognized the approaching vehicle. At the sight of his favorite human, Turbo executed a neat right angle, in full stride, effortlessly escalating his speed before stopping on a dime, inches from the fence. Nuzzling Josh's shoulder, Turbo accepted as his due the affectionate strokes that were accompanied by sugar cubes and soothing words.

"You don't seem worse for the wear, old timer. Thanks to your Arabian genes, we outsmarted some fool last evening. You deserve a vacation in a lavish layout like this!"

At that, Turbo took off on a pretend victory lap. His owner leaned on the fence, watching in proud amusement. The stunt signaled another "mood" day, free of saddle and rider. The great-hearted horse understood that his moment in the sun deserved pretentious reward. The pats, sugar, and words were accepted as his just due.

As show time approached, Josh felt determined to debut at the Chamberlain Playhouse with time to spare for the pending showdown with the crafty Marcus Brogan. Retreating to the jalopy, he called back reassuringly, "Bye, Turbo! I'll be busy for the next day or two and won't have time to hitch up the trailer to drive you home until the weekend. Have fun until then!"

The horse whinnied approval, reared heavenward, and raced away in a thunder of hoof beats and a shower of mud splatters. Intent on this display of horsepower, Turbo was beyond range before Josh shouted the promise, "I'll pick you up Sunday, old friend."

Tooling down the hill to the main campus entrance and then to the Chamberlain, Josh wished horses could live forever. A flash of irony crossed his mind. Life's beginning provided the overriding theme of *Monkey II*. Birth of a new life deserved celebration.  But what about life's end? Decline and death represented irreplaceable loss. Generations of long-gone humans had spent their lifetimes in quest of meaning. Now it was his turn to wonder. He was starting to understand fossil science, but the big picture—the meaning of life—boggled him. He wondered if anyone understood.

Refreshed but intellectually stymied by the gravity of the unknown, Josh mentally shifted gears. As he approached his favorite space at the far end of the Playhouse parking lot, he was impressed by the number of early arrivals. Intermittent autumn raindrops had failed to deter the milling crowd, swelled by the eruption of students who had seized the opportunity to party. Easing himself from the cramped cockpit of the Mustang, Josh plunged into the perimeter of the revelers without an inkling of the fiasco at the afternoon's press conference other than the heads-up from the lady broadcaster. Spectators spotted the reticent scholar as he picked his way through the fringes of the crowd.

Ear-splitting cheers blended in a cacophony of chants. *"We want Josh! We want Josh!"*

Josh hesitated, uncertain what to do or say. Seeing no alternative, he simply smiled and waved like a vote-hungry politician, accepting the boisterous backslaps from the guys and hugs and congratulatory kisses from the girls as he waded into a sea of grinning youngsters, most of whom he never met. Coming at him head-on, organizers Ira Lonergan and Marie Mackin were determined to recruit another party animal. As they strode forward, a pathway peeled open as though parting the Red Sea.

"Welcome to Action City, Josh Ryan! We've been waiting!" A hefty handshake from Ira and a square-on-the-lips buss from Marie sealed the red-carpet greeting. Grabbing his arms, they alternately pulled and prodded him through the mob until they reached the base of the bronze statue of General Chamberlain.

"Take a look," they said, pointing skyward.

The tip of the upraised sword carried a sign bearing the crudely scrawled paraphrase, *Support Your Local Darwin.* Hanging shoulder-high from the backside of the unprotesting general, a placard campaigned shamelessly, *Dar Ryan for President.*

Catching Josh's look of pained puzzlement, Ira and Marie tried to put his mind at ease, assuring him, "Even the general seems pleased! We know you need to joust windmills, but you've got time to join the party for a minute or two. Don't worry, Dar...before long, the drizzle will chase away everyone without tickets."

Having little choice, Josh offered no resistance. Gamely clutching his briefcase filled with proof that the fossil Archy existed, he joined the lawn dance, whirling on the grass a few times with Marie before being spun off to a half-dozen giggling coeds. The primitive throb pulsing from the synchronized efforts of student musicians lent cadence to the dance steps. Even a rambunctious Redondo Calizar, sans tie and jacket, mixed with the rowdy music-makers, borrowing, playing and then exchanging one instrument after the other in an unscripted recital.

Trapped in the melee, Josh Ryan's heretofore concealed lack of rhythm emerged for all the world to see. He attempted a perfunctory show of appreciation for the wacky demonstration, all the while weaving erratically through the undulating throng and plotting his escape. Finally breaking free, he felt a heavy-handed grip on his shoulder.

"This appears to be quite an auspicious outpouring in your honor, Mr. Ryan." The student-sensitive professor's voice sounded cautiously critical.

"They're just having fun, Dr. Morley. You know students. I'll bet even money that half these kids can't tell a fossil from a wishbone."

Morley laughed at the exaggeration. "I learned long ago that students possess far more intelligence than teachers are willing to recognize."

Pointing to a sign hoisted at the far side of the crowd reading, *Sam Houston Morgan Would Be Proud,* Morley added, "Look there, Mr. Ryan. The news seems to have filtered out that you, Sir, are the anointed heir to the illustrious founder of this university and the Morgan dynasty."

Josh shrugged. Having skipped Mid-Atlantic University as an undergrad, he usually took pains to diplomatically sidestep the connection. "I can't see that anyone's great-great grandfather is relevant...unless, of course, some genealogist insists that the great ape deserves ancestor status!"

Gus Morley laughed good-naturedly, gratuitously slapping the *Monkey II* witness on the back. "You're right about that, Mr. Ryan. I just wanted you to be aware of the depth of campus friendship you currently command."

After lowering his voice to a conspiratorial whisper, the professor confided, "As you can well imagine, that sympathy is not confined to students. Nevertheless, you will want to watch your step until you find a way to make peace with the division chairman."

Josh pulled away, muttering, "Thanks, I'll remember." Inwardly, he realized this self-appointed champion of student rights was a gutless wonder who talked big but carried a toothpick-sized stick.

Inside the playhouse foyer, no trace of the press conference debacle remained. But the outsized cartoons poking fun at Darwinian dogma still drew mixed reviews, ranging from high praise to contemptuous derision. Early arrivals streamed in to gawk, snicker, or snort at the irreverent depictions. Terhune, the opportunist, unabashedly exploited the cartoons to stoke the fires of controversy and boost box-office revenue.

Rumors flew that tickets for the ballyhooed evening performance were being hawked at premium prices. Giant TV screens looked down on SRO courtyard crowds. Unlike his desperate ride and skin-of-his-teeth arrival Wednesday evening, Josh Ryan burst through the playhouse stage door in ample time to dress and to groom himself. Feeling cool and at ease, he felt few qualms facing cross-examination by tiger-lawyer Brogan.

# 30

A week in the limelight served to invigorate rather than intimidate Josh Ryan. Having shed his soggy-clothes image, he sparkled on stage, ready for a duel in the arena of ideas. Tonight's theme dealt with a threshold issue: Does life progress upward to something bigger and better, or is the trend toward deterioration and decline? Does evidence support Darwin's claim that "both the parent and all the transitional varieties will generally have been exterminated by the very process of the formation and perfection of the new form"[99] or did a gargantuan cataclysm accelerate decline and cause extinctions of entire species?

Nothing today compares to the fingerprints of catastrophic intervention that once shattered Planet Earth's ecology. Josh remembered reading that *"The average thickness of the sediments on all of the continents is approximately 1,500 meters"* and that *"The average sedimentation rate measured over a period of one year is approximately 100 meters per thousand years."*[100]

He concluded that uniformitarian gradualism must exist only in myth.

As he pondered the issues, the curtain rose. Scanning the audience for familiar faces, he spotted a group of partying scholars packed in a block of seats toward the rear. Karl Striker, flanked by minions Gus Morley and Slade Lassiter, observed from a side aisle near the auditorium center. Seven rows deep in the center-front orchestra section, a radiant Lady Carrington presided serenely, belying her *Iron Pants* reputation. Josh noted with satisfaction that his father, Michael, sat immediately to the right of the Duchess, while Bishop Brock Daniels occupied the spot to her left. A plethora of cameras confirmed the presence of media reps hoping to follow up on the afternoon's fiasco.

Judge Stone extended a cordial introduction to the prestigious visiting barrister. Polite applause welcomed the guest. Acknowledging the acclaim, Marcus Brogan bowed in turn to judge, jury, and audience. Then, in a thinly veiled hint to the guest attorney, Judge Stone reminded the jury and audience of the less-than-rigid courtroom protocol.

Before the guest lawyer could sit down, the judge referred to the pending civil litigation initiated by Striker in which Cabot, Calvert, and Stone represented real-life defendant Josh Ryan.

"Is it fair to assume, Mr. Brogan, that your presence this evening confirms your stipulation that conflict of interest will not be an issue between us?"

"Certainly Sir! It is so stipulated!"

As the advocate took his seat, Judge Stone signaled JT: "You may call your expert witness, Mr. Daniels."

With a "Thank you, Your Honor," and a modest bow to the judge, JT rose with fluid grace to address the emotionally wired Josh.

"Last evening you quoted Charles Darwin's claim that *'improved...descendants...will generally cause the extinction of the parent-species.'*[101] So how can you explain a fossil record loaded with extinct species?"

"Bald eagles came close to extinction in this century, but I don't see any *improved descendants* flying around, taking the rap!"

"Then what was to blame? An ecosystem upset by human intrusion?" JT asked.

"Man destroyed the passenger pigeon and almost wiped out the eagles and the bison! Still, the hands-down champion for species extinction remains sudden, disastrous cataclysm—not superior offspring!"

"So unlike Darwin's slow, continuous gradualism over mega-chunks of time, you claim the evidence shouts discontinuity—sudden appearance of a full range of organic life forms, followed by sudden extinctions—correct?"

"Mega time chunks can't gloss over the impossible! The most muscular human athlete will never leap high enough to touch the top of the Empire State Building—despite billions of attempts launched over millions of years! Natural limits to change impose impossible odds!"

Intending to distract the witness, Brogan leaped to his feet, objecting loudly to the unimpeded flow of opinion. The judge gaveled him down. Anticipating the ruling, the courtroom tiger huffed and puffed in a show of disdain for the less-than-strict protocol. Judge Stone pretended not to notice, while JT continued to lead the witness.

"Can you define *impossibility*?"

"Anything beyond one chance in 1050 represents mathematical impossibility—that's the number ten with fifty zeroes after it. The odds against the 2,000 enzymes essential to the simplest cell life form appearing spontaneously at one time and in one place reputedly runs 1040,000 to one.[102]

"And you contend that Darwin's '*innumerable transitional forms*'[103] remain numerically sparse and elusive?" JT asked.

"Somewhere between bleak and blank!"

"Can you be specific?"

"The extinct ichthyosaur vanished from its '*adaptive zone...long before the porpoises and dolphins appeared...and during the interval the adaptive zone was simply empty.*'[104]

"Despite the fancy title, '*deferred replacement,*' no evidence suggests any *struggle* or *competition* in which dolphins and porpoises swam triumphantly into the sunset after theoretically exterminating the previous resident species. Nor can Darwinists point to transitionals confirming that porpoises and dolphins descended from ichthyosaurs."[105]

"So mass species extinctions don't prove evolution?" JT asked.

"Toss in the millions of missing transitionals; persistently surviving species; empty adaptive zones; and mega-evolution trips on its own petard!"

"Are you suggesting that prototype life forms live virtually unchanged through hundreds of generations?"

"I'm not suggesting, I'm asserting!"

"Name names, Mr. Ryan! What life forms from ancient strata resemble living creatures?"

"Everything from alligators to oysters: sea urchins, horseshoe crabs, bowfins, Australian lung fish, sturgeons, bats, opossums, and the platypus. Even the lowly cockroach and the gingko tree show up in antiquity and yet persist, virtually unchanged, in today's world.

" *'We can propose that the major body plans were separate and distinct from the very beginning...that living things appeared abruptly, that they were fully formed when they appeared, and that modern organisms were derived from these types...similar morphological distances of both Cambrian and present seem to argue that a process of phylum-level stasis has operated during the history of life...Clams have always been clams; brachiopods have always been brachiopods; fish have always been fish.'* [106]

"And don't forget the survival factor of so-called *common ancestor* life forms, be they bacteria or whatever. These micro-organisms continue to live, unconquered by alleged complex descendants, and apparently immune to being *'gradually rendered more perfect through natural selection.'* [107] Those less well-fitted microbes ignore evolution's warning that natural selection dictated their death. Instead, like the Energizer bunny, they keep going, and going, and..."

"Tangible proof, Mr. Ryan?"

"Take a look!"

The jury strained to see as Josh whipped out a fossilized chambered nautilus from his personal collection.

"This baby owns a pedigree going back to a time when dinosaurs thrived. Now take another look!" Showing his own excitement, the witness displayed a full-color photo of a twentieth-century creature, an exact replica of its fossil ancestor.

"Could be a twin, don't you think?" The jury's heads bobbed in agreement.

"That's because the living nautilus from the Western Pacific is a dead ringer for its long-dead ancestor! Its been alleged that *'In every way they are virtually identical to the living chambered nautilus. The creature that swims in our oceans today is the same one that was swimming around 100 million years ago.'* [108]

"The fancy title describing this biological reality is *stasis*, incompatible with mega-evolution without some serious hoop jumping."

Exploiting this threshold moment, the witness continued, "Where are the *short, sure, slow* steps producing change? What happened to those multiple mini-modifications progressing ultimately to an entirely new and more perfect prototype? Why is the fossil record barren of persuasive evidence of transitionals, which should exist in profusion?"

"Your point?" JT challenged.

"Seems as how the nautilus tribe failed to evolve sufficiently to read Darwin's books and undergo the radical face-lift that his dreams demanded!"

To underscore his point, with the court's assent, the witness hand-carried the marine trophy and photo over to the jury, who scrutinized as they would a display of exquisite diamonds. Josh took his time, encouraging the admiring jurors to trace with their fingers the curving coils of the ancient relic.

JT chose the interlude to ask another question. "Suppose Darwin's conjecture is correct, that across-the-board mega-evolutionary change bridges prototype-to-prototype? Why couldn't this random traffic pattern reverse direction and return to the beginning of its journey?"

Josh responded, "Darwin discounted *'the tendency to reversion to long-lost characters'* asserting *'any actually injurious deviations in their structure would of course have been checked by natural selection.'*[109]

"His *of course* supposition posed as incontrovertible fact, when, in fact, it was nothing of the kind. Why indeed couldn't random chance trigger reversion? If genetic chaos could produce random links bridging divergent prototypes, what's to prevent another detour, this time retreating all the way back to 'Go'? Or a lateral slippage to nowhere? Or even erratic hopscotch leaps back and forth between prototypes?

"If random chance is the controlling factor, rather than predictable rules of the genome, what's to prevent an across-the-board return to microbe status?"

Shuffling feet reflected spectator infatuation with the monologue and growing endorsement of the young scholar's logic.

Annoyed at Josh Ryan's uninterrupted testimony and its effect on the jury, Brogan seized the moment to showboat.

"Objection, Your Honor. The witness is subjecting this court to biased, intellectual wanderings. Seems to me the jury might be entitled to see some real evidence—if he has any!"

The bench overruled, with a warning. "This is the last time I intend to explain, Counsel, that while this proceeding enjoys the trappings of a courtroom, the primary purpose of the event is to entertain. Ideally, illumination may enlighten. Testimony introduced on this stage inevitably includes opinion. Be assured, you will have time to cross-examine. Is that clear, Mr. Brogan?"

"Very well, Your Honor." Anticipating the ruling, Marcus Brogan retreated to his seat. The furrowed brow of the lawyer who had built a flashy career as junkyard-dog lawyer betrayed no frustration, only satisfaction that his stunt had served his client as well as the judge's instruction.

Undistracted, JT obeyed Judge Stone's signal to continue.

"Darwin promised *progress toward perfection*. Does the fossil record confirm this thesis?"

Josh protested the question's limits. "The *progress towards perfection* package insists on simple-to-complex as well as bigger-is-better mentality. '*The larger and more dominant groups within each class thus tend to go on increasing in size; and they consequently supplant many smaller but feebler groups.*'[110]

"Fossils march to a different drumbeat! Show me the big stuff! Nothing today compares to the gigantic sizes that used to be!"

"Can you prove that life forms have *decreased* rather than *increased* in size?"

"Deterioration reigns: Surviving prototypes decline rather than increase in size; exquisite diversity is diminished; and the triple tyrants of cataclysm, loss of habitat, and disease unite to destroy entire species.

"Darwin's pal, Alfred Russell Wallace, observed that '*we live in a zoologically impoverished world, from which all the hugest, and fiercest, and strangest forms have recently disappeared.*'[111]

"Today's twenty-foot-long great white shark strikes terror, but an ancestor shark found near Oildale, California, weighed eight times as much, measured forty feet long, and sported a twelve-foot head. And '*the monstrous Carcharodon...possessing distinctive triangular teeth up to 8" long, may have had a 6-7ft wide jaw gape, and...a length of 80 ft.*'[112]

"Every kid knows that dinosaurs, some of which are believed to have been as much as 150 feet long, once roamed the planet at will. The Smithsonian displays a sloth as large as a pickup truck...rhinoceros-sized marsupials and giant kangaroos inhaled Australian air...foot-long trilobites occupied what is now Morocco...the Caribbean's Anguilla Island housed a three-hundred pound rat...oversized lemurs hung out in Madagascar...and a Denver museum exhibits the fossil remains of a cow-sized pig."

Josh warmed to his roll call of bigger-than-today animal types. Unimpressed, Brogan scribbled.

"A dragonfly with an inch-thick body and a two-and-one-half-foot wing span once hovered over pools of water. The list of jumbo-sized fossils includes a fourteen-inch tarantula...a two-foot scorpion...a millipede-like creature six feet long and a foot wide...a forty-foot crocodile twice the length of its living descendants...giant beavers...*eurypterids* (enormous crabs), ranked among the largest invertebrates ever...ammonites several feet in diameter...pterosaurs with wingspans reaching fifty feet...canary-size mayflies...and the enormous Australian bird, *Genyornis newtoni*, which makes today's emu appear puny by comparison.

"Imagine the pesky ancient cockroach, equipped with a four-inch body to better perform mischief; a tortoise stretching out to twenty feet; or an elephant reputed to sport tusks that were fourteen feet long and three feet in circumference!"

"And for an up-close inspection of giant floral life, don't forget to gaze up the trunk of the three-hundred-foot sequoia, with its hardy offspring still stabbing at California skies."

Josh continued in his mildly extravagant style. Stranded without his traditional legal tools to impede the monologue, Brogan adjusted to the unfamiliar role of restrained onlooker.

"Today's armor-plated armadillo missed Darwin's message that descendants must *go on increasing in size*. Instead it's a mini version of its nine-foot ancestor. Since when does runt-of-the-litter mean bigger? Maybe I missed something in Freshman English—or Neo-Darwinism!"

"And so another chunk of Darwin's dream goes out the window?"

"Right on!"

Student applause erupted. Judge Stone rapped his gavel to restrain any hint of orchestrated enthusiasm.

"So the fossils don't match the theory?" JT asked. "Then how do you account for size deterioration?"

"I've wondered if prior to the convulsion of a universal cataclysm or major shifts of tectonic plates, there could have been a superior system of interrelated underground aquifers that modified climate and balanced ground-level moisture and temperature—comparable to an underground sprinkling system with built-in radiant heating/cooling.

"Imagine a vapor canopy; screen out UV short-wave radiation; mix in a higher oxygen ratio; add a dash of $CO_2$ and that conjures an environment surpassing a balmy day in the Caribbean!" Remembering his dismissal of Darwinspeak for its lack of corroborating evidence, the witness admitted cagily, "Understand, this is *Joshua* Darwin Ryan speculation. I figure if the *master wriggler* can get away with it, why can't I?"

Laughter greeted the confession.

Josh continued, "Thanks to the Internet, the riddle of dinosaur extinction has been solved. Some wag postulates that by calculating the minute distance the moon moves away from the Earth each year and borrowing from uniformitarian methodology to measure the past by the present, a bazillion years ago, the moon allegedly orbited much closer to the Earth. As it dipped like a scythe to a tree-top altitude of thirty-five feet, this monstrous guillotine allegedly did in all the gullible dinosaurs browsing at night—at least the tallest ones!"

The audience cracked-up at the witness' misuse of mathematics; Judge Stone put his head in his hands to hide his laughter; and Brogan turned up his nose in disdain.

Biting his lip, JT struggled to suppress a grin. "Assuming these monsters ducked instead of trying to bite green cheese, what killed the big guys?"

"Cataclysm! Without overpowering violence having wracked the earth, snuffing out organic life and burying remains in moments, fossil finds would be rare. Normally, biological decay follows death!"

"Can you identify the primary culprit for catastrophe?"

"Easy! Hydraulic power!"

"Didn't Darwin contend that *'no cataclysm has desolated the whole world*?'"[113] JT challenged.

"Yup. Sudden ecological disaster didn't track with gradualism. His rush to judgment leaves a batch of questions: How could fossil cemeteries form without rapid, massive, hydraulic energy? How is it possible to explain dinosaur fossils forming if sediment accumulates gradually at the minuscule rate of a fraction of a millimeter annually when burial requires multiple tons? How could billions of barrels of petroleum and tons of coal be formed without the sudden submersion of flora, inundated by sweeping water action? How else can one explain mass burials that created fossil graveyards? What explanation other than hydraulic catastrophe better accounts for the preservation of Wyoming's giant palm leaf fossils—up to eight feet long and four feet wide—which came from a one-time temperate climate?

"There is pervasive evidence of past water intrusion on current land mass, worldwide. At least seventy, some say eighty percent of the planet's surface remains covered by water today.

"Although the date may be suspect, it's been said that, *'the world got soaked seventy million years ago, as sea levels rose five hundred feet.'*[114] A Darrow witness at the *Scopes* trial avowed that *'practically all of the earth has at some time or other been covered by water.'*[115]

"Test results presented to the National Congress of Sedimentologists at Brest in 1991 *'contradict the idea of the slow build up of one layer [of sediment] followed by another. The time scale is reduced from hundreds of millions of years to one or more cataclysms producing almost instantaneous laminae. These innocent-sounding words are the death knell of...the idea that the existence of thousands of meters of sediments is by itself evidence for a great age for the Earth...Today, there are no known fossiliferous rocks forming anywhere in the world.'"[116]

Encouraged by JT's prodding, Josh continued to spout opinion in lecture format. "Planet Earth originally boasted a pervasively mild climate; a thick carpet of lush forests and verdant vegetation; differently positioned land and water masses; a plethora of jumbo-sized organic life forms; and an atmospheric envelope conducive to ecological balance. But since that springtime of life, profound changes have engulfed the natural world.

"I suspect that nuclear equilibrium might have been disrupted by an explosive rupture of subterranean water systems, which sent jets of water gushing through Earth's crust from the fountains of the deep, matched by a massive collapse of the atmospheric canopy triggering universal deluge—all of which lack any present parallel.

"I believe the Earth convulsed in a prolonged tumult. Wave after wave of violent hydraulic action spread mega-tons of multi-layered sedimentation, demolishing life, ripping tectonic plates apart at the seams, upsetting magnetic fields, and redesigning geography.

"Meteorites tore jagged holes in Earth's face; volcanoes erupted with chaotic fury, spewing rivers of molten lava, clouds of climate-altering ash, and iridium-bearing gases; ridges below the ocean's surface shifted, altering currents; towers of surging tides smashed the coastlines as ocean levels shifted radically, pushed by hurricane-force winds and pulled by the tug of the moon; rampaging hydraulic force bit into land surfaces, building craggy chains of mountains; and resulting ice mountains strangled land surfaces, further scrambling temperate climates."

Brogan twisted restlessly in his seat. Though raring to object, he was reluctant to incur Judge Stone's displeasure. Eventually, he could not restrain himself any longer.

"With all due respect, Your Honor, I have to object to this dreary litany! Must we afflict the jury with pedestrian fantasies that flow like magma from the fertile imagination of the witness?"

"Sustained!"

The unequivocal response of Judge Stone surprised and encouraged Brogan. JT countered by asking Josh to corroborate his opinion with hard data.

"You've described a killer of a disaster, Mr. Ryan! This must be more than a private pipe dream? Or is it your version of *fact-free science?*"

"It's what I read in the collective message of the rocks. Darwin touted gradual, uniformitarian change, while I see a sudden, earth-shattering event! Extinction of entire species and ecological decline for survivors are joint casualties of cataclysm. Fossil graveyards confirm that dry land succumbed to ancient hydraulic power. To allege that a deluge never swept the Earth is to put your head in the multi-layered sediments deposited by the inundating water that built the fossil record in the rocks.

"Fossilized dinosaur tracks scale sheer mountain cliffs, which are tilted topsy turvy by some unseen, latent power—trademark testimony to the magnitude of cataclysmic force. Seashells and fossilized marine life litter bone-dry hilltops and mountain slopes, far above today's sea level. Chains of today's high-altitude, rugged terrain lay submerged underwater in the past, until sea beds awash in the currents of a cataclysmic deluge or powered by some convulsive thrust inside the Earth's crust pushed mountains skyward from the ocean's floor.

"A fossil fish has been unearthed 17,000 feet up the slopes of the Andes and marine fossil limestone has been spotted in the Himalayas at an altitude of 20,000 feet! *'Marine fossils are found on top of glacial deposits as in the case of the whale skeletons...covering glacial deposits in Michigan...Whale fossils have also been found 440 feet above sea level north of Lake Ontario; more than 500 feet above sea level in Vermont; and some 600 feet above sea level in the Montreal area.'*

"The Siwaliks, foothills to the Himalayas, which run for several hundred miles and are 2,000 to 3,000 feet high *'contain extraordinarily rich beds crammed with fossils: hundreds of feet of sediment, packed with the jumbled bones of scores of extinct species...the remains of terrestrial animals, not marine creatures.'*[117]

"*'Marine sedimentary rocks are far more common and widespread on land today than all other kinds of sedimentary rocks combined. This is one of those simple facts that fairly cry out for explanation and that lie at the heart of man's continuing effort to understand more fully the changing geography of the geologic past.'*[118]

"Every continent shelters caves crammed with graveyards of disparate fossils, stacked in jumbled piles—the remains of colossal deluge! Combine turbulent water with a gargantuan cosmic discordance that smashes the earth with some celestial force, and the ingredients fall in place for unimaginable cataclysm. The planet has endured violent land mass convulsions causing massive crustal displacement, collapse, subsidence, and upheavals."

"When do you think this cataclysmic deluge occurred?"

"The date is a matter of opinion, but if Carbon-14 dating methodology is correct, a cataclysmic deluge rearranged the face of Planet Earth in the recent past—no more than 11,500 years before the present."[119]

"Anything else?"

"Perhaps it's more than coincidental that Northern Hemisphere oak and pine tree rings with overlapping dates reach back a projected 11,000-plus years—then abruptly dead-end. Whatever the time before the present, evidence points to planet-wide, cataclysmic deluge!"

"Do you credit cataclysm for the creation of fossil fuels?" JT probed.

"Energy resources are available thanks to the pressure of sediments piled up by sudden cataclysm. Gradual decay won't do it. Show me a slow, evolutionary process that is now in progress manufacturing a replacement supply!

"Pools of petroleum flow deep below seas and desert sands. Except where large blocks of granite dominate Earth's crust, vast deposits of coal and petroleum lie deep below the surface, underground and underwater—where living plants and animals once thrived. Coal fields represent mass collections of ferns and trees, many extinct, with occasional animal fossils interspersed in the seams."

"So, we'd have no traffic jams without fossil fuels, the footprints of cataclysm?" JT asked with a smile.

"Nope! Of course, a million or so horses and buggies could tie up the Beltway at rush hour!

"When there was ecological balance, Earth's ecosystem recycled naturally. Marine invertebrates characterized the Cambrian period because the oceans provided the water suitable for fossil creation. That changed when hydraulic cataclysms struck the land. Sudden inundation by water-borne sediment wiped out entire species and created fossil graveyards. Even those tiny fossil embryos discovered in China were '*most likely buried alive one day in a sudden catastrophic overflow of sediment.*'[120]

"Thousands of fossil dinosaur eggs have been discovered strewn about over a parched square mile of layered mudstone within the Argentine badlands at a site dubbed Auca Mahuevo. '*Every evidence shows that the embryos may have perished in a flood that quickly buried the eggs in a layer of silt and mud. This made it possible for the soft tissues to fossilize before decaying, an extremely rare occurrence.*'"[121]

"Is it conceivable that somewhere in the residue of that cataclysm, fossil bones of extinct species resembling modern skeletal structures qualify as the missing intermediates predicted by Darwin?" JT asked.

"Lots of extinct species—absolutely! But transitionals linking two distinctly different life forms? No way! Morphological resemblance of bone design doesn't prove anything without corroborating DNA. A fossil of an extinct life form, standing alone, does not confer ancestor status or link the defunct species to a living prototype! Instead, it proclaims a genetic dead-end—informing us about a permanent loss of information and providing a memorial to the dazzling diversity that used to be.

"As for more recent, post-cataclysm years, the evidence reveals additional decline and deterioration. There is no sign whatsoever of living transitionals spawning links to new and different prototypes."

"Then what about the evolution of new prototype life forms as the result of interbreeding within small, geographically isolated populations?" JT asked.

"Negatory!" Josh exclaimed. "The diminished genetic base can set in motion a '*decline to oblivion for small populations...called the 'extinction vortex.'...Reduced genetic variation has been shown to reduce population growth and increase probability of extinction.*'"[122]

Josh wrapped up his testimony with a well-worn anecdote. "It reminds me of the kid trying to hawk George Washington's hatchet. When the customer inquired if the hatchet was really an *original*, the youngster offered this guarantee: 'Dad says the handle may have been replaced several times along with one or two new ax heads.' The kid went on to allege, wide-eyed and straight-faced, 'No doubt about it, Sir, this hatchet belonged to George Washington!'

"That hatchet job of a myth describes the extinction of Darwinian dreams."

# 31

During the intermission, Marcus Brogan glowered, pacing the stage like an angry bull. Whatever the jury thought of the charade, Josh Ryan remained relaxed and unimpressed.

Once upon a time, Brogan had cut a lean, mean figure, viewed by some as dashingly handsome. But his sex appeal had long since eroded. He was now stocky, ruddy-faced and rapidly aging, though uniquely skilled at crafting his reputation by clinging to past glories. However, he was by no means a has-been. His bluff and bluster and perpetual scowl camouflaged remnants of rapier wit and photographic memory. To compensate for his eroded prowess, Marcus Brogan exploited high drama and pitched public temper tantrums.

Brogan's ego craved a flashy public image and private approbation. He bragged that he had accepted the *Monkey II* assignment gratis. However, the gesture did not glow with magnanimity in light of Brogan's promised $50,000 retainer to litigate against the Ryan scion, and his expectation of gaining additional dividends from the public exposure and publicity of a trial.

Marcus Brogan neither understood nor particularly cared about Charles Robert Darwin nor his controversial theory of evolution. Time was when he would have crammed into the wee hours to prepare for the *Monkey II* appearance. But advancing age had dulled his competitive edge.

To his credit, the old warrior had shown up, three nights running, to get a read on his adversaries. Also, he had scanned the *Scopes* trial transcript, searching for clues to Clarence Darrow's strategies. But beyond that minimal investment of time and energy, the battle-worn tiger arrived woefully under-prepared.

Staring over half-cut lenses balanced precariously on his nose, Brogan eyed Josh with studied indifference. He confidently expected to impress his patron with a brilliant performance. If this meant impugning the reputation of the star witness, so be it.

Immediately after the curtain rose, and Judge Stone had given approval to proceed, the gravelly voice unrolled like a thunder clap, unleashing verbal lightning bolts at the faux Darwin. Sharp-edged rhetoric enveloped the stage with reckless fury, honed by a lifetime of cunning confrontation.

Traci Kilburn winced at the first belligerent volley. The steam-roller style intrigued JT. The jury sat bolt upright. Edward Anthony Stone tapped his left index finger in amused syncopation—he had presided over similar spectacles countless times.

Strutting and stalking, Brogan approached the stand. Barely concealed glee marked his deeply lined face. Words spiked with arsenic dripped as sweet as molasses.

"What a comfort to know that a successor *Darwin* walks the planet, dispensing scientific interpretations to the public—free of charge, no less!"

Josh absorbed the demeaning innuendo without a blink. Brogan got down to business, like a tiger circling its prey.

"Earlier this week, Mr. Ryan, I understood you to allege that you had discovered a 1969 letter from Chancellor Brooke Fielding that you deemed timely and supportive of your testimony. Is that not correct, Sir?"

"Absolutely correct!"

The slashing assault caught most observers by surprise. An instant later, another attack ensued, sugar-coated with sarcastic innuendo.

"Then, Mr. Ryan, are you able to explain—bright young man that you seem to be—just how a renowned research scientist such as Dr. Karl Striker reported that the mysterious document was non-existent, only hours after your phantom encounter?"

Josh stood his ground, refusing to back down. "Are you inviting my own theory or a Darwinian-style conjecture?" he asked, determined to return tit for tat. He needn't have bothered. JT cut him off, objecting strenuously.

"It appears that Counsel prefers to badger the witness rather than to address the substantive issue before the jury. I can assure the court that Mr. Ryan plans to address further the matter of his own credibility with admissible evidence and without irrelevant argument from Mr. Brogan."

"Sustained!"

The ruling took the acclaimed lawyer to task.

"You know better, Counsel. Let me remind you again that although this is a make-believe courtroom, the bench's discretion is real—and absolute!"

Marcus Brogan pretended to obey, taking elaborate pains to show due respect. Since the seed of doubt had been planted, the wily lawyer tried another tack.

"Is it not true, Mr. Ryan, that you are a washout from a doctoral candidacy program on this campus?"

"Well, I, uh..."

"Come, come, Mr. Ryan. A simple 'Yes' or 'No' will do!"

JT jumped to his feet. "Objection, Your Honor. Counsel continues to badger the witness."

JT had warned Josh to respond at a deliberately slow pace, using simple answers and not volunteering information beyond each question's scope.

"I was just testing witness credibility," countered Brogan.

"Overruled," pronounced the judge. Looking at Josh, he instructed, "Feel free to summarize your scholastic credentials for the benefit of the court if you would, Mr. Ryan."

A genuinely modest Josh admitted to an undergraduate double major in both geology and paleontology, with a math minor. "For the past couple of years I've been pursuing a masters here at MAU with a doctorate to follow in the same double major."

Prodded further by the judge, the embarrassed witness allowed as how his GPA had never fallen below a straight "A" average, at both undergraduate and graduate levels.

Brogan commented with saccharin-coated acerbity, "Obviously this book brilliance equips you to second-guess a learned major professor? Or even a Charles Darwin?"

Unmoved, Josh said nothing. Tasting blood, the ferocious barrister went for the jugular! "Let's get specific, Mr. Ryan. Is it not true that earlier this year you were dropped from the doctoral program at this university as unqualified, if not unfit?"

"Yes, but..."

The cross-examination raced on, leaving no leeway for explanation.

"Then, is it not possible, or even probable, that the testimony you have placed in the record this week has been motivated by personal pique—the desire to embarrass and humiliate your former major professor?"

"No way! Rather..."

"Just answer the questions, young man. Try a simple 'yes' or 'no.'"

Josh shrugged in resignation while Brogan pressed the attack. "Do you seriously expect this jury to believe that without so much as a doctorate in sight, your biased amateur blarney offers more weight than the experienced views of Mid-Atlantic University scientists who have earned their spurs in laboratory and field research?"

The unflappable witness threw Brogan a curve by answering with a question. "Time was when experienced scholars taught that the Earth was flat. Would science have been better served if truth had not been pursued by questioning erroneous dogma...however orthodox it may have seemed at the time?"

The question twisted the tiger's tail. The cat pounced with a threatening roar. "Don't forget, young man, I'm asking the questions. I'm the lawyer here—unless perchance you claim credentials in the legal as well as scientific disciplines!"

The admonishment drew audience snickers while Brogan bored in for the kill. "Now then, let's talk about your experience, Mr. Ryan. Have you ever written a scientific treatise published in book form under your own name?"

"No." The answer came without apology.

"Perhaps you've authored an article published in a respected scientific journal?"

"No one has requested my literary services yet!"

Brogan pretended surprise. "Well then, no doubt you've seriously undertaken a scientific research project providing the basis for such a treatise?"

Josh blurted out a naïve, "I've been excavating a bird fossil in Tennessee..." JT cut him off, bounding to his feet to object, in an attempt to cover the witness' vulnerable flank. "The status of Mr. Ryan's scientific research project hardly seems relevant to the issue as to whether or not *evolution is fact.*"

"Sustained! It strikes me as diversionary to debate the professional credentials of the witness. May I suggest, Mr. Brogan, that you proceed with your examination of the evidence previously introduced into the record by this witness."

Brogan offered his best fake smile before continuing the pursuit. "Very well, Your Honor." Then, turning toward Josh, he continued, "Mr. Ryan, the world eagerly awaits your ballyhooed wisdom. Perhaps you can provide us with your alternative explanation for the origin of organic life?"

Josh refused to snap at the bait. "The world will have to continue to wait. I'm not equipped to propose an authoritative alternative."

Not to be denied, the innovative Brogan shifted to a collateral attack. "But you are quite willing to poke fun at the likes of Charles Darwin, are you not? He at least showed the courage to stick his neck out for his beliefs."

"My testimony focuses on the total lack of empirical evidence supporting Darwin's philosophical conjecture that mega-evolution is fact."

"Do you mean to tell me you profess to be both scientist and creationist?"

"I mean to tell you I'm a student of science and I've never studied creationism."

"Can't say that I've given much thought to the subject myself," Brogan interjected, in an attempt at levity. "It's my understanding that creationism attributes all organic life to the all-powerful command of a Supreme Being. Given your hostility toward mega-evolutionary theory, presumably you must believe in God?"

"I've never carefully considered the matter. Logically, there likely is an all-knowing Being with unfathomable intelligence and infinite power...but I've never met Him...Of course, I've met a couple of professors who like to think they are gods."

The quip drew laughter and a smattering of applause from the students at the back of the auditorium.

Before Attorney Brogan could interrupt, Josh took the initiative, adding, "Quite frankly, I would have a hard time believing in a tyrannical Being that tortures victims eternally as a penalty for seventy or eighty years of bad personal choices."

Then with a knowing look at the Duchess and his dad, he alluded to his own anguish. "Perhaps someday an enlightened theologian can explain to me why God allows a child to be partially orphaned at birth or a parent to experience the anguish caused by the death of a child."

"Surely you've been to church?" Brogan challenged.

"Only if you count weddings and funerals."

"Do you believe the Bible?"

"I've never read the Bible."

"Surely you own a Bible?"

"No, but once I saw one in a wedding, carried by a Bible boy! Does that count?"

Audience laughter implied that many endorsed his impertinence.

"You've never read the Genesis account of Creation?"

"No."

"Apparently you don't share the creationists' belief that Adam and Eve were the first human parents?"

"I'm not going to quibble about names, but don't mega-evolutionists recognize that somewhere in time there had to be a first-ever male and female *Homo sapiens* pair?"

Stumped, Brogan scratched his head, choosing "not to quibble" with this clever twist. Instead, he reverted to mimicking the Clarence Darrow script.

"Have you ever read the Genesis account of Noah and the Flood?"

"I've heard allusions to Noah and his ship, but I've never read the account."

"Then you dispute the existence of Noah's flood?"

"Are you asking me to debate the label of an ancient event? Whatever title given to hydraulic cataclysm, the planet's land mass became immersed in the colossal power of surging water, with crustal convulsions, a few thousand years ago."

Brogan left that one alone, choosing to chase another Darrow tactic.

"Impressed as you are with water power, what do you think of the story of the gigantic whale-sized fish reputed to have swallowed Jonah in a stormy sea only to cough him up on the beach, alive, three days later?"

"I think nothing about it except that the runaway Jonah must have caused a whale of a case of indigestion!"

Rolling his eyes heavenward in frustration, the increasingly impatient counselor-at-law clung resolutely to the interrogation of William Jennings Bryan made famous by Dayton, Tennessee's *Scopes* trial.

"Now, I know you must be familiar with the biblical name Joshua if I understand your given name correctly. Per chance you have been informed of your namesake and his military exploits?"

"Per chance I have, Sir, and I am right proud to have been named after a distinguished military officer the likes of General Joshua Lawrence Chamberlain!"

Raucous hee-haws in the playhouse prompted Judge Stone to tap his gavel for order, commanding, "Please continue, Counsel!"

Brogan's growing agitation showed. He knew little Civil War history and even less about the hero of the Twentieth Maine. He surmised that the witness knew more about the biblical general than he acknowledged.

"Joshua Ryan, do you believe it is astronomically possible for a divine force to cause the sun to stand still to ensure a battle victory for the biblical Joshua?" The attorney stepped back, arms folded, waiting to pounce.

"My knowledge of biblical history, to say nothing of Divinity, is pathetically inadequate to respond intelligently. Will an uninformed opinion suffice?"

"Try me, young man!"

"Well, Sir, if we ordinary *Homo sapiens* have figured out how to flood Baltimore's Camden Yards with enough candlepower to play an Orioles baseball game at night, it seems to me that a Divinity wise and powerful enough to hang the sun in space might be able to devise a way to reflect enough light to help the good guys win a battle without disrupting gravity or cosmic orbits."

The answer brought down the house. Brogan recognized the futility of revisiting the issues raised by Darrow in the 1925 *Monkey I* trial. A stubborn Joshua Chamberlain Ryan refused to play the role of William Jennings Bryan, adeptly sidestepping the intellectual minefield laid by the cunning Clarence Darrow. Marcus Brogan vowed not to let his own over-reliance on *Scopes* go to waste. He diverted the testimony to time measurements.

"Earlier this week you admitted skepticism about conventional geochronology. Tell us, Mr. Ryan, do you think the Earth is 6,000 years old?"

"Are you asking my opinion?"

"Of course! Do you need the question repeated?"

Josh responded with matter-of-fact impertinence. "Since I was born in 1973, my opinion is short on eyewitness authority...But yes, I think Earth is at least 6,000 years old...and probably older!"

Uncomfortable at the laughter at his expense, and recognizing the shortcomings of the question, Brogan snorted.

Josh lapsed into lecture mode. "My opinion is that a sterile, inorganic sphere could be very ancient. I also believe organic life began abruptly and recently."

"What do you mean by *recently*?"

"Atmospheric lead deposition in a peat bog in Switzerland's Jura Mountains has been Carbon-14 dated at 12,370 years before the present.[123] Radiocarbon tests at Chile's Monte Verde site have been interpreted at 12,500 years before the present.[124] Both measurements relate to elevated mountain terrain locations and share post-glacial, Pliestocene dates."

"So you claim that life on Earth is only a few thousand years old?"

"I claim only that life on Planet Earth originated comparatively recently—not many millions of years ago!"

"Surely, being the scientist you claim to be, you must enjoy access to some evidence that inspires such grandiose conjecture and rejects a conventional three point six billion-year estimate for the gradual evolution of life?"

"You must have read my mind, Mr. Brogan. Earlier you heard me refer to radiometric methodology and the interpretation of Willard Libby's data concluding that Earth's atmosphere is somewhere in the range of 10,000 years old. Last I heard, most organic life can't survive without an atmosphere."

Brogan started to snort his displeasure, but Josh pressed on, rattling off a litany of stats. "Darwinism expires at the starting gate without a sustaining atmosphere in place for endless eons of time! And other booby traps threaten the concept's viability!

"*At the present rate of erosion, the North American continent would have been eroded away about 250 times in 2,500 million years...Why are the earth's continents still here if they are so old?...The earth's volcanoes release an average of about four cubic kilometers per year...extended over 2,500 years, there should be 74 times as much volcanic material as we now find.*' And at the present rate of average mountain uplift, calculated at five millimeters per year, mountains would rise '*500 kilometers high in just 100 million years.*'[125] Measured at 3,280 feet per kilometer, mountain altitudes would stretch upward an impossible 310 miles—more than fifty-five times the height of Mount Everest."

"That's absurd, Mr. Ryan!"

"Correct, Counselor—you and I agree! But the absurdity is mega-evolution's out-of-sync irreconcilability with scientific reality, not mine! An avalanche of related questions baffle:

"Given the quantity of the sun's neutrinos detected on Earth, can the sun be older than ten million years? Is the sun's rotation rate decelerating at a rate that limits its age to one million years? Did Earth's magnetic decay begin more than 11,500 years ago? What about the atmosphere's ratio of helium 3 to helium 4, suggesting a young earth?"[126]

"And you accuse Darwin of conjecture? Surely you can do better!" Brogan snorted.

With his hold on the *Scopes* dialogue slipping, a fuming Brogan confronted the nebulous unknown, a trial lawyer's worst nightmare. But he ignored the most important axiom of the profession: *Never ask a question without knowing the answer in advance!*

"Conjecture may be one thing I learned at the feet of the *master wriggler*. I simply said that questions baffle, undercutting mega-evolution's refuge in vast time periods," Josh said.

"Are you claiming that Planet Earth is less than 11,500 years old?"

"I claim only that I don't know! Still, I believe life began recently on a lifeless, formless, inorganic planet. I wasn't present as an eyewitness and neither were Darwin or Darrow—or any living lawyer, Mr. Brogan!"

Brogan shuddered in disgust at the blizzard of data. The blunt candor following the bewildering array of hard facts left him stymied and annoyed, wandering through an information jungle of his own making. Skilled at covering his vacuity, he slogged on, he knew not where. He decided to retreat from the confusion of radiometric dating mysteries to borrow more of Clarence Darrow's pet themes.

"Surely you've had occasion to read the record of the *Scopes* monkey trial, Mr. Ryan?"

"Yes, Sir."

"And were you not impressed by the written testimony of distinguished scientists introduced into that record by Attorney Darrow?"

"Yes, I was impressed. Unfavorably impressed in some cases!"

"So you find the 1925 scientific establishment distasteful—just as you thumb your nose at today's conventional wisdom...Is that not right?"

"Objection!" JT enjoyed jousting with the bombastic Brogan. "Counsel is arguing with and leading the witness."

"Overruled!"

"Remember, Counsel, leading the witness reigns supreme on this stage. Surely an artful lawyer-in-training like yourself must have noticed," Judge Stone scolded.

JT accepted the pointed reminder and took his seat without protest.

"Shall I repeat the question, Mr. Ryan?" Mock deference masked Brogan's inquiry.

"No, Sir. I recall the question. My answer only speaks to fallible bias.

"'*One part of the scientific mythos states that two groups of scientists, using the same information and knowledge of physical laws and principles, will come to identical conclusions. This notion is based on the premise that scientists analyze and process information in a dispassionate, rational and bias-free manner, unlike their counterparts in the arts, humanities, industry and government.*'[127]

"That's why yesterday's *fact* can become today's *heresy*."

"Then perhaps you can honor this court by sharing an example of such heresy?" Brogan challenged.

"Yes, Sir! Dr. Fay-Cooper Cole, University of Chicago anthropologist and Columbia University Ph.D., touted the Sussex, England, Piltdown man, endorsing it as *'distinctly human...an approach toward man in very ancient strata.'*[128]

"Cole's written testimony, introduced in the 1925 *Scopes* trial, attributed big-hit status to the 1912 alleged discovery. Darwinism promised ancestral linkage between hominids and ape-like creatures. After scads of primate fossils had been unearthed, finally, bingo: the *missing link*—discovered in England, no less!

"Big problem, though: Some headline-hungry entrepreneur had concocted Mr. Piltdown from a relatively recent human cranium and an orangutan's jawbone. Both the cagey Clarence Darrow and Dr. Cole had been hornswoggled by their own preconceptions and bias...so they swallowed fraud as *fact*.

"The Piltdown *discovery* hyped a calculated fraud—bogus from the beginning, but it took forty years for the news of the hoax to reach the unsuspecting public."

"Are you accusing Attorney Darrow of conspiring with Dr. Cole to submit fraudulent evidence in the *Scopes* trial?"

"Far from it! I'm only reporting on the arrogance of ignorance. Who knows, maybe Dr. Cole's identification of *very ancient strata* is no more credible than his *distinctly human* assessment!"

The retort struck broadside. Still, the exasperated Marcus Brogan didn't flinch, given his own skimpy review of the *Scopes'* written exhibits. Instead, he shifted gears, addressing Judge Stone:

"Your Honor, the witness himself has introduced the matter of his Tennessee research project. It seems fair to assume that the subject is now fair game for examination."

"You may proceed, Counsel. But don't play games that will affect the civil action currently pending against this witness!"

Brogan thanked Judge Stone with studied courtesy, then turned back to the witness. "You mentioned some research project in Tennessee that reportedly received generous funding thanks to Mid-Atlantic University's Science Research Foundation. Is that not correct?"

"Yes, Sir, and I'm grateful for the support."

"And I presume that this project enjoyed some grand purpose? At least that is your claim, is it not?"

Edward Anthony Stone interceded with a stern, "Careful, Counsel."

The attorney muttered a respectful "Yes, of course, Judge Stone," as Josh continued.

"The project involved the discovery of what I believed to be a fossil bird. The study had been intended as a source of data for my doctoral dissertation relative to the origin of birds. Regrettably, this find mysteriously vanished."

"If it ever existed..." shot back the doubting interrogator. Judge Stone sustained JT's insistent objection.

Brogan pressed more hot buttons, satisfied the point had struck home. "Perchance, Mr. Ryan, could this alleged bird-like discovery of yours have been a rare *Archaeopteryx*?"

"Don't I wish!"

"Maybe even a transitional proving dinosaurs are ancestral to birds?"

"No way does Archy qualify as an intermediate linking dinosaurs to birds— no way! Gossamer-winged hummingbirds from dinosaurs? Now there's a confused family tree! Pity the hummingbird tribe trying to concoct a menu for a family picnic in which theoretical dino ancestors showed up as guests!"

The audience enjoyed the impertinence. Brogan challenged: "How can you be so sure?"

"Show me the DNA evidence confirming dino-bird linkage! Dinosaur or bird, *Archaeopteryx* emerged from Bavarian limestone as a paleontological Rosetta Stone for Darwinists seeking transitionals to pump up evolution."[129]

"Come now, Mr. Ryan, would it pain you to show evidence rather than spout opinion?" Brogan asked sarcastically.

"Birds appeared abruptly in the fossil record and shared the landscape with the first-known dinosaurs—as contemporaries, not descendants. Bird fossils much older than *Archaeopteryx* have been discovered—and they resemble modern birds. The bird-from-dinosaur scenario doesn't fly any more than *Tyrannosaurus rex* could have rivaled California's condor for air time. Since when do offspring precede ancestors?"

"Evidence, Mr. Ryan! Evidence!" Not put off, Bulldog Brogan pursued.

"Two bird fossil specimens dubbed *Protoavis* turned up in Texas rocks, reputedly seventy-five million years older than *Archaeopteryx*.

"*Protoavis* shows '*a keel-like breastbone, or sternum, and hollow bones*' in addition to other modern bird-like features. The crow-sized *Protoavis* claims antiquity '*as old as the oldest fossil dinosaur.*'"[130]

"So what is *Archaeopteryx*—dinosaur or bird?"

"Does it matter?"

"Then what's your point, young man?"

"Simple! Birds neither descended from dinosaurs nor shared a common ancestry! *Archaeopteryx* could not have been an extinct transitional linking modern birds to ancient dinosaurs. It's '*not an ancestral bird, nor is it an 'ideal intermediate' between reptiles and birds. There are no derived characters uniquely shared by Archaeopteryx and modern birds alone.*'"[131]

With the questions crowding the edges of the cross-examiner's knowledge, Brogan countered with a feather gambit.

"Don't you admit that feathers could evolve from reptilian scales...given enough time?"

The witness explained patiently: "The bird's incredibly complex feathers are key to its flight. Asymmetrical feathers display unique shafts, barbs, and barbules essential for aerodynamics and insulation—unlike the symmetrical feather shafts of flightless birds. Acrobatic flight is possible thanks to the alula, the thumb-like joint at the end of the wings. In contrast, dinosaur scales resemble armor plate. Bird feathers and reptilian scales are produced from different skin layers. No one has seen reptile scales elongate when exposed to excessive solar heat.

"Imagine a pathetic transitional having to make do with a hide coated with 'sceathers'—part scale, part feather. Pity the critter spending a lifetime trying to fly by mounting thousands of fruitless leaps from a tree, only to crash to the ground in pain."

The caricature drew hoots from the audience.

"You can spare us make-believe nonsense, Mr. Ryan. Instead, perhaps you would be kind enough to enlighten the jury as to the ancestry of birds—if neither dinosaurs nor *Archaeopteryx*, then what?"

"Birds, Counselor! Birds! There's no way any bird—hummingbird, eagle, woodpecker, or cardinal—can trace its ancestry to a dinosaur, crocodile, or other reptile. Pardon the pun, Counselor, but that dog won't hunt—and that bird won't fly!"

Brogan searched for an escape while Josh held him at bay with a clincher.

"By the way, dinosaurs were cold-blooded beasts. The fossil clue: They lacked turbinates! '*Cold-blooded animals like crocodiles have narrow, hollow nasal cavities. Warm-blooded animals...have wide cavities housing sheets of bone or cartilage called turbinates.*'[132]

"So?"

"The *so*, Mr. Brogan, appears obvious—birds are warm-blooded!"

Wandering far from his reserve of biological knowledge, Brogan struggled, resorting to bluster. "Turbinates, smurbinates, whatever—don't try to snow us with fancy phrases! Do you expect this jury to believe that a theory long-held by respected scientists lives or dies by the shape of some nasal cavity? Excuse my own pun, Mr. Ryan, but do you honestly believe your testimony passes the smell test?"

The barrister snorted at his anemic jest.

"The quick and simple answer to your question, Mr. Brogan, is an emphatic, 'Yes!' There is more supporting evidence—if you would care for me to proceed."

Brogan shrugged, "We can hardly wait for your incandescent enlightenment, young man—by all means, proceed."

The groping Brogan got more than he bargained for.

"Paleontologist John A. Ruben has examined photos of a fossilized dinosaur from China, so remarkably preserved that its guts could be identified. Heart and lungs were separated from other organs by a diaphragm—and birds don't have diaphragms! Warm-blooded birds without diaphragms coexisted with cold-blooded dinos with diaphragms. Dr. Ruben concluded that the dino's *'lung was like a crocodile's'* and was *'not capable of exchanging enough oxygen in the air for carbon dioxide in the blood to accommodate the needs of an active, warm-blooded animal.'*[133]

"Ruben's subsequent examination of a *Scipionyx samniticus* in Salerno, Italy, fortifies his finding that the *'hepatic-piston diaphragm, in theropod dinosaurs rules out the possibility that they breathed with a sophisticated bird-like lung.'*"[134]

The attorney was baffled at the scientific terminology, but he knew he had allowed enough bird testimony. "Surely, Mr. Ryan, you can lower yourself from that high-and-mighty pedestal to find an accommodation with some fragment of the dinosaur-to-bird theory!"

Josh ignored the challenge, retorting with a flip, "Of course. Common ground exists on two basics: We agree that most birds fly and that reptilian dinosaurs can be found exclusively in extinct fossils!"

The bird issue had flown Brogan's coop. *Archaeopteryx* offered no clues to aid mega-evolutionists' quest for intermediates. Belatedly, the veteran litigator grasped the inadequacy of his superficial scan of the *Scopes* trial testimony and the futility of his furious note-taking prep during previous *Monkey II* performances.

Desperate to redeem his reputation and to assuage his client, Dr. Striker, Brogan tried a final round of Clarence Darrow tactics.

"Previously this evening, you asserted you had no personal knowledge of God. Is that not so?"

"Yes. But my ignorance on the subject is a source of personal regret!"

Brogan ignored the confession. "Apparently, you've never seen God?"

"Not to my knowledge. However, as I said, I've met a couple of professors who like to think they're gods."

The attorney plunged ahead, despite the echoing mirth in the audience. He could not allow this wisecracking kid to make the tiger appear toothless. Judge Stone rapped for order.

"Then, as a scientist in the making, is it a fair assessment that lacking evidence, you refuse to believe in someone you can't see?"

"That's not true! I could believe if shown persuasive evidence!"

251

"Oh?"

Brogan's eyebrows arched, feigning astonishment. Energized, he pursued. "Perhaps you could share a specific example of your belief in the unseen?"

Convinced that his quarry had been cornered and was certain to raise the white flag, the attorney smirked confidently. But again, the unruffled witness responded with wide-eyed aplomb.

"Certainly! I believe in my mother—and I've never seen her."

"And your evidence?"

"I'm here today, thanks to my mother. Perhaps you've heard...She died just after my birth!"

Brogan sighed; he hadn't heard. The academic underdog was beginning to emerge as nearly everyone's hero. Both Iron Pants Carrington and seat-mate Michael Ryan wiped their eyes.

Gamely, the courtroom tiger renewed his assault on witness credibility, desperate to salvage some hint of momentum. "From your earlier comments, I recall you acknowledged that you don't possess Ph.D. credentials in science, nor are you presently a valid doctoral candidate?"

"Correct...but hope springs eternal."

Another round of applause greeted the comeback. Brogan trudged on, unfazed by the retort.

"And is it fair to say you don't represent Mid-Atlantic University, officially? Nor, for that matter, do you speak for any of the various departments of science affiliated with the university in this mock trial?"

"I don't recall making that claim." The snappy response came devoid of guile.

"But nevertheless, given the limited status of your formal academic training, you do consider yourself qualified to dispute the thinking of a revered science spokesman like Charles Darwin, do you not?"

"Without being presumptuous, let the record show that it's quite likely that I've benefited from more formal scientific education than Mr. Darwin."

"You mean to tell this jury you don't consider Darwin a credentialed scientist?"

"Let me put it this way. If Darwin had replaced the Wright Brothers at Kitty Hawk, we'd likely still be flying hot air balloons..."

Audience mirth rippled, then crescendoed, intruding on the punch line.

"...and navigating by the seat of our *random* pants."

Explosive laughter rolled through the auditorium. Red-faced and glowering, Marcus Brogan challenged, "How dare you suggest to this jury that Charles Darwin should not be ranked as a great scientist!"

"Darwin answered that himself by admitting in a letter to Asa Gray that his *speculations* ran *beyond the bounds of true science.*"

Judge Stone rapped his gavel announcing, "We're out of time for this evening, Counsel. Unless you have any further questions, we should adjourn until tomorrow's final session, scheduled for 10 a.m."

Brogan made one last attempt to shatter the poise of the sometimes flippant, sometimes stoic, but always unshakable witness.

"With the court's permission, I have a significant matter to pursue that shouldn't take more than a few moments."

"Very well," intoned the judge. "You may proceed...if you can wind it up in five minutes—or less!"

The fuming attack dog lawyer raised his voice to a booming crescendo as he struck at the core of witness integrity. Brogan had no intention of being outmaneuvered by a pipsqueak student who had so easily evaded the cage that had trapped William Jennings Bryan.

"Mr. Ryan, your testimony is loaded with self-serving opinions about the lack of evidence in support of evolutionary theory. And yet in your own professional life, you repeatedly make reference to a supposed ancient fossil you claim to have discovered in Tennessee. You have the audacity to imply that these bones might belong to the rare *Archaeopteryx*. But where is the empirical evidence confirming your so-called find? Do you expect the court to believe that this mysterious prize, lost since pre-history, has somehow, hocus-pocus, vanished once again? Or is it all a figment of your fertile imagination?"

Listeners reacted to the verbal barrage with hushed silence. The repeated crash of the gavel telegraphed Judge Stone's bluntly negative reaction. Almost forgetting that he presided in a pretend courtroom, he threatened Attorney Brogan with contempt, announcing in clarion tones his intent to summarily terminate the cross-examination and bring down the curtain.

"I warned you, Mr. Brogan, at the outset of this evening's proceedings, that the subject of the civil action you have brought against this witness in the federal court is off limits for any theatrical discussion on stage."

The chastened cross-examiner churned. "I'm sorry, Your Honor...It's just that..."

Before the attorney could complete his apology, the incredulous judge saw the witness he was trying to protect waving his hand like a student in a classroom, hoping to speak.

"Very well, Mr. Ryan, what is it?"

"This may be out of order, Judge Stone, since I know you're trying to act judiciously. Nevertheless, I'd like the privilege of answering Attorney Brogan's question..." He finished by tacking on some Perry Mason-style legal jargon, adding, "...if it pleases the court!"

Perplexed by the provocative request, Judge Stone motioned Josh, JT Daniels, and Traci Kilburn to approach the bench. After a full minute of indecipherable stage whisperings, a grinning JT and an amused Traci returned to their seats. The star witness swaggered back to the witness stand, emanating confidence.

Judge Stone displayed resignation, cautioning, "Mr. Ryan, you may proceed to answer the question—with the understanding that my advice would be that you defer doing so."

Ignoring the protective suggestion, Josh snatched up his briefcase and removed the box of CDs delivered that morning by his father. Without speaking, he whipped out an LCD portable projector, placed it on the lawyers' table, and aimed it toward the white space on the wall behind the judge. When he touched a remote control, an automatically focused, full-color picture of the fossil Archy flashed across the wall, just as it was being lifted from Tennessee strata. Still without comment from the witness, another thirty action sequences followed within a minute's time, recapturing the instant of discovery.

Returning to the witness box, Josh sent a faint smile toward his father and the Duchess, then beamed at Traci, JT, and the jury. Finally, he sought the eyes of Marcus Brogan.

"Mr. Brogan, I'm no hired gun. My integrity is not now, nor ever has been for sale. I've been known to make a wager on the horses, but I've never tried to fix a race. My meager winnings and somewhat larger gambling losses confirm that reality."

Brogan stared, blinking blankly, uncertain what to expect.

"These pictures address the matter you have just brought to the attention of this court. What you see are unposed pictures of the fossil bird as it appeared at the time of its excavation. I can assure the jury that there are many more pictures similar to these that will authenticate my Tennessee project as significantly more than the figment of a fertile but fraudulent imagination.

"While my scientific credentials are admittedly meager, I can assure you, Mr. Brogan, that my integrity stands unimpaired. I'd like to think these pictures will successfully pass muster as evidence confirming fact in harmony with the lofty traditions of the legal profession."

The sound of a pin dropping could have been heard. A defused Marcus Brogan stared fascinated at some unidentified spot on the floor. Finally, the commanding voice of Judge Stone probed for a finale. "Any further questions, Counsel?"

"No, Your Honor. No further questions."

Again, the barrister sighed with frustration, creating an effect akin to a slowly deflating balloon. The reverberating boom-box of a voice shrank to a whisper. The flames of oratory had been doused, leaving only smoke and mirrors.

"Then this court stands adjourned. It will reconvene tomorrow promptly at 10 a.m., Friday, September 6."

As the curtain descended, the roar of applause muffled the echoing rap of the gavel. The audience rose as one person, applauding the evening's high drama.

A split second before the curtain hit the floor, Josh spotted a center-front contingent spring to their feet, leading the enthusiastic acclaim. In the mind and heart of a proud Lady Carrington, a quarter century of alienation lay buried in fossil history.

# 32

Chaos reigned the instant the curtain touched the stage deck. A dignified Edward Anthony Stone returned the gavel to its resting place atop the judge's bench, then rose to shed his custom-cut judicial robe. The jury engaged in lively conversation. An animated Pace Terhune savored success befitting a Ziegfeld. Jessica Saunders sauntered soberly toward Joshua's side as if to shake his hand before spontaneously reaching up to plant a congratulatory kiss on his cheek. She whispered, "General Joshua Lawrence Chamberlain and the biblical Joshua would both be proud."

As the witness reassembled his evidence and equilibrium, the two lawyers-to-be rushed to his side. JT enthusiastically pounded his back and shoulders, proffering endorsement. "Sure beats scratching Tennessee stone, Dar. Appears to me you intercepted for the home team and ran back a hundred big ones for the score!"

Josh grinned sheepishly, shifting his attention as Traci moved in. Slipping hands behind his neck, she pulled him close and kissed him full on the lips, affirming with fierce pride, "What a man!"

Still seated at the Counsel's table, the ignored Brogan smoldered, stunned and embarrassed by his own lackluster performance. However, thoughts of the $50,000 retainer in the pending civil case worked like a balm to salve his psychological wounds.

The battling barrister glumly reviewed the inadequacy of his perfunctory preparation. Mentally he kicked himself for his blind overconfidence and lethargy. Too late, he recognized the appeal and brilliance of a youthful adversary who could not be bullied, bought, or sold.

Haunted by the humiliating memory of *Monkey II*, he rose to take a last ill-tempered shot at his nemesis before leaving the stage. Catching the jovial student perched beside the witness box, chatting with a bevy of admirers, Brogan elbowed his way in. The tiger's snarl had the searing force of a match touched to a cloud of methane.

"I'm not through with you, Mr. Ryan. I'm looking forward to meeting you in federal court in the near future. That's a real court, you know...with tough rules of law...strict procedure...all the things that assure a lawyer sweet dreams! Not that I want *you* to lose any sleep, but don't forget to bring your checkbook when you show."

Having hurled this malignant threat, Marcus Brogan swaggered through the exit with the same brand of disdain displayed earlier in the week by his client. But vitriolic innuendo could not rain on Josh Ryan's parade this day.

Surfing the wave of audience enthusiasm, the jocular crew filtered outside to mingle with the chorus of celebrants. Reluctant to leave, reveling students fed the chaos. The last of the cast to emerge from the playhouse, Josh soon succumbed to the spell.

In the midst of the roiling push-and-pull, he felt a firm tug. He looked down on the radiant face of an uncharacteristically patient Regina Ann Carrington. Clasping the arm of her reclaimed grandson, the socialite towed him to an open space near the crowd's edge.

The Duchess gushed while Josh blushed. "Your father must have done something right these past twenty-three years. I only wish your mother could have seen her offspring in action."

"I'll second that!" Michael had plowed through the crowd in time to add fatherly endorsement. With a twinkle, he quipped, "I might even take time to drop by for tomorrow morning's performance."

"You'd better, young man!" The Duchess invoked her Iron Pants voice to issue the command.

Josh beamed. "Thanks, guys. But I can't wait to escape this shtick and abandon my infamous stage career forever. This week has been a non-stop blind stagger."

At this, the exuberant Duchess locked her arms around her grandson's neck, snuggled her face close to his and nestled there. The uncomfortable Josh looked imploringly in the direction of his dad, pleading for deliverance as he awkwardly patted Lady Carrington on the back.

Still holding his shoulders in a vice-like grip, the Duchess glowed. "It's true you are a Ryan, and a fine one at that! And never forget, you will always be a Morgan...and a Carrington!"

"And a Magruder," reminded Michael.

They laughed heartily, completing their farewells. After delivering the Duchess to her limousine, Michael headed home to Fort Ryan to banish lingering jet lag with, he hoped, about 12 hours of refreshing shut-eye.

A gaggle of starry-eyed students, sporting a tattered placard touting *Josh Ryan for President,* immediately latched onto the beleaguered candidate. Intuitively he knew that the happy exhibition had as much to do with natural instincts to party and to protest establishment authority as with plaudits for his own outspoken behavior.

Having no clue how to respond to the outlandish camaraderie, Josh grinned and gave thumbs-up salutes. Courteous to the core, he took care to respond to every greeting. All the while, he kept working his way doggedly through the crowd to the edge of the parking lot.

The crowd's giddy hoopla couldn't compare to the lingering glow from the passionate kiss by the girl of his dreams. This was all her doing. Without Traci

Kilburn's catalyst role, no way would he have consented to participate in *Monkey II.*

Eventually Josh emerged from the slowly dissipating throng, striding across the macadam in quest of his vintage wheels. Not ten steps past the edge of the lawn, he froze. For a split second he thought he saw a menacing tongue of fire licking at the rear window of the Mustang.

It couldn't be! He didn't smoke. The car's electrical connections were maintained religiously. Perhaps it was just a reflection of the street lights.

Dropping his briefcase, he broke into a limping run. The spreading orange-yellow blazed brighter by the moment. He felt sick in the pit of his stomach.

No mistake! It was no street light reflection! No mirage!

Bolting desperately toward the inferno, Josh searched his mind for emergency methods to douse the flames. The powerful force of a explosion snuffed out that thought. A cascading shower of orange-red sparks flew upward, while debris crashed around him. He stopped in his tracks, numbed by the grim sight of molten metal. All the firefighting equipment in Montgomery County could never salvage the flaming mass that had served as his primary means of transport since he had first learned to drive.

Though deterred by the force of the blast and the scorching heat, Josh felt a chill go down his spine at the sight of a masked figure sprinting from the inferno, heading in his direction. No doubt about it—the car had been torched and the villain was now trying to escape! Josh charged instinctively as fast as his limping left leg would allow, determined to intersect the fleeing phantom. The question flashed: Could the arsonist and the harasser of the week be the same person?

Josh crouched, reviving skills honed years before at Sherwood High. The crippled leg ruled out a high-speed sprint, but Josh knew that a disabling, open-field tackle could give him one chance to subdue and unmask the culprit. In desperate retreat, the arsonist hadn't noticed anyone coming his way until he had almost brushed the outstretched arms. At the last second he lunged away.

Josh plunged ahead recklessly, throwing his weight headlong at the elusive runner. Despite the disadvantage of beginning from a standstill, he dropped the enemy to the pavement, where the arsonist struggled to escape the steely embrace, kicking violently. Slowly, inexorably, Josh subdued his attacker, tasting victory in a battle matching brawn with brain.

Before he could exult, the attacker slashed at him with a razor-sharp instrument. Josh saw its glint against the glow of the eerie inferno. With a single swath of the switchblade, the contest's odds shifted. Josh's long-dormant athletic skills empowered him to dodge, duck, and weave, but each slash of the knife had him struggling for his life! While he could never subdue the elusive prey without moving closer, he knew doing so put him at deadly peril.

Launching a kick toward the attacker's groin, he grabbed at the knife-wielding wrist. The arsonist twisted frantically, swinging the blade in a wild arc.

At first Josh felt nothing! Then a warm trickle down his neck announced that the steel had struck home, barely missing the jugular. Josh grabbed again, this time managing to break the assailant's grip on the weapon and flipping him to the ground in a heap, thanks to leverage gained in jujitsu training.

The weapon clattered to the pavement while the prone aggressor gulped for air. Before Josh could retrieve the knife, the enemy recovered his bearings and with one gymnastic jerk, jumped to his feet. Despite the blood flowing down his neck and across the collar of his coat, Josh attempted a final plunge to corral the masked man. Weakened and dizzy, he lacked the strength to hang on. The antagonist wrenched free and escaped.

Josh held nothing but a jeweled Rolex and the abandoned switchblade as souvenirs of the struggle. Rising slowly, he reached down and numbly retrieved the lethal weapon. Only then did he remember to search his pockets for a handkerchief to staunch the flow of blood.

Before his senses returned, he heard whistling zings, along with two sharp cracks from the adjacent woods, shattering the flaming darkness. Not until he spotted the ragged tear in the left sleeve of his jacket, only a few inches from his heart, did he realize the fleeing phantom worked with a partner—intent on murder. Again his body chilled as his wildly beating heart pumped ice water.

The event had consumed no more than sixty seconds, start to finish!

A contingent of curious onlookers flowed en masse from the playhouse lawn to the parking lot. JT charged ahead, the first to join his friend.

"What happened to you, Dar? You OK? Trying to start World War III?"

"Yeah, I'm OK! But some loony tune torched my jalopy—and someone else with shaky aim mistook me for a turkey!"

Josh pointed at the billowing acrid smoke while fingering his jacket's unsightly rip at the left shoulder. JT faked controlled calm. "And tried to slit your throat in the process, it appears!"

Yanking a fresh handkerchief from his pocket, JT put pressure on the wound to staunch the flow of blood, cautioning, "We'd better get you to the E.R.—just in case!"

His tone betrayed urgency. Brushing aside Josh's insistence that they hang around to report to officialdom, he took command with brotherly concern.

"First things first, Dar. Your account of the crime won't amount to much if you bleed to death trying to do your civic duty!" Determined to rush his charge to Montgomery Memorial's emergency room, he ordered, "Grab your briefcase, and let's get you outta' here—now!" With tires squealing, the two tore out of the playhouse parking lot, JT steering the makeshift ambulance like a protégé of racing great Barney Oldfield.

By the time they were outward bound, the fire had begun to subside. Pungent smoke from burning rubber tires curled skyward in lazy circles. University fire department personnel eventually dragged up some hoses to douse the embers, creating hissing steam and burgeoning gray-black clouds. Campus security searched the woods for clues, methodically taking notes. Dumbstruck spectators stood by, witnesses to the aftermath of a hate crime.

Once hospital-bound, JT lightened up. "You must think I'm your permanent lifetime ambulance!"

"Just this one last time, if you don't mind, Mr. Daniels."

"I suppose you need a ride to the home place after we visit the E.R.?"

"I thought you'd never ask!"

An exhausted Josh had been too emotionally drained to laugh. Now, as the color returned to his cheeks, his wit and good humor returned. Off-beat humor between the Magruders Mill neighbors had never felt so good.

"This transportation thing looms as a larger problem every day. With a toasted jalopy and a moody mount interested only in occasional, cross-country jaunts, I may be doomed to a future of *shanks mare* unless you unleash your political clout and lobby for a Metro subway line to Fort Ryan!"

Before JT could respond in jest, Josh remembered the spoils of war—the Rolex watch and the switchblade. "By some fortunate fate, this watch may carry more cash value than my old car. As for the knife, maybe I can learn to whittle on the front porch."

The lawyer-in-the-making proposed a better idea. "This could be hot stuff, Josh! We need to check the fingerprints. Wrap it all in a Kleenex so as not to smudge the evidence. Better yet, why don't you let me take them to Judge Stone—and let the legal beagles do the grunge work? Isn't that why you pay those big bucks?"

"You're a genius lawyer, Mr. Darrow. Or should I say Mr. Brogan? You're worth every penny I don't pay you."

The zany pronouncement from the flat-broke client produced another round of boisterous guffaws. Still, the carefree mood did not detract from the mission at hand.

Emergency Room physicians acted rather unimpressed with the severity of the neck gouge. After skilled patching to close the cut, they pronounced the patient fit and good to go. Nonchalantly, they described the gash as "at least an inch from the jugular."

The balance of the ride to Fort Ryan proved uneventful. The duo lapsed into silence, consumed in thought.

Josh broke the reverie first. "The line is thin and fragile."

"Busy as it is, this highway could use a freshly painted stripe every six months or so."

Concentrating on navigating the winding ribbon of road leading deeper into the rolling countryside, JT missed completely the philosophical implications.

"No, JT. Not Maryland's state highway system...I'm talking about the road of life! Without that fragmentary bit of dodging and broken-field running I learned at Sherwood, I might have been too slow to duck that knife...Barely an inch, JT...but for that single inch, by now you might have been moanin' and groanin,' planning my funeral...all the time saying nice mournful things like, 'Oh, but that Dar Ryan was a good dude!'"

"What makes you so sure I'd be saying good things? I'm trying to learn to be a good lawyer, always telling the truth—like my star witness."

"Get serious, JT. Life-spans can be no more than a blip on the screen. Look at your dad! And my mom! She barely made it to twenty-five and your dad flew to a hero's death at twenty-seven. Makes you think. Up till now, I'd always expected to go on forever!"

JT nodded affirmation. Josh Ryan had just walked through the *valley of the shadow* and had lived to tell about it. He deserved a chance to think out loud.

"Look at the Duchess! Seventy-eight feisty years, but downhill every day since her prime. If anyone seems able to go on forever, immune from the grim reaper, it's her. But just like any seventy-eight-year-old, she's deteriorating— imperceptibly, relentlessly. Sure as shootin' the day will come when she'll ride over the top of life's escalator and disappear from sight forever!"

JT joined in priming philosophy's pump. "Granddad Daniels is no spring chicken, either...It's hard to imagine a world without his overshadowing presence."

Not wanting to drift into morbidity, Josh let the matter rest. Soon he resurfaced to touch on an ultimate issue.

"Darwin never offered an explanation for the origin of the first-ever spark of organic life or the prospect of life after death. The 1851 loss of his ten-year-old daughter, Annie, crushed the poor guy's spirit. No matter how he longed for progress toward perfection, reality trashed his dreams and taunted him with the opposite—deterioration, decline, decay, and death.

"If claims of a gradual continuum propelling organisms from the simple toward the more complex by the random good fortune of *natural selection* had merit, the question jumps out: Why doesn't the same chaotic process eventually increase longevity?"

Intrigued by the question, JT invited the punch line by asking, "In the words of the distinguished Marcus Brogan, *explain*, Mr. Ryan."

"I see seasons and rhythms in nature. For some extravagantly marked butterflies, it's all over in a few days. The smartest, most loyal dog in the world is lucky to be barking for fifteen years. Turbo could be around for thirty years. We *Homo sapiens* run on borrowed time after seventy or eighty years.

"Take the Duchess. She rides horses, works like a Trojan, watches her weight, and eats nutritiously. Already she's beaten the law of averages."

"Meaning?"

"Looks to me as though we all exist as part of a balanced cycle involving some finely tuned cosmic sequence! No way is the big picture one of unplanned chaos—otherwise we could never predict the seasons, the sunrise, the tides, or astronomical positions in time and space. Humans are central to this rhythm of life, possessors of a predictable life span."

"I declare, you almost sound like a religionist, Josh."

"If that means I believe in some universal big picture, count me in! For now, one thing is certain. If asked to choose between producing a literary masterpiece thanks to the haphazard dance of a kitten on the keys of a computer or by the handwriting of a creative author, it seems to me that only a fool would give odds favoring a random tap dance on a word processor. Every gambler I know would place his bet on the sure thing.

"If religion requires faith in the unprovable, then the ultimate zealot is the mega-evolutionist who places bets on a dice throw loaded to lose!"

Josh's philosophical musings had run down. JT signed off with an agreeable, "Makes sense, Dar."

For the remainder of the homeward trek, the duo sank into silence. The ride was punctuated only by an occasional oncoming headlight probing the shoulder of the rural highway. Within minutes, the Fort Ryan entry drive loomed.

As the car rolled to a stop near the mill house, JT returned to his role as intuitive lawyer-to-be. "Do you know anyone who wears a Rolex, Josh?"

Stepping out of the car, Josh hesitated. "Funny you should ask, JT. For the life of me, I'm stumped. Most folks I know wear waterproof throwaways. Still, there's the nagging thought I've seen this watch somewhere, sometime...I just can't..."

At the moment, Josh's normally reliable memory failed!

This Thursday night, no relentlessly ringing telephone intruded on his sleep. Instead, his racing mind kept him awake as it wrestled with the puzzle.

Keyword clues accessed cerebral search engines, asking the identity of the watch's owner. He dozed off and on, figuratively thumbing through the computer files of his multi-megabyte mind. From time to time, almost asleep, he would waken with a start, with a sense of proximity to the elusive answer. Then, drawing a blank on his memory screen, he would drift away again, only to awaken to reruns of the exhausting episode.

Three hours into the night, he awakened suddenly. The search had paid off! He saw a three-dimensional image in his memory. He had observed this Rolex watch on a single occasion, earlier in the year. He recalled with precision the time, the place, even the date. The identity of the owner came through loud and clear. The subliminal thought erupted, "Eureka!"

Josh Ryan exhaled slowly, savoring the moment. Satisfied he had survived another day, he rolled over and drifted into peaceful, uninterrupted sleep.

# VI
## Day Six: Friday

# *The Human Edition*

## RANDOM CHANCE GENETICS

**"DON'T BE RIDICULOUS CHARLES!
'PREBIOTIC SOUP'
DOES NOT EXPLAIN MY CONDITION!"**

# 33

Charred metal chunks lay strewn in haphazard disarray across the singed surface of the playhouse parking lot, marking the site occupied by a vintage automobile the day before. This Friday morning, a swath of yellow tape quarantined the crime scene. Curious onlookers shuffled around the perimeter. Inside the boundary, law enforcement and insurance agencies measured, jotted, and took photos.

Michael Ryan's jaw tightened at the sight.

"Looks like someone's trying their best to make a pedestrian out of you, Son."

Josh ignored his dad's attempt at humor. Now derailed was his plan to restore the car to pristine beauty. Emotions churned. Swallowing bitter disappointment, Josh assessed stark reality.

"As long as I have Turbo and the mountain bike, I'm not reduced to what Grandpa Ryan used to call *shanks mare*. Ever since last April's camper theft from Center Hill Lake's Camp Ryan, something sinister keeps doggin' my steps. Last night it dawned that someone would like to kill me. I'm not sure *who, what,* or *why,* but I'd have to be an idiot not to recognize I don't rank very high on some rat fink's popularity poll."

Michael mulled his remarks without comment. After parking the car, he offered fatherly counsel as they hiked to Chamberlain Playhouse.

"For what it's worth, Son, I've canceled all business trips for the immediate future. I'm going to hang out with you at Fort Ryan until these mysteries are solved. The legal team working your corner consists of heavy hitters. Sooner or later the jumbled clues will form a clear picture.

"In the meantime, let the good guys do the worrying while you knock 'em dead for the *Monkey II* finale."

"Yeah. Makes sense."

Pausing at the main courtyard entrance, Josh took off for the stage door waving a, "So long, see ya."

Michael stopped and turned, offering one last admonition. "Son? I know this sounds stogy, but you're my one and only. You're smarter than most anyone I know, but don't take anything for granted until this sad episode is behind us! Be careful! OK?"

Josh smiled and nodded as they headed in opposite directions.

With barely ten minutes to go before curtain time, the preoccupied witness plunged through the rush of late arrivals, only to inadvertently smack head-on into Dr. Karl Striker. The force of the impact jarred Striker's glasses a notch or two down his nose. Reacting to the severity of the jolt, Josh instinctively offered an apology before recognizing the victim.

Straightening his horn-rims, Striker would have none of it. Straining to exert every ounce of discipline at his command, he grumbled a graceless reply. Only a tin ear could miss the menacing tone.

"You are notoriously reckless, young man. It would pay you to watch your step."

The cheerful but gutless wonder, Gus Morley, reluctant witness to the embarrassment, walked in lockstep with his demanding master. The normally effusive assistant, uncomfortable in his role as peacemaker in the wake of Dr. Striker's wrath, tried stammeringly to smooth things over.

"I'm sure the young man didn't jostle you intentionally, Karl...I can't see that any damage was done."

Morley laughed nervously. Other than directing daggers at all comers, Striker didn't change his sauerkraut expression.

Josh piloted his escape with an end-run, surprised that the bitter professor and his number-one underling seemed intent on showing their faces at the final act of the disdained event. While the offending student remained in earshot, the professor managed one last caustic barb while readjusting his half-lenses and recapturing his dignity.

"The word on the street, Mr. Ryan, is that you're in the market for a new means of transportation."

Unsure whether the retreating Josh could still hear, Karl Striker raised his voice to deliver a final ill-tempered taunt.

"Good luck shopping! They say credit is hard to come by for the unemployed."

A florid-faced Morley hustled the apoplectic division chairman through the playhouse entry, knowing that the big boss faced two final hours of teeth-gnashing frustration. Striker's remaining hope for intellectual vindication rested with a jury verdict that might repudiate the obstreperous student.

Inside the playhouse, early arrivals were treated to a gravel-voiced Louis Armstrong swinging out Satch classics on recorded stereo. Moments before curtain time, the moving strains of "What a Wonderful World" greeted listeners' ears. Whether by coincidence or in a subtle stroke orchestrated by Pace Terhune, the rendition just happened to be a Lady Carrington favorite.

Due to scheduling conflicts, the climactic presentation of *Monkey II* had been allocated the unusual 10 a.m. time slot. Pace understood show biz and the premium placed on timing. He covered the glitch by boasting to the media that, "The morning setting is ideal for the summation and the verdict scenes since it more closely approximates the daytime setting of the *Scopes* Tennessee trial."

Unimpressed, a wiseacre shot back, "Sure, and if we hurry, you'll sell us tickets to last year's Super Bowl!"

News people roared. Pace grinned like a Cheshire cat. With or without this thin rationale for the off-beat timing, the public clamored to know the verdict, undeterred by wind, rain, or unconventional scheduling.

The curtain rose on a standing-room-only crowd. Congratulatory applause reassured Terhune.

An exceptional trio occupied Row G orchestra seats. Front and center, socialite Lady Carrington ruled in serene grandeur. The Duchess' heretofore estranged son-in-law flanked her left elbow, while Bishop Brock Daniels, in town for a churchman's conclave at Washington's National Cathedral, sat to her right. Josh beamed at the sight. Except for the telltale patch on his throat, he appeared in fine fettle, unruffled by the previous evening's brush with death.

As Judge Stone intoned the come-to-order ritual, Josh nudged JT to scan Row G. At sight of his distinguished grandfather, JT flashed an ear-to-ear grin, signaling thumbs up. Traci Kilburn exchanged knowing winks with the Magruders Mill chums, satisfied the selection of the talented two would have done a casting director proud.

An invigorated JT wasted no time in launching Friday morning's entertainment menu. The sight of his mischievously snapping eyes and eager demeanor led Josh to surmise that his protector at the bar had cooked up something more than prebiotic soup. JT's first words to the bench confirmed his guess.

"Your Honor, if it pleases the court, before calling Mr. Ryan to the stand for this final day's testimony, I respectfully request the privilege of introducing two witnesses prepared to respond to the credibility issue raised by Slade Lassiter this past Wednesday."

Josh sat erect, intrigued. Judge Stone eyed Traci. "Have you been made aware of this request and the identity of the proposed witnesses, Ms. Kilburn?"

"Yes, Your Honor, earlier this morning."

"And do you have any objection?"

"None whatsoever, Judge Stone, but of course I'll reserve the option to cross-examine."

"Very well then, Mr. Daniels. You may proceed, provided you confine the presentation to matters relevant to Mr. Lassiter's previous testimony."

"Thank you, Your Honor. First, I'd like to call Ms. Jessica Saunders."

A buzzing undercurrent echoed through the playhouse!

Jessica Saunders' credentials fairly shouted credibility—CPA; FBI-trained, computer whiz, and small-arms sharpshooter; even a karate black belt for good measure! The witness abandoned her court clerk post for the witness box. Immaculately attired in a stunning black business suit, Ms. Saunders radiated dignified authority as she took the oath, swearing to tell the whole truth and nothing but.

JT flaunted her flashy résumé for the jury's benefit as though hoisting battle flags for a victory parade. Satisfied that the jury had gotten the message loud and clear, he got right to the point! He pulled out a copy of Dr. Striker's single-page letter dated Wednesday, September 4, 1996, in which the professor reported finding no sign in the Science Research Foundation's library computer index of any 1969 correspondence from Chancellor Brooke Fielding to Regina Ann Carrington.

"Ms. Saunders, you have seen this exhibit admitted to the record of this case and have read its contents?" asked JT.

"Yes, I have, Mr. Daniels."

Her demure manner personified assurance. The lady understood the drill, having spent many a day as a seasoned witness in real-life courtrooms.

"Have you had occasion to check the library's computer index yourself?"

"As a matter of fact, in the presence of campus security, I entered the library early yesterday morning...a few minutes before 6:00 a.m. I spent the better part of two hours reviewing the index and checking the archives."

"Did your research confirm the findings reported by Dr. Striker as of 2:33 p.m. the previous afternoon?"

"As a matter of fact, Dr. Striker's analysis appears correct...as far as it goes!"

Josh groaned inwardly, pleading in his thoughts, "What kind of stunt is this?"

JT exuded placid confidence. "Can you explain what you mean by *as far as it goes*?"

"Certainly! Dr. Striker's analysis accurately describes what he likely found—at the time he looked. My check of the index confirms that the Fielding letter did not appear on the computer screen as of 2:33 p.m. yesterday."

"So, you agree with Dr. Striker?"

"Yes, but only to a point. A more thorough analysis disclosed that reference to the letter still resided in the computer's recycle bin, waiting for retrieval."

"Can you interpret *recycle bin* for us computer illiterates?"

"Most sophisticated computer systems have backup files to prevent inadvertent erasures. Whoever tampered with the library's computer index forgot to delete the backup."

"Then, what about the onionskin? Were you able to find that fragile copy of the Fielding letter in the library's archives?" probed JT.

"No. The onionskin carbon of the unseen original is obviously gone. But the index reference to the letter still lurks in the computer's recycle bin, as confirmed by this printout."

She handed a copy to JT, who in turn showed it to Traci and Judge Stone before offering it into evidence. Pausing for dramatic effect, he asked a final question.

"Your conclusion, Ms. Saunders?"

"Someone stole the letter from the archives and attempted to erase the reference from the index, but, due to inexpertness or haste, botched the job! There is every reason to believe Chancellor Fielding's 1969 letter to Regina Ann Carrington did in fact exist!"

"That's it?"

"One more thing, Mr. Daniels. I've heard Mr. Ryan quote a piece of wisdom from his father that fits the moment: *Truth stands in three dimensions; it needs no defense. Falsehood collapses on itself, melting at the whim of the sun.*"

JT allowed the devastating conclusion to speak for itself.

"No further questions, Your Honor."

Fascinated, Traci Kilburn shook her head, declaring, "I have no questions of this witness, Your Honor." Relieved, Josh heaved a huge sigh, "JT Daniels, you're an all-fired genius...and as for you, Jessica Saunders, may we always walk the same side of the fence and *whup* the cowboys together!"

Judge Stone quipped, "Thank you, Ms. Saunders. Feel free to return to your clerk's post of duty that you abandoned so willingly. Seriously, Ma'am, the court appreciates your testimony."

Looking again at JT, he added, "Are you ready with your second witness, Mr. Daniels?"

"Ready, Your Honor. It gives me great pleasure to call Lady Regina Ann Morgan-Carrington to the stand."

The introduction created a stir, with muted conversations erupting in a wave. Josh could hardly believe his eyes, but there she was, stepping regally to the isle from Seat 14, Row G, bound for the crimson, carpet-clad steps leading to the stage. No shrinking violet, this lady. Judge Stone gaveled for silence several times. By the time the Duchess mounted the witness box, the decibel level had subsided to faint mumblings.

Traipsing across the stage, the Duchess affected a modest bow aimed at Judge Stone; blew a kiss in the direction of Traci Kilburn; delivered an approving pat to the shoulder of Jessica Saunders; shook hands with a startled, but reciprocating Pace Terhune; and as a finale, exerted her considerable charm to beam at the entranced jurors.

Bewildered, Josh first put his face in his hands, elbows resting on the counsel table. Then he slowly raised his head, deciding to enjoy the show. JT Daniels knew what to do, and Judge Stone smiled his approbation at the bold stroke. Once the oath and foundation data reached the record, JT headed direct for the bottom line.

"Tell us, Mrs. Carrington, as a Mid-Atlantic University trustee, can you recall ever meeting Chancellor Fielding in person?"

She threw up her hands in mock astonishment. "Land sakes, young man, you know me well enough to know the answer to that question! Of course I knew Brooke Fielding. My dad, Joshua Chamberlain Morgan, and he were great friends. The chancellor acted as godfather at my christening; I audited several of his lectures on genetics; he even presided at my several-days suspension from the university, thanks to my unsavory involvement in a minor fracas, the details of which have long since escaped my mind.

"You bet I knew the chancellor, Mr. Daniels. Right honorable man he was too! Straight arrow! Even though my dad functioned as a university trustee, it made no difference to Brooke—he enforced my involuntary vacation from class brandishing authority he asserted to represent the *unanimous consent of the faculty.*"

"I take it that is a 'yes' answer as to your recollection of meeting the chancellor in person?"

"Of course it's a 'yes,' Mr. Daniels. Didn't your granddaddy the bishop teach you to listen?"

The audience reveled at being treated to *Iron Pants* at her fabled best—or worst, depending on your point of view. The aging widow's stiletto tongue lashed friend and foe alike. She could purr like a kitten one minute then casually cut you off at the knees the next. Turn on the showbiz spotlight and the charming wit blossomed. On command, the Duchess could portray any emotion in the thespian book. Instinctively she knew how to reach inside the heart and touch the soul of a listener. Her skillful stroking of audience reactions matched the artistry of a Kenny G coaxing melodious strains from a saxophone.

For this one golden moment, the vivacious socialite enjoyed the opportunity to play-act, to perform, to sway the audience—but most of all, deep down, she welcomed the chance to circle the family wagons in the finest Morgan tradition.

Eyes dancing, tongue-in-cheek, the Duchess volunteered a confession of sorts, guaranteed to titillate. "Oh, by the way, Mr. Daniels, perhaps my familiarity with Dr. Fielding will be confirmed with my admission that I had occasion to sit on the chancellor's lap from time to time..." Her blue eyes twinkled at the innuendo.

"He was a right friendly kind of man...if you know what I mean!"

Unabashedly stealing JT's measured flamboyance, she paused for effect before adding hastily, "Of course, this was during family visits...I would have been four or five years old at the time!"

The audience roared while the grand lady of Morgan Manor feigned innocence. JT hastily interjected another query, determined to short-circuit her wandering reverie. "Stipulating that Chancellor Fielding was a close friend of the Morgans, are you, per chance, also familiar with Doctor Fielding's signature?"

"You lawyer types insist on asking the obvious, consuming time, to keep the billing meters running, no doubt!"

She chided and charmed so good-naturedly that even Judge Stone neglected to rap the gavel to still the laughter. Undeterred, the Duchess meandered on.

"His name is written everywhere but on my body parts! He signed my diploma from MAU; he wrote to congratulate me on my marriage to Lord Carrington; his letter announced my appointment as university trustee; and best of all, he had nice things to say upon the birth of my daughter, Leslie Anne."

"Did you receive a letter from the chancellor written in August 1969?"

She didn't answer immediately, instead staging a ponderous fumbling through her purse, deliberately building high drama. Finally, she looked up, clutching a wrinkled memento.

"I'm sure you know about women's purses, young man. Carry everything, find nothing."

JT welcomed the stall, deliberately exploiting the moment. Basking in the shared glow, he made no effort to intervene and risk breaking the spell.

"I don't save everything, maybe I should. For some reason I saved this. But it took me all day yesterday to find it in the attic, with many other things I'd completely forgotten."

Nonchalantly she fished out a sheet of paper tied with a red ribbon explaining, "Why, look here! My MAU diploma—in case you were wondering if my awful suspension became permanent! See! It carries Dr. Fielding's signature—in ink—and the official gold seal of the university, no less!"

She brandished the diploma with impish glee.

"And?" JT brought her back on track, massaging her thoughts without pushing.

"Oh yes, of course. You asked about the August 1969 letter. Well...this is it! Eureka!"

Exulting at the discovery, she waved several pages of crisp stationery high above her head.

"Does it carry the original signature of Chancellor Brooke Fielding?"

"Why no, not at all. Remember, he long since had retired and no longer served as chancellor..." Then, after making the most of the moment with another interminable pause, "...but of course it carries Brooke's original signature...Who do you think signed his correspondence, Charley McCarthy?"

Here she tried a touch of huffy, self-righteous indignation. "I already told you I can identify his signature...and Morgans don't prevaricate...and neither do the Ryans, I might add!"

"Prevaricate?"

"Don't be impudent, young man. Any Daniels knows full well that a prevaricator is a liar! And the Morgans, Ryans, Carringtons, and Magruders never lie! Of course, I'm not prepared to vouch for other matters of lesser consequence such as gambling, and..."

Once she drew ripples of laughter from the spectators, her voice trailed away. JT next handed her a photocopy of the battered onionskin document Josh had found in the science library's archives.

"Tuesday evening, on this stage, Joshua Ryan produced this copy of a letter allegedly written to you by Chancellor Fielding in August 1969. Do you hold the original to that copy?"

After taking the print, she ceremoniously adjusted reading glasses for eyes intimates swore functioned perfectly at a 20/20 level without the props. Inspired by the show and hype, she set about comparing original and copy with less-than-deliberate speed. JT knew better than to prod. He waited patiently. Finally, with authoritative flair, the Duchess spoke. The mock astonishment wreathing her face would have done Ethel Barrymore proud!

"Wouldn't you know, it's the exact same letter! I had no idea he kept a copy for the science library!

"Brooke was the sweetest man. It's always bothered me that he passed away before I ever saw this last letter written to me. Leslie Anne and I were touring jolly old England at the time...It's a shame...The grand old man up and died...been buried a week before we even heard he had passed."

The Duchess shed a genuine tear or two at thought of the long-departed man of eminent stature. JT sensed when enough was enough, and quickly moved to thank the witness and offer the original Fielding letter as evidence in the court record.

"Any questions for this witness, Ms. Kilburn?"

"No questions, Your Honor."

"Very well then, you may step down, Mrs. Carrington. Thank you for the courtesy of your voluntary appearance."

The Duchess rose with grace, flaunting her best prima donna pose, bowed to the audience, and departed the stage waving and winking in all directions as though winning the Y2K Miss America crown. For the first time in the week, playhouse spectators exploded in a standing ovation as she made her triumphant return to center-front Orchestra, Row G, Seat 14.

Captivated by the drama, Judge Stone ordered an unscheduled recess to allow all parties to savor the once-in-a-lifetime performance.

Virtually lost in the clamor was the game of wits that had been played with the playhouse stage serving as symbolic chessboard. The queen had just knocked off the adversary's pawn, jeopardizing the would-be king. Left with no intellectual escape route and fully aware that *Monkey II*'s final act threatened

274

checkmate, Karl Striker longed to slip unobtrusively out the auditorium's rear exit. But he gritted his teeth and hung tight. Never a quitter, the tough professor grimly determined to fight back another day, in another way.

# 34

The curtain didn't drop during the unscheduled recess. Players on stage milled about informally. Aroused by the intriguing testimony of the Duchess and Jessica Saunders, the audience clustered in small discussion groups, disdaining the opportunity to exit. Aware of the excitement level, Judge Stone interrupted the recess to preserve continuity, calling the court back into session in exactly five minutes.

Confident that his credibility was no longer a major issue, a poised Josh Ryan waited impatiently for one last opportunity to provide fodder for a verdict. JT Daniels obliged with his first question.

"Darwin uses analogy to suggest that '*both animals and plants may have been developed*' from some lower algae and '*all the organic beings which have ever lived on this earth may be descended from some one primordial form.*'[135]

"Excuse my making this personal, Mr. Ryan, but is it conceivable that the lofty Ryan/Morgan genealogical family tree stretches its roots back to a Great-great grand pappy Primordial Algae?"

Josh shook his head violently. Although he tried to sound objective and professional, the answer nevertheless came tinged with a hint of scorn.

"A story in today's *News Press* identified Sam Houston Morgan as my honest-to-goodness great-great grandfather. That foxy old wildcatter provided some of my genes. That family-tree thing, speculating that humans and apes share a common ancestor, is dead as a dodo bird and dry as a desert fossil! It's one thing to draw an illustration on paper, but it's something else to corroborate superstition with empirical evidence! That fictional nineteenth-century human genealogical-tree thing is as imaginary as Jack's beanstalk."

"I take it that's a 'no'?"

"With a capital 'N'! Throughout post-glacial history, *Homo sapiens* ancestry continues exclusively human. Today's four blood types correlate precisely with the types discovered in Egyptian mummies; human organs remain in identical body placements and perform identical functions."

"And DNA codes?"

"The same exclusively human identity applies—as far back in time as test specimens can be found. To conjecture otherwise equates fact with fabrication. The arrogance of ignorance never ceases to amaze."

"No man from molecule, then?"

"The last I heard, it's all in the genes—human genes produce *Homo sapiens* and fruit fly genes produce fruit flies. Luther Burbank cited the law '*of the Reversion to the Average...I can develop a plum half an inch long or one 2½ inches long, with every possible length in between, but I am willing to admit that it is hopeless to try to get a plum the size of a small pea, or one as big as a grapefruit...there are limits to the development possible, and these limits follow a law.*'[136]

"Where is the evidence that plum genes can produce a grapefruit-sized plum? Or better yet, a grapefruit from a plum? How can random chance breach those genetic boundaries even with mutations and millions of years? When does mega-evolution kick in to produce an entirely new prototype animal or plant?

"*Escherichia coli* bacteria can reproduce a new generation every three-and-a-half hours. Twenty-four thousand generations later, genetic adaptability responding to scientifically modified environments assures variety—but all descendants remain *E. coli* bacteria![137]

"What then about the odds for some ape-like ancestor evolving into a *Homo sapiens* in 24,000 generations averaging 25-years each?"

"How do you define *reversion,* Mr. Ryan?"

"Luther Burbank described genetic homeostasis as '*a pull toward the mean.*' If descendants carry the original diversified gene pool, they can revert to original prototype."

"What does reversion have to do with *Homo sapiens* and mega-evolution?" JT asked.

"That's just the point! Laboratory genetic experiments demonstrate a '*pull toward the mean*' rather than a push away from the '*mean,*' contrary to evolutionary theory.

"Listen to Darwin's fantasy. '*I can see no difficulty in a race of bears being rendered, by natural selection, more and more aquatic in their habits, with larger and larger mouths, till a creature was produced as monstrous as a whale.*'[138]

"Excuse me? Where's the proof demonstrating a jump to an entirely different prototype? What is to prevent lateral slippage to the side or reversion back to the original? What barrier keeps *Homo sapiens* from reverting to primordial slime under Darwinism? No wonder the guru of chaos deleted the *race of bears* rhetoric from subsequent editions of *Origin*!

"Bottom line: Genetic defects: Yes! Entirely new animal prototypes: Never!"

JT switched topics to the *Homo sapiens* family tree. "If you claim the root of the human family tree remains forever human, do you still disagree with Darwin's *survival of the fittest* mentality among disparate human cultures?"

"The time has arrived to welcome Charles Darwin to the ranks of the politically incorrect as well as the scientifically obsolete."

"Meaning?"

"Darwin's *survival of the fittest* suggests boot-strap elitism!"

"Maybe some of that *reflecting the attitudes of his time* whitewash?" The normally ebullient Jonathan Thomas Daniels delivered the question in icy tones. He bridled at the least hint of prejudice.

"When Darwin's first edition of *Origin* hit the book markets with its demeaning cultural stereotypes, human beings were being routinely bought, sold, and broken in bondage. Nearly a century and a half after Sam Houston Morgan's *ride to glory* took him from Texas to Gettysburg's bloodbath, racism still lurks in the minds of the insecure."

Josh Ryan shared disdain for the bias virus that plagues the human race, presenting a blunt response by exposing George Bernard Shaw's take on Darwin's nineteenth-century English environment.

"'*Never in history...had there been such a determined, richly subsidized, politically organized attempt to persuade the human race that all progress, all prosperity, all salvation, individual and social, depend on an unrestrained conflict for food and money, on the suppression and elimination of the weak by the strong, on Free Trade, Free Contract, Free Competition, Natural Liberty, Laissez faire: in short, on 'doing the other fellow down' with impunity.*'"[139]

"Consequently, the human edition came in for some tinkering and manipulation by self-perceived superior types?" JT fired off the question with jaw flinty and eyes flashing.

"Driven by *survival of the fittest* rationale, Darwin's cousin, Francis Galton, and a tribe of followers determined to categorize *Homo sapiens,* eugenics saw the light of day in 1883—or the dark of night if, you prefer. Galton basked in the privileges of upper-strata British society.

Galton's view represented the '*science of improving the stock...to give the more suitable races or strains of blood a better chance of prevailing speedily over the less suitable.*'[140] This recipe for conflict offered plenty of ammunition for the spouters of super-race propaganda."

"Can you pin the racist label on Darwin?"

"The guru of *survival of the fittest* painted a grand-scale epic of contemporary humanity in which *savages* and *barbarians* occupied the lowest level while unapologetically, European Caucasians pre-empted pinnacle status. Darwin tipped his hand in the full title of his book: *On the Origin of Species by Means of Natural Selection, or the Preservation of Favoured Races in the Struggle for Life.* Let his own words speak.

"'*It is chiefly through their power that the civilized races have extended, and are now everywhere extending their range, so as to take the place of the lower races.*'[141] '*At some future period, not very distant as measured by centuries, the civilized races of man will almost certainly exterminate and replace throughout the world the savage races.*'"[142]

"Any boldly pristine examples?"

"Darwin was not bashful in citing the J. Barnard Davis idea alleging the *'mean internal capacity of the skull in Europeans is 92.3 cubic inches; in Americans 87.5; in Asiatics 87.1; and in Australians 81.9 inches.'*"[143]

"Just because Darwin said it, does that make Neo-Darwinists racists?"

"Not likely! Sweeping condemnations are as mindless as the malady. Today's mega-evolutionists shun cultural *survival of the fittest* labels as they would a blood transfusion donated by an AIDS victim. But it's a catch-22 for Darwin devotees—another arena of the master's junk science to disavow.

"But hang on, ladies and gentlemen of the jury! Darwin delivered further inspiration to the authors of eugenics and master-race philosophies, suggesting with stiff-upper-lip mentality that *'we must bear without complaining the undoubtedly bad effects of the weak surviving and propagating their kind...but there appears to be at least one check in steady action, namely the weaker and inferior members of society not marrying so freely as the sound.'*"[144]

Seeing audience eyebrows raised, the witness continued his brash assault.

"Darwin's pronouncements didn't match those of a pompous politician claiming to feel your pain. Without apology, he bemoaned the fact that *'vaccination has preserved thousands, who from a weak constitution would formerly have succumbed to small-pox. Thus the weak members of civilized societies propagate their kind...It is surprising how soon...care wrongly directed leads to the degeneration of a domestic race.'*[145]

"Come again? Condemnation of vaccines that can save the lives of little kids? A far cry from a sensitive social conscience or the declaration echoing across the Atlantic that...*all men are created equal!*"

The audience was hushed. Pausing to allow listeners to savor the implication of Darwin's sentiments, Josh breathed a disgusted sigh and hammered away.

"Married and a prolific parent, the writer left little doubt that he intended the Darwin family to be identified with the *we* of the equation. The savvy naturalist betrayed no reticence in embracing the 'Rule Britannia' complex."

"How do you interpret Darwin's *civilized races...extending their range to take the place of the lower*'?"

"My take on his *civilized* reads 'Europe'; and *extending* translates to empire exploitation! The American colonies had broken free, but other places in the world still felt the heel of unchecked social Darwinism!

"What would you expect from a guy who believed that *'the early progenitor of all the vertebrata must have been an aquatic animal provided with branchiae with the two sexes united in the same individual, and with the most important organs in the body (such as the brain and heart) imperfectly developed.'*[146]

"Since he knew nothing about DNA, Darwin blithely envisioned a hermaphrodite human ancestor that could answer simultaneously to *Grandma*...or *Grandpa*! So much for sex appeal in the Darwin family tree!"

"Do you blame the molecule-to-man scenario for world war?" asked JT.

"It certainly didn't provide a charter for peace. It did provide a veneer of justification for greed enforced by power—compliments of a contagious *survival of the fittest* rationale.

"William Jennings Bryan drafted a proposed address for the *Scopes* trial that he never delivered. Written days before his death, the document was eerily prescient of a future holocaust. Bryan cited *The Science of Power*, published the year World War I ended.

"'*Within half a century the* Origin of Species *had become the Bible of the doctrine of the omnipotence of force...Nietzsche's teaching represented the interpretation of the popular Darwinism delivered with a fury and intensity of genius.*' Nietzsche '*gave Germany the doctrine of Darwin's efficient animal in the voice of his superman...military textbooks in due time gave Germany the doctrine of the superman translated into the national policy of the super state aiming at world power.*'"[147]

"Do you blame Darwin and Nietzsche for plotting World Wars I and II?" JT's blunt question pulled no punches.

"I didn't live during *the war to end all wars* or the *great war*. Nor do I claim a historian's credentials. It's enough that Darwin's *survival of the fittest* doctrine didn't earn him a posthumous nomination for the Nobel Peace Prize."

Trained in the art of trial preparation, JT had spent the summer boning up on paleoanthropology. His lexicon nailed down familiar nicknames like *Lucy* as well as fancy titles such as *Australopithecus*. By skillfully exploiting this knowledge, the savvy law student's questions revealed more than abstraction.

"Let's take a look at that *Homo sapiens* genealogical tree, Mr. Ryan. Are those fancy graphics anything more than make-believe? Or is there hard proof that human and simian branches sprouted from a common trunk?"

"Pure palaver, Mr. Daniels! Fanciful as Jack's beanstalk! Fossil record gaps dominate. Missing linkage is the rule, not the exception. Mega-evolution's tree enthusiasts swing from non-existent limbs."

"Then what's the inside scoop about the fossil record? Do humans share a common ancestor with apes and chimps?" asked JT.

"No way! Genus *Homo* has coexisted with apes from the beginning. Extinctions have since diminished the diversity of both groups. The phantom *common ancestor* allegedly linking *Homo sapiens* with apes simply never existed."

"You'd better explain for those of us who rarely see a tree stripped naked of leaves and limbs."

Titters from the audience acknowledged the need for clarification.

"Evolutionists acknowledge that '*Prior to 4.5 m.y.a. the hominid fossil record is a virtual blank for ten million years.*'[148] It takes brassy guesswork to concoct a family tree within evolution's time frame that admits to a ten-million-year fossil gap before the tree began to grow!' Josh exclaimed."Just how do you explain a tree planted with seeds pictured as distinctly ape-like *Australopithecus* creatures; which sends up shoots the likes of lady *Lucy*, a diminutive pee-wee of an ape that stood three-and-a-half feet tall and weighed in at 64 pounds; and eventually blossoms with a *Homo sapiens* crop? Last I heard, even Luther Burbank didn't try to graft a citrus branch onto a coconut tree!"

"So bones of extinct species don't prove ancestry?" JT challenged.

"Fossils of extinct apes such as *Australopithecus anamensis* or the slightly smaller *Australopithecus boisei* (translated *southern apes*) share ape-like characteristics. Skulls, forelimbs, and femurs—obviously!

"But do bone relics confirm common ancestry? Hardly! There is no junction on the chart to graft the limb confirming Lucy's hook-up with humans, nor is there a whit of evidence supporting the connection.

"It's one thing to conjecture that an ape-like ancestor parented the human family tree, but something else again to prove it with nothing but dry fragments of fossil ape bones! As I said earlier in the week, homology does not equate to genealogy."

"What about *Neanderthals* as human ancestors?"

"Contemporaries, sure! But ancestral...no chance! '*Anatomically modern humans existed in Africa and elsewhere well before the Neanderthals.*'[149] '*Between the classic Neanderthals of 100,000 years ago and the earliest Homo fossils in Africa, there is a gap of around 1.5 million years that is sprinkled with just a few.*'[150]

"*Homo sapiens* left their footprints '*on the sandy shore of a South African lagoon after a violent rainstorm some 117,000 years* ago.'[151] "*Neanderthals* disappeared forever, some say as recently as 5,710 years ago. Since those rugged fellows appeared on the face of the earth after *Homo sapiens,* lived as our contemporaries, and left the scene before us, it makes no sense to nominate them as ancestral candidates!"

"Could *Neanderthals* pass as *Homo sapiens?*" asked JT.

"Many believe *Neanderthals* to be an extinct race of fully human beings. One reason: A large-sized brain with a hypoglossal canal feeding nerves to the tongue suggest power of speech and point to *Homo sapiens* identity!"

"Is there another view? Don't *Neanderthal* skulls show bulging brows and receding craniums?"

"Ian Tattersall and Jeffrey Schwartz '*saw two triangular bony projections jutting into the front of the nasal cavity from either side. They have not found these features in any modern human skulls.*' The scientists believed that their '*discovery of yet another basic difference in Neanderthal anatomy supports the view that Neanderthals and modern humans are separate species.*'[152]

"But other respected paleoanthropologists contend that *Neanderthals* qualify as fully human—a unique *Homo sapiens* race: '*We have no reason to assume that biological differences between them determined different intellectual capabilities...The Neanderthals were cultured human beings.*'"[153]

"And your personal opinion, Mr. Twentieth-Century Darwin?"

"The big guys supposedly did stuff most humans do—certainly genus *Homo*, arguably a robust but extinct branch or race of *Homo sapiens*."

"Then what about Java man as missing link—*Homo erectus*?"

"Again, just another extinct member of genus *Homo* that co-existed with *Homo sapiens*—contemporary, yes! Ancestral, no way!"

"'*For the evolutionist, Homo erectus is the major category bridging the gap between the* australopithecines *(which everyone recognizes as nonhuman) and the archaic Homo sapiens.*'[154]

"'*According to conventional theory, Homo erectus first appeared in Africa about 1.8 million years ago and spread through Eurasia until it vanished around 200,000 BC displaced by slender, handier, brainier Homo sapiens...new research...sets an upper age range for putative Homo erectus fossil remains found on the island of Java at only 53,000 years. And the lower estimate overlaps by some 20,000 years the period in which Homo sapiens inhabited that part of southeast Asia.*'[155]

"*Homo erectus'* coexistence with *Homo sapiens* demolishes any *Homo erectus* ancestry aspirations!

"A couple of Cal Berkeley scientists, Garniss Curtis and Carl Swisher, assign even more recent dates to *Homo erectus* fossils. The radically younger '*age estimates...range...from 56,000 to only 21,000 years ago...By 20,000 years ago...humans around the world were essentially indistinguishable from people living today...some archaic Homo populations seem to have lingered beyond their time, alongside more modern-looking people—implying that the two types weren't interbreeding and therefore must have belonged to different species.*'[156]

"'*Homo erectus was an essentially Asian species. Contrary to long-held belief, we are not descended from it.*'[157] Some respected scientists pitch the *Homo sapiens* genetic tent large enough to shelter *erectus* as a unique race within the human family rather than a distinct species of genus *Homo*.

"Either way, without *Homo erectus*, the imaginary linkage connecting today's humans with some ancient, less-than-human species vanishes, leaving nothing but a yawning fossil gap, a trail to nowhere. It's Gap City—major league!"

"What about *Homo habilis*?"

"I call that one *Hopeless habilis*. Like its pal *rudolfensis*, that ape can't even qualify for the genus *Homo* when fossil specimens are subjected to a wide range of comparative tests.

*"'H. habilis and H. rudolfensis are more similar to the type species of* Australopithecus *and* Paranthropus *than they are to H. sapiens...should be removed from Homo'* and *'for the time being...should be transferred to the genus Australopithecus.'"*[158]

Josh turned toward the jury, anticipating their unanimous confirmation.

"Since you insist on dismissing *habilis, erectus*, and *Neanderthals* as links in the genealogical chain of *Homo sapiens*, let's get back to *Lucy*, the tiny hominid described as *Australopithecus afarensis,"* said JT.

"Should we address that little lady as Great-grandma? Not likely!

"A population of 20,000 bonobos survives in Congo.[159] I imagine them as *Lucy*-like because standing erect they reach a height approaching four feet; weigh in close to a sack of cement; and seem capable of bipedal walk. Their DNA is no match for the human genome, with ratio variations comparable to the difference between chimpanzees and *Homo sapiens*.

"Clearly, bonobos are contemporary and not ancestral to humans, although some evolutionists claim the species shared a common ancestor with us six million years ago—well back into the time of that virtual blank primate fossil record. From the pictures I've seen, they're a far cry from human! Big toes jutting outward off the sides of their feet...Need I go on?

"Imagine, if these guys were extinct, found only as fossil fragments in a geologic column. Some committed Darwinist might be tempted to trim the theoretical tree of human ancestry with the bonobo. But it's a scientific non-starter. Not only are bonobos living contemporaries of humans, but also their different genetic template poses an insurmountable obstacle to the relationship—with or without mutations."

"Bottom line: You believe *Homo erectus* to be a now-extinct former neighbor of *Homo sapiens,* but not an ancestor, much as we coexist today with bonobos and chimpanzees. Is that not correct?" JT asked.

"That's what I believe!"

"In that case, if humans don't share a common ancestor and a genealogical tree with ape-like creatures, just where did they hang out before leaving footprints on the sand 117,000 years ago? How do you explain the dearth of *Homo sapiens* fossils during earlier times?"

"Honestly, I don't know."

The disarming response startled listeners.

"Surely you must have an opinion?"

"Yup. But that's all it is, opinion. I can't prove it any more than mega-evolutionists can conjure up provable family trees that link diverse animal prototypes."

"Then would you be generous enough to share the benefit of your own unprovable opinion with the eagerly awaiting jury?"

Jonathan Thomas Daniels grinned, pleased that his witness continued to protect his credibility by refusing to crawl out on a limb.

"Sparse human population, for one thing!"

"What does that mean?"

"Under any scenario, there had to be a first-ever *Homo sapiens* couple. Human reproduction rates hardly match the lightning speed of bacteria—or rabbits. Quantitatively speaking, the species must have been comparatively rare at its beginning—a single couple!

"Estimates of world population fifty years into the future are 9.8 billion humans. A thousand years ago, Planet Earth accommodated 265 million inhabitants. In the year 1 A.D., humans totaled 170 million. These extrapolations suggest a two-millennia population burst fifty-six times over![160]

"Computing reverse geometric progression based on observable rates of births and deaths, it would take a time frame of only a few thousand years to span back to the first-ever handful of human couples, ice-age survivors existing under primitive conditions. Hardly a fertile geological rock garden conducive to prolific production of *Homo sapiens* fossils!"

"Any other fine-spun logic to bolster those extrapolations?"

"Ah, yes! Unlike apes, dignified burial rituals activated biological decay of human remains rather than fossilization. Also, I like to think that *Homo sapiens'* brain power inspired recognition of cataclysm and avoidance of onrushing sediments that trapped and fossilized species endowed with less-acute mental skills. Can you imagine a horde of humans wading into the teeth of the deluge, begging to qualify as fossil candidates?

"Fine spun logic perhaps, but what about evidence?" JT pushed the witness, anticipating the response.

"Will seriously debated evidence suffice?"

"You're the expert, Mr. Ryan."

"Excavators claim to have uncovered giant-sized human footprints tramping a Texas trail alongside dinosaur tracks! I don't have a clue whether or not the data is valid, but committed Darwinists shout 'foul.'"[161]

"Why would that be?"

"Because dinos were supposed to have died out 65 million years before humans allegedly evolved! Either the prints are not what the discoverers believe them to be—or else conventional time measurement separating humans from dinos must be erroneous and mega-evolutionary theory dissolves in shambles. Darwinists have no choice but to pooh-pooh the findings or to go back to the drawing board!"

"And your take?"

"I'm betting that all life on Earth debuted more recently than presently postulated, with convulsive cataclysm shattering the Earth and creating fossil graveyards sometime between the beginning of life and the present!"

"Evidence, Mr. Scientist? The jury waits!"

"*Res ipsa loquitor*—'the thing speaks for itself,' as you lawyer types say. A meager scattering of *Homo sapiens* fossil specimens suggests a mere handful of survivors following a worldwide cataclysm—primitive tools; hostile climates dipping 10 to 12 degrees Celsius; radically reshaped coastlines with massive ice shelves causing ocean levels to drop drastically."[162]

"Does this fancy reverse-progression formula of yours carry sinister implications for Darwinism?"

"Seems to me if *Homo sapiens* left footprints on the sand as long ago as 117,000 years before the present, Planet Earth would have long since run out of land for a population explosion reaching the tens of billions!"

"Just what are you saying, Mr. Ryan?"

"Darwin's multi-million-year scenario for organic life doesn't compute. Inorganic Mother Earth may be ancient, but the appearance of organic human beings could not have preceded the appearance of an atmosphere—some say as recently as 10,000 years ago!"

Worried that the audience might drift off in response to his geometric numbers game, Josh introduced a tongue-in-cheek diversion.

"Hey, maybe this DNA thing can be the litmus test of the Darwin myth with Charles Robert himself a test case?"

"What do you mean?"

"Why not put Darwin's own DNA on the line? Just think, since his bones repose in Westminster Abbey, a simple exhumation and laboratory examination of the guy's genes might surprise genealogists! What a smashing way to cut through fancy big talk and replace it with some honest-to-goodness evidence!"

As Josh expected, the off-the-wall quip roused the jury's interest.

"Evidence of what, pray tell?"

"Ancestry, of course! Imagine this knockout headline: *Darwin's Own Bones Prove Theory a Fact; DNA Links Aristocrat's Ancestry to Apes!*"

"You've got to be kidding, Mr. Ryan!"

"Now that's a fact—sure, I'm kidding! The old philosopher's ancestry is *Homo sapiens* from the git-go—no trace of a Lucy, or bonobos—just people like us. Of course, his wife's grand pappy was the well-fixed Josiah Wedgwood—an unlikely success story if left to *lucky mutations* and *natural selection!*"

Theater patrons exchanged whispered pleasantries, convincing Josh that his diversion had worked. JT pushed his witness back to serious issues.

"Tell the court, Mr. Ryan, whether you are prepared to introduce documented fossil evidence suggesting that *Homo sapiens* coexisted in antiquity with now extinct primates?"

"Even in the context of conventional dating, the evidence shakes mega-evolution's mythical tree. For starters, 32 fossil corpses have been discovered in Spain, reported to predate *'the last time Earth's magnetic field switched directions, around 780,000 years ago. Before then, the magnetic field had negative polarity: it pointed south instead of north.'*[163]

"Curiously, *'all date from the same killing season, and probably from no more than two or three years'* with one of the excavators expressing the belief that the victims perished from an *'ecological catastrophe.'*[164]

"The discoverers reported *'what we found was a totally modern face...it is so surprising, we must rethink human evolution to fit that face.'*"[165]

"Anything older?" The prober persisted.

The star witness obliged. "In 1972, Richard Leakey found a femur in Kenya *'dated at 1.9 m.y.a...virtually indistinguishable from modern human leg bones.'*"[166]

"Anything else?"

"Sixty-nine human-like footprints marking a trail through fresh volcanic ash conventionally dated at least 3.6 million years ago, were uncovered by Richard's mother, Mary Leakey, in 1978 near the Olduvai Gorge in Tanzania. She shared her personal impression that this trail of footprints appeared *'remarkably similar to those of modern man.'*[167]

"For sure, the foot did not match the stand-up bonobo! Instead, *'the best-preserved print...shows the raised arch, rounded heel, pronounced ball, and forward-pointing big toe necessary for walking erect.'*[168]

"'*What do these footprints tell us? First, they demonstrate once and for all that at least 3,600,000 years ago, in Pliocene times, what I believe to be man's direct ancestor walked fully upright with a bipedal, free-striding gait. Second, that the form of his foot was exactly the same as ours.'*[169]

"A University of Chicago scholar concluded in 1989 that *'the Laetoli G prints are indistinguishable from those of habitually barefoot Homo sapiens...If the G footprints were not known to be so old, we would readily conclude that they were made by a member of our genus Homo.'* The problem cited *'is that to ascribe those fossil footprints to Homo does not fit the evolutionary scenario time wise.'*[170]

"The 3.6 million-year date ascribed to the human-like tracks fails to raise the eyebrows of devout mega-evolutionists. Nor does the coincidence of the presence of a plethora of tracks laid down by animals clearly ancestral to and identifiable with species walking today's world impress them: elephants, hyenas, hares, rhinoceros, ostriches, guinea fowl, giraffes, gazelles, antelopes, baboons, and pigs.

"*'Fossilized whistling-thorn leaves appear identical as today's...Oddly—or perhaps not so oddly, given the geologic continuity of East Africa—we find the same type of wildlife in roughly the same proportions as exist today.'*[171]

"Then why not recognize the fossil hominid footprints as *Homo sapiens*? The Chicago scholar's 1989 answer deserves an A+ for candor: *'to ascribe those fossil footprints to Homo does not fit the evolutionary scenario time wise.'*[172]

"Such cage-rattling observations don't seem to fluster Darwin loyalists! When evidence conflicts with mega-evolutionary doctrine, its easier to dance with who brung ya!"

The insight generated audience murmurs. To underscore the point, Josh offered an explanation for the paradox of the Laetoli footprints.

"'Evolutionists concede that a giraffe must have made the giraffe prints, an elephant must have made the elephant prints, but *'their preconceived ideas about evolution and the age of these formations do not allow them to concede that a human made the human prints.'*[173]

"Still, if it looks like a duck, walks like a duck, quacks like a duck, and even leaves duck tracks—could it possibly be...a duck?

"Recognize the conventionally dated, 3.6 million-year-old footprints as *Homo sapiens* and a watershed paleoanthroplogical dilemma emerges: Either sacrosanct conventional dating techniques become suspect or mega-evolution's tottering genealogical tree collapses in a heap of splinters."

Acquiescing nods encouraged the star witness to believe that his picturesque speech had scored with jurors.

JT prodded for more. "Let's get back to fossil bones. Anything arguably human older than these footprints?"

"The fossil fragment of a humerus awarded a conventional date of 4.5 million years before the present *'could not be distinguished from Homo sapiens morphologically.'* Despite recognition of an ancient human bone, the assessment bowed to mega-evolution, admitting *'that time allocation to Homo seemed preposterous, although it would be the correct one without the time element.'*[174]

"KP 271, a well-preserved, left upper-arm bone, *'indistinguishable from modern Homo sapiens'*[175] was discovered in 1965 at Kanopoi, Kenya. The conventional date of 4.5 million years before the present places this human humerus at the edge of the time boundary closing the alleged ten-million-year gap during which hominid fossils evade discovery—a pre-history moment even before *Australopithecus anamensis* swung in the trees!

"'*The lower end of a left upper arm bone...becomes virtually the oldest hominid fossil ever...older than "Lucy."...True humans were on the scene before the australopithecines appear in the fossil record...*

*"'Homo erectus demonstrates a morphological consistency throughout its two-million-year history. The fossil record does not show erectus evolving from something else or evolving into something else...Anatomically modern Homo sapiens, Neanderthals, archaic Homo sapiens, and Homo erectus all lived as contemporaries at one time or another...As far as we can tell from the fossil record, when humans first appear in the fossil record they are already human."'*[176]

"And you contend that evolution's missing links remains AWOL?" JT prodded.

"Without DNA corroboration, there is no absolute methodology that confirms *Homo sapiens* ancestry linkage with any species other than *Homo sapiens! Neanderthal* appeared *after* humans, then disappeared. *Homo erectus* coexisted with humans, hardly qualifying for ancestral status. *Homo habilis,* alleged progenitor of *Homo erectus,* doesn't fill the bill either: It reputedly coexisted with *H. erectus* for 500,000 million years before going extinct.

"And if you claim the small ape-like creature, *A. afarensis,* affectionately dubbed 'Lucy,' acted as the ancient custodian of human DNA, that dog just won't hunt! No evidence exists that mutations ever added brand-new information to Lucy's DNA—nor to the DNA of any other extinct primate allegedly ancestral to *Homo sapiens!*"

"Is it fair to conclude that you don't believe anyone in this auditorium shares common ancestry with an ape?"

"Sounds fair to me. Neither do I believe we can link human ancestry to amphibians, fish, or bacteria!"

"Your testimony may impress us non-scientists, Mr. Ryan—but does that mean anything? Just the other day I heard that as many as twenty-seven percent of Americans still believe that the sun orbits the Earth! However plausible your views, why don't more scientists rush to support your assault on mega-evolutionary traditions?"

"Because tradition binds minds to dogmatism rather than inspiring objective inquiry!"

The unreconciled Dr. Karl Striker internalized the message and struck a power pose. Gritting his teeth, the top-gun academic raised his nose in disdain. Sensing the hostility, JT steered the witness to philosophical mode.

"In the last analysis, isn't the issue of human origins little more than personal whim and interpretation?"

"No, it's a lot more! The weight of evidence must control—unless bias subverts objectivity."

"Do you equate scientific research to jury evaluations?"

"That's just the point—neutrality promises the best results—but it's tempting to spin data to reinforce pre-determined opinions."

Feeling somewhat worn down, Josh sighed. "After a week in this hot box matching wits with Charles Darwin, I'm beginning to understand how bias builds!"

Ensuing guffaws confirmed the obvious: Theater spectators had embraced the appeal and integrity of Joshua Chamberlain Ryan, irrespective of his take on pre-history. Jonathan Thomas Daniels seized the moment of rapport to wind up direct examination.

While JT expressed a, "Thank you, Mr. Ryan," the curtain descended to enthusiastic applause. By the time it touched the floor, all *Monkey II* players had reached their feet, stretching in prep for the last act.

Jessica Saunders took Josh aside for a confidential briefing relating to the off-stage events plaguing the star witness' personal life.

"Just wanted you to know we're zeroing in on a prime suspect," she whispered with an insider's authority. "With luck, there could be an arrest late this evening or early in the morning."

"Can you clue me in?"

"The less said, the better. Put it out of your mind for now. Don't talk to anyone about it, but for pity's sake, be careful."

With that, Jessica ended the conversation.

# 35

Stuck on the stage for the final act of *Monkey II*, Josh Ryan felt trepidation at the prospect of facing further cross-examination by the one woman in his life he had fallen for unreservedly. The last thing he wanted was to be trapped at the mercy of this all-time charmer who possessed the power to lure him into a verbal sparring match at the wink of an eyelash.

He saw Traci as embodying an intelligence surpassing that of the acerbic, bar-hardened cross-examiner of yesterday. Strange as it seemed, he much preferred the bellicose flailings of the shopworn Brogan.

Instead of diatribe, Traci flashed disarming smiles. Vulnerable to her spell, Josh felt helpless, yet overcome by passion.

Only a few steps into the cross-examination, Traci telegraphed the message that she had no intention of bashing or bruising the hero of *Monkey II*. Her questions came across short and perfunctory, devoid of the killer instinct characteristic of real-life litigation. She struggled to conceal her affection for the handsome witness. With barrister's pride she confronted, then almost succumbed to her emotions. Finally, she opted for blank salvos with token touches of humor and occasional sarcasm.

"Let's get serious, Mr. Ryan! Given your scholar's fetish for evidence, just what do you call the array of hominid fossils discovered in the 20th century? Any reason some of those old-bone fragments can't qualify as pieces of missing-link transitional*s* connecting primitive primate to modern *Homo sapiens*?"

"Existed, certainly! But ancestral, no way! Square pegs can't fit round holes. Every creature with a leg bone isn't related to every other walking critter!"

"Then how do you account for homological resemblance? Surely even a skeptic like yourself must recognize some sort of progression from ancient primate skulls to present human craniums?"

"Of course apes and humans have skulls. They also have arm and leg bones. Trees and plants have leaves. What does that have to do with ancestry? Sequence can be neither implied nor imposed by functional resemblance!"

Josh thought the answer sufficient, but Traci kept probing. "Let's stay with the skulls, Mr. Ryan! Even you must acknowledge that there is remarkable similarity?"

"Touché, but *Homo sapiens'* characteristic differences stand out: imperceptible brow ridge; chin jutting forward; vertical forehead protecting a proportionally larger cerebral thought-center; 'U'-shaped jaw; a decidedly gracile structure. And don't forget the brain's speech processing site, Broca's area inside the cranium. The human edition reflects one-of-a-kind skull design—*Sui generis*, as you lawyers say!"

"Never mind what we lawyers say. What about small-brain to big-brain progression?"

"The size gap between human cranial brain capacity and perceived capacity of ancient non-human primates may be greater than previous estimates.

"'*Recognition that no australopithecine has an endocranial capacity approaching, let alone exceeding, 600 ccm, and that several key early hominid endocranial estimates may be inflated, suggests that current views on the tempo and mode of early hominid brain evolution may need reevaluation.*'[177] Computed tomography technology may not only expose inflated overestimates of extinct hominid brain capacity but also could upset evolutionary time scales."

"So you rule out the possibility of non-human ancestry for *Homo sapiens*?"

"It's not me that rules it out—DNA rules it out...and I'm unaware of any genetic evidence that rules it in! Extinct apes don't match the human genome but coexisted as separate species—just like chimps and bonobos coexist with us today...Although admittedly, I wasn't there and can't prove a thing."

"Your 'I don't know' is refreshing, Mr. Ryan. Does this guise of humility leave room for you to buy into the mitochondrial DNA argument for a common *Eve* ancestor?"

"It's no big whoop of a revelation to stipulate, 'Sure, somewhere in time the first-ever female *Homo sapiens* lived!' But where's the evidence confirming her less-than-human ancestry? It's inconsistent to accept the mitochondrial DNA code factor controlling several billion descendants, thanks to the gene pool conferred by the mother of all humans, while claiming simultaneously that *Eve* descended from less-than-human parentage, compliments of unproven, unidentified multiple mutations!"

"Do you buy into the 'out-of-Africa' scenario for Great-grandma Eve?"

"The flimsy effort to link humans to ancient African apes through a common ancestor is at best a tenuous tangle of genealogical roots. A recent study comparing PDAH1 gene variants found in geographically diverse, present-day populations could reveal the 'out-of-Africa' Eve to be anthropological fiction."[178]

Abruptly, Traci switched focus to the witness. "You do claim to be a scientist, do you not, Mr. Ryan?"

"A scientist in the making...at least until recent days!"

"All right then, *scientist in the making*, you claim to be committed to scientific method, do you not?"

"Yes, Ma'am!"

"Is there some universal definition of *scientific method*?"

"Research and discover evidence; assemble and collate data; analyze objectively; and reach conclusions based upon fact, not fancy."

"Sounds like high and mighty muckity-muck stuff to us average citizens! I take it you're committed to this so-called scientific method?"

He knew where she was going, but he stood his ground. "To the best of my ability."

"Are you implying that among all scientists, both real and aspiring, you alone champion the scientific method?"

"That would be arrogant and false. But be assured, I'm no isolated wacko. A growing legion of brilliant scientists doubt Darwinism!"

"Then who is right, Mr. Ryan? What is truth?"

He winced at the twist.

"The truth would be that two honest observers examining identical evidence independently can reach opposing conclusions."

"But how can this be if both claim to be honestly pursuing truth?"

"Easy! It takes a strong, open-minded person to consciously abandon the burden of culturally imposed traditions and taboos. Brainwashing from infancy can create a Hitler youth mentality so firmly entrenched that the victim willingly dies for insidious fraud. To imperialize thought mocks academic freedom. Clarence Darrow railed against '*bigotry*' and power wielded '*to inhibit learning.*'[179]

"Peer pressure plays a role, too. Otherwise, why would young guys wear baseball caps with backward-facing bills and oversize jeans bunched at the ankles, barely hanging on their fannies? There's nothing like good old peer pressure to homogenize and impose. Once an idea is published, it can calcify and ego resists retreat.

"Some lack the guts to stand tall for truth at the risk of losing a job...or squandering a shot at an academic degree. Mr. Daniels speaks of a wise USC law professor who warns law students that *moral choice is not without cost*."

Traci noted mentally that the witness had handled the hot potato with astute persuasiveness. But he hadn't finished.

"It's easy to swallow hype posing as truth. Erroneous notions can be perpetuated with the aid of widely published, impressively designed graphics. Make-believe comes alive when embellished and marketed in living color.

"New discoveries surface daily, shaking establishment's status quo and shattering obsolete conventional thought."

"Didn't Darwin upset the status quo of his day?" Traci challenged.

"The establishment deserved some major-league upsetting!"

"Was Darwin boldly sincere in his pursuit of truth?"

Caught by surprise, the witness struggled to be objective. "Certainly bold...but biased by the diet of evolutionary dogma fed to him through the writings of his granddad, Erasmus Darwin." Pausing in thought, Josh concluded with a frown. "In fairness, I doubt Charles Darwin would have settled for much of his conjecture if he had been exposed to today's technology."

"Clearly you see Darwinism as obsolete, and reject the assertion by his admirers that *evolution is fact*. Now, if humans didn't evolve from microbes, how did we get here? Did some magician wave a magic wand?"

Instantly JT jumped to his feet, objecting to the relevancy of the question. He reminded the court that the parties stipulated to confine the scope of evidence presented to "whether or not evolution is a fact" and not "an open discussion of alternatives explaining the origin of life."

Traci argued that her query was relevant, asserting that "the witness should be allowed to share his expertise on a subject that appears to be a consuming passion."

Looking at Josh, Judge Stone ruled "Sustained," reminding the jury that the focus of the drama had been narrowly drawn from its inception, and it would be unfair and inappropriate to introduce an entirely new body of evidence that should be reserved for a *Monkey III* trial.

Judge Stone followed with some folksy, off-the-cuff bench wisdom. "Finite minds lack the capacity to understand—much less explain—the infinite. The quest for a belief system and a rational basis for faith exceeds the scope of this trial's stated parameters—namely this witness's interpretation of evidence relating to whether or not *evolution is a fact*. I doubt even a bright scholar like Mr. Ryan claims inside knowledge of the infinite."

Nodding benevolently, he said, "Feel free to proceed, Counsel."

Thanking the judge respectfully, Traci Kilburn softened the inquiry by offering Josh a chance to address the marvels of the human race.

"Earlier this week, Mr. Ryan, you cited impossible mathematical odds against the human edition descending from a microbe ancestor. Given the amazing complexity of the one-celled life form you artfully described last Monday evening, what makes humans so special?"

He welcomed the question. "Darwinism postulates incongruity: Human intelligence derives from unintelligent cause. Hello? Hot-wired intelligence thanks to unintelligent blind luck? Intelligence spawned haphazardly by unintelligent coincidence? Impossible!

"The heart pumps something like 2,625,000 pints of blood every year. Healthy blood has a clotting mechanism that knows when, where, and how to plug up a wound and prevent the body bleeding to death—automatically! Blood that fails to clot because of a defective gene or that clots in the wrong place at the wrong time can kill.

"Vision is three-dimensional thanks to two eyes protected by six bony frames. We hear in stereo because of two middle ears featuring the tiniest bones in the body; olfactory power registers the full range of scents; taste buds alert the brain to sensations of salty, bitter, sweet, and sour, decoding menus acceptable to personal palates; and the protective power of touch, so lightning fast that the least sensation of pain will alert the brain to danger, triggers a micro-second muscle reaction and withdrawal from a too-hot stove.

"The human body is a living symphony of 500 trillion cells functioning in concert, marching to the beat of the same drummer. Nervous system, skeletal system, muscle system, endocrine system—all blended in cohesive beauty as a work of original art."

Lowering his voice, the witness reflected, "Incidentally, the package can pack a powerful wallop of personal attraction." This afterthought carried its own subliminal code intended for the lawyer-to-be!

At this hint of things to come, Traci Kilburn arched an eyebrow, asking innocently, "Are you about to favor us with a lecture on sex, Mr. Ryan?"

His face flushed. "Well, no...not exactly...that is..."

"What, then, did you have in mind?" Her eyes danced, eyelids fluttering in expectation.

"I was just going to say that unlike the alleged hermaphrodite common ancestor concocted by Darwin, humans come in two genders: male and female."

"So?"

Sensing his face turning crimson, Josh began to stumble, unable to maintain his cool. "Well, it's a great idea. You know what I mean...There's a downright chemical attraction...more like a jolt of electricity...that can run between a man and a woman..."

Folding her arms in amusement, Traci flaunted her control of the exchange. She toyed with his testimony, teasing relentlessly. "Is there a chance, Mr. Ryan, that when you fling words of profound insight citing chemistry and physics you are really talking about *love*? Could it be possible Sir, that your scientific expertise extends to the lofty realm of romance?"

The audience tittered. His white-flag surrender emerged.

"No,...not that I...however,...love is...!" Madly scrambling for words, he mumbled a series of unintelligible phrases.

Finally, through sheer willpower, he managed a semblance of coherence. "What I'm trying to say is that a woman can be beautiful and downright attractive to a man. And I believe the reverse is likely to be true as well...or so I've heard..."

The audience roared with glee. The brilliant scholar who had cut the battle-scarred Marcus Brogan off at the knees was now wrapped around Traci Kilburn's little finger. The exercise proved great sport for everyone except Josh. Even pal JT was howling his head off. But this was nothing compared to Traci's last zinger.

"I recall the first day of this staged spectacle, you raising your hand, solemnly declaring to tell the truth, the whole truth, and nothing but the truth. Is that not correct?"

"Absolutely!"

"Nothing like straightforward old-time integrity. Isn't that right, Mr. Ryan."

Josh replied meekly, "Yes." He shuddered at the coming slam-dunk. The lady's pursuit left no place to run or to hide.

"Then, Mr. Ryan, since you happen to be a man and I happen to be a woman, would you please tell the jury, this court, and this eagerly awaiting audience the *whole truth* as to whether you are swept away in electric rapture when you gaze at a human edition such as myself?"

Eyes telegraphing mischief, she delivered the question without qualm. A roar swept the playhouse. Judge Stone could barely hear JT's frantic objection.

"Your Honor, the witness claims the Fifth Amendment!"

"Or even the Fifth Commandment, if that will help," chimed in the witness.

At that, the normally composed Edward Anthony Stone nearly slid off his judge's chair in helpless laughter. Jessica Saunders convulsed at the mindless plea. The unflappable Pace Terhune hovered in the background, crowing like a banty rooster at the success of the spectacle he had engineered. Traci swung around toward the audience, smilingly seeking the plaudits of the Duchess, Bishop Daniels, and Michael Ryan—at which, all three signaled thumbs-up endorsement of her bold stunt.

Joshua Chamberlain Ryan alone sat in befuddled silence, not knowing whether to laugh, sputter, or stand and pledge allegiance to the flag. Eventually, he regained his composure, but his mind still raced to craft an acceptable comeback that might capture the fancy of the audience but would avoid embarrassing the woman he adored.

Judge Stone rapped for order, and the crescendo of appreciation subsided.

"With your permission, Your Honor, I hereby embrace the counsel of Mr. Daniels and claim the protection of the Fifth Amendment."

The judge endorsed the move, to a backdrop of audience laughter. Temporarily assuaged, the witness pressed on, hopeful the light bearing down at the end of the tunnel did not signal an onrushing locomotive.

"However, it seems as though I've heard something in legal proceedings to the effect that I can reserve the privilege to address the question again at some more convenient and appropriate occasion."

Again there were ripples of friendly laughter, this time with spontaneous applause. Josh sensed that this was a prime time to unleash a bombshell exploiting the politically incorrect sayings of the late philosopher Charles Robert Darwin.

"With the court's permission, perhaps I can muster the sympathy of my interrogator by referencing an unheralded observation of Charles Darwin with which I am in total disagreement."

"Feel free to proceed," intoned the bemused judge.

"Charles Darwin didn't hesitate to describe his own private niche of exclusivity for the *fittest* who successfully struggled and survived. His self-serving traditions gave no quarter to illness, poverty, or disadvantaged cultures. Even classification as a healthy, wealthy English citizen didn't cut it with Darwin if the person happened to be other than the masculine gender.

"Married, with a houseful of children, and pampered by a domestic staff of eight, Darwin excluded his wife from his *fittest* definition. Sounding like a cheerleader for the macho mantra that regards women's niche as 'barefoot, pregnant, and in the kitchen,' Darwin unabashedly pronounced men mentally superior to women.

"*'The chief distinction in the intellectual powers of the two sexes is shown by man attaining to a higher eminence, in whatever he takes up, than woman can attain...the average standard of mental power in man must be above that of a woman.'*"[180]

Josh rolled his eyes at the explosive quote. He could not help but believe that he and his cross-examiner stood in 100 percent agreement.

He continued, feeling more confident again. "The English language contains as many as two hundred thousand words. Linguists say there are at least 800 distinct languages and dialects on Planet Earth, with unknown others lost to time.

"Communication by speech is taken for granted. Yet, conversation involves a reception of a barrage of sound waves; translation into meaning; with the brain framing a split-second reaction. Millions of neurons in the brains of both communicants work at breakneck speed to share thoughts, ideas, and emotions. *Homo sapiens* alone, of all species, enjoy the power of creative speech, thanks to awesome intellect, reasoning power, and a memory rivaling state-of-the-art computers.

"Humans understand justice and moral choice. Their language communicates it all—from the most innovative ideas to the richest emotions!"

Fixing his eyes squarely on the face of the beautiful lawyer-to-be, he tested his own communication skills in a voice loaded with emotion.

"Take it from me, Counselor, humans are born to love!"

His eyes glistened as he delivered a last punch line in a stage whisper intended primarily for Traci's ears. "And that's the truth, the whole truth, and nothing but...!"

The gutsy, let-the-chips-fall witness left the witness stand, looking ahead to the dawning of the first day of the rest of his life.

# 36

Prior to Day One curtain time, Pace Terhune had stage-managed the event to omit summations by either Traci Kilburn or JT Daniels. All participants had endorsed the choreography. As Josh rejoined JT at the counsel's table, Judge Stone first instructed then dispatched the jury to an adjacent room to discuss and deliberate. With fifteen minutes to go till noon, time remained for a pre-planned filler—an innovation concocted by Pace Terhune as frosting on the showbiz cake.

As per instructions, Josh had assembled some of the classic but more outlandish observations of Charles Darwin, matched with wacky questions engineered for comic relief. An opportunistic Jonathan Thomas Daniels suggested that the compilation be enshrined as part of the evidentiary record of *Monkey II.* Judge Stone refused, but did agree that JT could entertain the audience with several samples of manufactured monologue, once the jury began deliberations.

JT Daniels, legal-eagle turned showman, took center stage, and with the jury out of earshot, faced the vacant witness chair and read the out-of-context questions in a sober, lawyerly demeanor. After each question, he would seat himself in the witness box with exaggerated stiffness, posing as the nineteenth-century Darwin delivering answers. To complete the spoof, he substituted a reedy tenor for his own baritone and affected a fractured English accent. The fictional inquiry conspired to poke uproarious fun at Darwin's quite real, but out-of-context words.

Q. "Tell us, Mr. Darwin, did your family tree spring from a hermaphrodite?"

A. "'*The early progenitor of all the Vertebrata must have been an aquatic animal...with the two sexes united in the same individual.*'"[181]

The audience responded with derisive shrieks. The reaction egged on the thespian instincts of the law student-turned-performer. He returned to his feet to again address an empty witness chair.

Q. "Tell us, Sir, if a bear spends too much time swimming in surf, does it risk its descendants evolving into seagoing creatures?"

A. "'*I can see no difficulty in a race of bears being rendered, by natural selection, more and more aquatic in their structure and habits, with larger and larger mouths, till a creature was produced as monstrous as a whale.*'"[182]

Alternating roles between interrogator and witness, JT jumped up and down like a jack-in-the-box, demonstrating his physical as well as verbal alacrity. Somehow he juggled competing lines and accents without missing a beat.

Q. "With all due respect for your capacity for the *high intellectual work* you ascribe to the male gender, and recognizing that the cerebellum's *neodentate* essential for cognitive skills exists exclusively in human heads, just how does your own *high intellectual work* explain the origin of this addition to the cerebellum's *'dentate nucleus'*?"[183]

A. "*'Man is descended from some less organized form.'*"[184]

JT took his cue from the audience guffaws and pressed on with the parody. The make-believe questions, cleverly matched with eyebrow-raising quotes, presented a gloves-off rehash of Darwin at his worst.

"*'Man is descended from a hairy quadruped, furnished with a tail and pointed ears.'*[185]

"*'We must bear without complaining the undoubtedly bad effect of the weak surviving and propagating their kind...vaccination has preserved thousands, who from a weak constitution would formerly have succumbed to small-pox.* '[186] *'Civilized races of man will almost certainly exterminate and replace...the savage races.'*[187]

"*'If two lists were made of the most eminent men and women in poetry, painting, sculpture, music, science, and philosophy, with half-a-dozen names under each subject, the two lists would not bear comparison...the average standard of mental power in man must be above that of woman...man has ultimately become superior to woman.'*"[188]

Feigning shock at the naturalist's out-of-whack speculations, JT surmised correctly that serious advocates of mega-evolution, including Karl Striker, felt embarrassed by the blunt rhetoric of the long-deceased guru. Regardless, the contrived English accent and mismatched quotations proved a hit.

Seeing the jury filing back onstage, the judge interrupted, bringing down the curtain on JT's solo performance. Tongue-in-cheek, he advised, "You've been quite enlightening, Mr. Daniels...Serious science may be hard-pressed to recover from your outrageous manipulation of the written record."

Jonathan assured the judge, "It's been my pleasure, Your Honor."

In benediction, he turned to the vacant witness chair, affected an exaggerated bow and pretended to award Darwin's legacy the triple-crown in the politically incorrect sweepstakes. He pompously thanked the long-absent Charles Robert Darwin, ending his spiel with the solicitous, "Thank you, Sir. No further questions!"

Wearing a grin as wide as mega-evolution's fossil gaps, he acknowledged the audience with another two or three deep bows and ceremoniously sat back. Spectators acclaimed the burlesque, rewarding the law student's stunt with wild applause.

Judge Stone added plaudits as he chuckled approval. "It appears you are a man of many talents, Mr. Daniels. It is conceivable that in one fell swoop you have succeeded in causing not only Mr. Darwin, but also William Shakespeare, to roll over in their graves at this astounding display."

Still grinning, JT quipped, "Thank you, Your Honor—I think!"

Amidst the frivolous banter, the jury took the center-stage seats facing the bench. Judge Stone inquired if deliberations had produced a verdict. The foreman stood, announcing, "Yes, Your Honor, we have."

Six jury ballots were delivered officiously to the outstretched hands of Court Clerk Jessica Saunders. The top half of each sheet carried printed findings to be marked. The bottom half left space for juror comments.

First, jurors were to check *"Guilty"* or *"Innocent"* after the question: *Did Charles Robert Darwin resort to fact-free science in his explanations of the origin of organic life?* The second question asked for a *"Yes"* or a *"No"* response to the core inquiry: *Is Evolution a Fact?*

Turning to the bailiff, the judge ordered, "You may proceed to read the verdict, Ms. Saunders."

She responded with crisp authority, "I've been given six sheets of paper duly signed by members of the jury. Each document carries supplementary handwritten comments, which appear to be personal observations. Would the court prefer that I simply tally the vote or that I also read aloud the individual insights?"

Looking at his watch, Judge Stone waved her on. "It's five minutes until scheduled adjournment. A bonus minute or two after six days of heavy-duty trial time could be worth it. My guess is that the patrons would count it a treat to hear those personal comments. Please read them with the verdict, Ms. Saunders."

"Juror No.1 finds Charles Robert Darwin guilty of fact-free science as charged and rejects the notion that evolution is fact. The handwritten comment volunteers this non-judicial insight: *'The star witness was very convincing...I thought he was kinda cute! Is he married?'*"

The audience tittered. Josh winced, turning his usual crimson. He assumed this vote of confidence sprang from the heart of the Sherwood High coed. Traci bit her lip to avoid laughing.

"The second juror also voted *'Guilty'* but with a caveat interpreting the true and false. *'Let's keep the semantics straight! If diversity, variety, and change resulting from combinations within the broad limits of the gene pool of a prototype life form can be labeled evolution, that's provable, scientific fact. But if that reality is stretched beyond the limits of genetic information to suggest microbe-to-man mega-evolution—no way, not in a billion-plus years!'*"

Josh noted with satisfaction, "At least one juror listened!"

The next ballot took a neutral stance. An *"I don't know"* coupled with *"Some mysteries cannot be fathomed"* answered the two written questions. A candid confession explained the equivocation: *"I was educated to be an evolutionist. Most of the material presented this week provides challenging food for thought. But Mr. Ryan proposed no alternative. In all honesty, I just don't know."*

The star witness never changed expression. Content with honest conviction, he rationalized, "Nothing wrong with old-fashioned integrity!"

The fourth juror created a stir by marking an emphatic *"not guilty"* as to Darwin's resort to fact-free science with a capitalized *"yes"* declaring evolution to be fact. The juror doffed his hat to academic establishment by admonishing. *"Accepted traditions endorsed by esteemed faculty leaders deserve public and student respect."*

Unimpressed with the verbal incense, Dr. Striker's rigid face didn't crack. Josh awarded the comment a "Big whoop," reasoning, "That kind of obeisance preserves flat-earth mentality." With two more ballots to go, the witness listened politely, attempting to be impassive.

The reaction of Juror No. 5 reflected views precisely the reverse of No. 4. Bold, hand-scrawled capital letters, superimposed on the boxes to be checked, pronounced Darwin *"GUILTY"* and decreed the *Evolution is a fact* proposition to be *"FALSE."* As with Juror No. 4, the sentiments expressed had been held long before exposure to *Monkey II*.

*"Some mysteries defy human comprehension!"* To remove all doubts as to prior commitments, the juror admitted bias: *"William Jennings Bryan championed religious faith. I will always believe he outperformed Clarence Darrow in Scopes."*

Old-timers in the audience were familiar with the religious stance on evolution. Many nodded their heads in agreement, but others dissented. Jessica Saunders grinned in amusement at sight of the last of the six ballots.

"It says here that Charles Darwin was *'guilty of preaching fact-free science and that evolution is not a fact.'* However, a political endorsement rides the coattails of the vote. *'I'm ready to vote Josh Ryan for Congress when he's old enough! Right or wrong, the guy exudes integrity.'"*

A still-smiling Ms. Saunders took her seat.

That was it! Joshua Chamberlain Ryan's frontal assault on hallowed tradition emerged with a 4-1 triumph plus the *I don't know* vote from one neutral juror. Any sense of vindication was mitigated by realization that except for the exclusion of both Slade Lassiter and golf pro, Lyle Grant, the decision could have been a slim 4-3-1 victory margin.

With a minute to go, Judge Stone took the floor before gaveling the case to a close. First he thanked the jury for its patience and understanding.

"Each of you has expressed his or her opinion honestly and openly in rendering a verdict regarding an issue scrutinized and debated with emotion for the past 150 years."

He peered into the audience with the commanding demeanor of a learned senior judge confidently presiding in a Federal District court.

"As indicated last Sunday, *The People v. Charles Robert Darwin,* staged in this make-believe courtroom, has been created for your entertainment. Hopefully, you have found it also to be thought-provoking. Whether or not you agree with the sentiments expressed or the verdict rendered, I commend you for being attentive and responsive to a cast of volunteers who unselfishly contributed their time and talent in celebration of the centennial of this magnificent university.

"Pace Terhune invited me to share the good news that thanks to the generosity of loyal supporters, MAU endowments have been enriched by nearly a million dollars from tickets purchased for this week's event as well as supplementary contributions and pledges. On behalf of the trustees and the administration—thank you!"

The judge ended with a flourish. "Here's a valedictory salute to Mid-Atlantic University; here's to the great adventure of scientific discovery...and here's to our quest for truth, the whole truth, and nothing but..."

A rippling stir confirmed the collective judgment that *Monkey II* delivered its money's worth. The curtain dropped on the audience's verdict of prolonged, standing ovation.

*Monkey II* belonged to the history books!

# 37

The instant the playhouse curtain touched the floor, Josh Ryan's mind raced ahead to the afternoon's rendezvous with Traci. Ever since Wednesday's encounter at the Hampshire Club's aquatic center, the eager suitor had found it difficult to concentrate on anything but Friday afternoon's plans. However, he remained trapped in the milling throng of well-wishers.

Anticipating the audience's natural instincts to reach out to touch and congratulate favorite actors, Pace Terhune had replaced the spotlights with soft, simulated daylight. The haunting recorded sounds of Michael Bolton's "When a Man Loves a Woman" played through the building's sound system. Following the example of Judge Stone, the jurors, together with the rest of the cast, remained in place exchanging pleasantries with the remnants of the audience who pressed forward to glad-hand and to chat.

Josh's frustration at being detoured dissolved at the soft touch of Traci's palms on his cheeks, accompanied by a full, on-the-lips kiss, which he returned with only partially restrained enthusiasm. Pulling back, she laughed easily, "See, I told you I would make you a star."

Without giving him a chance to reply, she disappeared into the crowd.

Recognizing that much of the attention and good wishes were directed his way, the reluctant ex-star offered a string of gracious "Thank you's."

"Tremendous, young man"; "You done good"; and "Couldn't agree with you more," typified the outpouring of verbal approval. Friendly but aloof detractors ventured restrained comments such as, "No one has all the answers, do they?" and "Keep an open mind!"

After dozens of handshakes, hugs, and backslaps, Josh took advantage of a brief opening in time and space to sidle over to JT. Mesmerized by an ivory-skinned beauty with dancing black eyes, JT ignored the approach of his colleague.

Josh barged in, nudging JT. "Sorry to intrude, old man, but I'm here to sign you on as my lifetime permanent mouthpiece."

"Oh, sorry, Josh...but I've been offering some legal advice to this young lady you should meet...uh...uh..."

Never having met his admirer previously, a flustered JT looked at her imploringly. She was delighted to take him off the hook. Suppressing a giggle, she introduced herself. "Natalie...Natalie Sherman." She couldn't resist spoofing the retired star witness, quipping, "I'm honored to meet this celebrated junk-science expert!" With a twinkle, she added, "I may need a great lawyer someday myself. You may have to wait in line, Mr. Make-Believe Darwin."

Josh took the hint, teasing, "Glad to meet you, Natalie. But before you wait in line too long, you better talk to me first. I've known the guy for a lifetime. I can tell you anything you want to know."

JT regained his poise, retorting, "Can't you see the lady deserves my undivided attention? I thought you scientist types had evolved more horse sense."

Before Josh could break away, Pace Terhune showed up with an immaculately groomed Redondo Calizar in tow. The producer personified Hollywood executive fiefdom.

"My friend, Redondo Calizar, from California, produces movies. As a courtesy to me, he's been present for all six *Monkey II* acts."

Terhune gushed on through introductions to Josh and JT. Responding, a smooth-talking JT identified the glowing Natalie Sherman as a newfound accomplice.

"Cool dude," thought Josh. "My pal may end up king of Planet Earth."

Eyeing JT Daniels, Calizar got right to the point. "Mr. Daniels, it's true that I produce movies. Pace promised a visit to Chamberlain Playhouse this week would be worth my while. As usual, he was right. I want you to contact my office immediately next week. As of Monday, my studio will offer a contract for your signature."

A stunned JT smelled a Pace Terhune prank, so tried to be flip. "Sounds great. Just call me Denzel Washington—or Sydney Poitier. I'll fly west on Mary Poppins' umbrella first thing Sunday."

The unfazed stranger didn't crack a smile but wasn't offended. An old hand at discovering hidden talent and used to having his way, Calizar zeroed in. "Those names are taken, Mr. Daniels. But I'm sure we can find an easily recognizable alternative to match your flair."

Natalie gazed with renewed admiration at the prize waiting at the end of her rainbow.

Josh alone ventured a witty punch line. "Think about it, Bro," he said. "Already a star on stage, it looks like you're bound for new fame on the silver screen, with television lurking close behind."

Drawing no response, Josh pushed his friend to explore all possibilities.

"Go for it, Bro! Besides, I need a job. I'll be your agent—for 15 percent."

"Fifteen percent of zilch equals goose egg, Dar!" Taken aback by the bolt from the blue, JT tried to play down the lightning from the west. But Josh enjoyed toying with the idea. He resorted to a last, humorous crack.

"I can see the headline: *Darwin Basher Abandons Bar, Goes Hollywood.* Mr. Terhune can dress that up for you first-class! But before you become too rich and famous, how 'bout an autograph, Denzel Poitier?"

For the first time in a week, Jonathan Thomas Daniels stood speechless. His new friend Natalie hovered in adoring silence. The moment belonged exclusively to JT as Redondo Calizar completed the pitch!

"Mr. Terhune reports that cable rights to a condensed version of *Monkey II* have been sold. Once that package hits the market, Mr. Daniels, expect your face to earn autograph requests. Depending on ratings, my studio plans a movie version. I've ordered a screen title that should appeal: *Ride to Glory!*

"Might even find a role for you, Mr. Ryan. Take my card, gentlemen!" In a single, sweeping motion, he handed business cards to both young men, followed by a round of handshakes.

That was it!

Proud as a strutting peacock, Pace Terhune waved "so long" to the California-bound Calizar, already walking briskly to the exit and a ride to the airport for his westbound flight. Without slowing, he pivoted to bark an executive command.

"Pleasure meeting all of you. Don't forget, Mr. Daniels, my studio expects your Monday call. It's never my custom to accept 'No' for an answer."

Josh took the cue and followed suit with good-bye rituals of his own. Bolstered by thoughts of the ensuing one-on-one with Traci, the retired witness made a clean getaway, mimicking the exit of the flamboyant movie mogul.

The normally articulate JT stood transfixed, jaw agape, staring first at Pace then Natalie. His effusive vocabulary had been swallowed by astonishment. Searching for appropriate words, he sported a bewildered, ear-to-ear grin. Seconds later, he mustered an abstract colloquial cliché resurrected from boyhood memories.

Apropos for all occasions, the folksy expression said everything and meant nothing. "Well, I wish I may never!"

# 38

Josh Ryan's romantic aspirations soared. As 2:30 p.m. approached, he paced the mill-house floor nervously. He hoped fervently that the threatening sky would not rain on his painstakingly plotted parade—and that nothing would deter this morning's cross-examiner from accepting his invitation to a walking tour of the picturesque pathways crisscrossing the rural acreage encircling Fort Ryan. Thoroughly smitten, the erstwhile witness felt hard-pressed to ask for the hand of Traci Kilburn if nature's elements conspired to drown his plan.

Early that morning, he had conceived an invitation format he deemed appropriately clever. Using his desktop computer, he did his best to piece together a formidable-looking document that he hoped would pass as a judicial order. Scavenging words from his summons, he spliced legal jargon into an official-looking instrument ordering her to join him for a 2:30 p.m. adventure. Despite its clumsy composition, it did have a certain charm.

Immediately following the final curtain, he had delivered the sealed pseudo-order to Traci while commending her "stellar performance." He didn't wait for her to read and respond. Without wheels, he couldn't offer to drive her to Fort Ryan.

His heart skipped when the knock at the door announced Traci's arrival at the targeted time. She waved the faux-legal document.

"While I confess to a touch of curiosity, the official power of the court order demanding compliance left me no choice but to appear."

"Whatever it takes," rejoined the host.

"While thinking that no scientist in his right mind would seriously consider a walk in the woods in today's abominable weather, I wore boots and blue jeans and brought rain gear—just in case."

He struggled to be casual, matching her flip attitude. "You've heard of mad scientists...Now you know what one looks like."

Josh's strategy had been fathered by desperation and expediency. He had convinced himself that a two-hour interval in mid-afternoon following the verdict and preceding the evening's university trustee centennial reception, would offer a rare window of romantic opportunity. Seizing this moment to propose to Traci Kilburn trumped all other priorities.

Before Josh could usher Traci out the door for the promised grand tour, sheets of rain threatened to drown his scheme. Raindrops ricocheted on the red aluminum roof in hypnotic rhythm. The repetitive rat-tat-tat reminded Josh of a flicker's desperate drumming on copper downspouts to attract the ear of a prospective mate. Listening to the hammering of his heart, Josh could identify with the frantic flicker. Undeterred, he conjured an interim plan.

"So that you won't think the mad scientist is completely nuts, let's postpone the hike in the woods...at least until the rain eases to less than a downpour.

"There's some stuff here I've been wanting to show you, anyway. How about some old-time ballads and country rhythms from before we were born?"

Glowing with pride, he spread out a collection of thirty-three-and-a-third records wrapped in pristine dust jackets. The rarely heard hits of another time and place had once belonged to Grandmother Ryan.

"Maybe some of these old-timers will grab you!"

The sweet sounds of the Forties, Fifties, and Sixties added syncopated beats to nature's background serenade. Her reaction reassured.

"The music is beautiful. I like the rain, and I like this old mill house. I'm flattered you bothered to take the time to show me around."

Then she spied the fireplace mantel, recently crowned with the coveted oil portrait of Leslie Anne that had been given to his safekeeping by the Duchess.

"What a beautiful painting! Is this a picture of some glamorous girlfriend you've neglected to warn me about?"

Too late, Traci's antenna warned that she had bumbled. Almost before the light-humored remark escaped, she bit her lip to choke it back. To her relief, her host betrayed no offense.

"I'm surprised you haven't seen the portrait before, seeing as how you pal around with the Duchess. That's my mom...Leslie Anne Carrington-Ryan. The stallion she's riding sired Turbo. She does look like a spectacularly beautiful woman. I wish I could have known her...even seen her...a time or two..." His voice trailed off.

"I'm sorry, Josh. I was trying to be clever. I wish I could have known her, too! You're right, she was a smashing beauty...I see a striking resemblance...especially in your eyes..."

He let it go, anxious to get on with his scripted show-and-tell. "You've got to see some of the other stuff the Duchess gave me this week!"

He led her to the candlesticks. Igniting her curiosity, he placed one in her waiting palms. She sheltered it as though it were a delicate bluebird.

Gently caressing its twin, he recounted tales of generations past, memorialized by the stalwart trees gracing Morgan Manor's manicured landscape. He puzzled at the irony of Leslie Anne's tree being struck down by lightning, mere days after her untimely passing, but his face brightened at mention of the tree planted to honor his birth.

"When I hold these, I think of nature's umbrella shading my mother's childhood. I imagine my parents' discovering love, wrapped in the approving embrace of that tree."

These reflections reconfirmed Traci's assessment that a sensitive and caring guy lived inside the host's rugged frame.

"...And here's my mom's diary." He traced the diary's spine with an index finger, knowing it was premature to reveal the ring box tucked safely behind it.

"The Duchess marked some pages of Mom's handwriting that promise to reveal fascinating family history. I'm going to reserve an hour or two Saturday evening to get better acquainted with my mom."

"You are a sentimental guy, Mr. Ryan."

He smiled bashfully.

As the stack of records spun lazily, the sultry voiced Patti Page caught his attention with the hypnotic "Tennessee Waltz." Josh remembered the tune as a Grandma Ryan favorite. Impulsively, he took Traci's hand and guided her to a patch of weathered oak flooring near the center of the room. Caught up in the mystical power of the refrain, he forgot for the moment he couldn't dance worth a whit. Instinctively he compensated for the shortfall by venturing short, measured steps. She slipped easily into the mood, edging close, mesmerized by a magnetic attraction.

Josh Ryan trod unexplored turf. Lost in rapture, he encountered a sensation akin to free-floating on clouds. Not wanting to break the spell, he stole a quick look to see her eyes flutter, then close, sharing the sublime journey into romantic space.

As the haunting strains of the ballad faded, the torrent outside relented. Josh seized the moment. "Looks like the clouds are lifting. It's time to introduce you to the deer trails hereabouts."

Still touching her hand, he led the way to the kitchen. Dominating one wall, a crude Montgomery County map displayed pre-Civil War geography.

"I copied this from a Montgomery County Historical Society 1850s map. In those days, the Alan Bowie Davis plantation spread across Westminster Pike—today's Highway 97."

Pointing to the northern boundary of the property, he explained, "Over here is the Fort Ryan acreage. And just down the slope, deep in the woods, runs the Hawkings River—and here, at the bend in the river, is my Shangri-La." His finger traced an arc to a spot on the antiquated map where the rambling river spilled into a bend resembling a small pond.

"X marks the spot where slave labor built a dam using slabs of granite. The rock barrier created a lake to channel water down the millrace to power a mill next to the river, just west of the pike. Fragments of the dam still exist, carving out a swimming hole that spawned bluegill and buckets of childhood memories."

Josh spoke with authority. She responded with the thought, "This guy packs more than a hi-fi scientific mind." She saw him as a connoisseur of history, but with all the signs of an old-fashioned romantic.

Donning rain gear, she beat her host to the door, teasing, "I'm at your mercy, Dr. Darwin. Take me to your hideout."

In one giant stride, he joined her, opening the door while bowing and waving her through with all the chivalry of a Walter Raleigh. Sloppy gray clouds shed intermittent drizzle. In no time, they had clamored up the hill behind Michael's manor house. Michael waved his blessing, having spotted the passing pair from the glass-walled sunroom. He shared not the faintest idea as to their destination or purpose, but aware of the vivacious charms of the young lawyer-to-be and the overheated blood surging through the arteries of his son, it didn't take a computer scientist to calculate the possibilities.

Michael Ryan mentally gave them his blessing. "It's time...I hope he'll be as lucky in love as his dad."

The giggling couple pressed on without a sideways glance. Josh pushed back limbs to reopen ancient, overgrown deer paths. Century-old trees, towering like haughty aristocrats, swayed to the rhythm of blustery breezes. Bowing in the wind, branches and brush welcomed the intruders, beckoning them to pass. Still the musty terrain belonged to primeval serenity, with humans as interlopers.

Attired in blue jeans and yellow slicker, with rain hat securely covering her carefully styled hair, Traci felt raindrops trickling down her forehead and tickling her nose. As the firm hand of her host steered her toward an unseen goal, her mood turned jocular.

"Definitely, you're neither a mad scientist nor certifiably insane. But you are one wild and crazy guy!"

He paid no heed, but quickened the pace through the Maryland jungle Her words slipped into oblivion in the dank underbrush.

Even in the darkened blue-green depths of the soggy forest, natural beauty triumphed at every turn. The journey led up and down gullies and crevices, moving relentlessly to the river, a course marked by water trickles swelling to tributary streams. Finally, as they reached the crest of a hill, a gurgling Hawkings River came into view, ambling eastward and carving out the floor of a narrow valley.

"There it is," he exclaimed, reporting the obvious with the all the fervor of a Meriwether Lewis reaching the churning mouth of the Columbia.

"Follow me," he commanded, releasing her arm to better dig in his heels while balancing the downward trek. Beating her by seconds to the base of the slope, he encouraged, "Don't worry, I'll catch you if you stumble."

Ignoring the oblique reference to their first meeting, she picked her way with deliberate caution. Every fiber of her being concentrated on avoiding another inglorious pratfall like the one that had memorialized her Tennessee arrival the previous April. After safely navigating the steep drop, she stood with Josh, savoring the sight of the millrace channel, which measured at least five feet deep and ten feet wide.

Once they turned upstream, it proved easier pushing through the hollow of the eroding earthworks. Still they dodged and ducked to escape the tentacles of the scratchy underbrush and protective overhang. Within minutes they reached the base of a pile of sculpted rocks jutting into the river in disarray. The rough-hewn boulders anchored the remnants of a monument to primitive engineering. Scrambling closer to the river and then up the scarred slope of the derelict dam, they were greeted by an astonishing view.

The heart of the dam had been destroyed either by the hand of man or some long-forgotten natural catastrophe. A rippling fish pool remained where a much larger lake once served the power needs of the mill. Aged oaks anchored the rubble on both sides of the river, challenging the granite rocks with their relentlessly penetrating roots. Humans had once cut a swath here, but the scene had long since reverted to its maker.

"It's gorgeous," she exclaimed.

"It's the capitol of my world," he corrected.

"Hey, I don't see any White House, and the Hawkings is a mighty poor excuse for the Potomac."

Patiently he explained the hidden meaning of his announcement.

"You see, JT and I selected this site as the capitol of *Arikmica*—the imaginary nation we founded as kids. Believe it or not, the White House is a mere twenty-three miles away by helicopter."

The disclosure approached surreal trivia, since this forsaken spot on the Hawkings lacked any trace of human habitation except for the wreck of what used to be.

"You founded your own country?"

"Of course. Doesn't every kid?"

She giggled at the absurdity. "Darwin the dictator, I presume!"

Her laughter dislodged a shower of raindrops, from her slicker. Undistracted, the erstwhile dictator continued his recitation of oral history.

"Bishop Daniels owned the two-hundred acre spread next door, purchased before the Civil War by his freeman ancestors. JT and I played founding fathers and merged the Ryan land with the Daniels' to carve our empire!"

"And?"

"We concocted the weird name, Arikmica, so we could pirate lyrics from patriotic melodies like 'God Bless America.'"

"Did you have a flag?"

"Couldn't find a Betsy Ross!"

"Sounds tyrannical!"

"No way! Democracy reigned supreme. A unanimous vote of two was required, but we never got around to legislating—we were too consumed with the power and the politics. Besides, the adults in our families refused to pay taxes.

But we did elect a president—we took turns every other day, always with a unanimous vote."

She rolled her eyes. "Other than boycotting taxes, what did your families think of your revolution?"

"Mostly good-natured embarrassment. They were wise enough to talk us into keeping the coup secret from the public to 'protect Arikmica's sovereignty'—or so they claimed."

Almost choking with laughter, Traci managed to mutter, "You were an absolute, certifiable nut...But no doubt a cute one or the Ryans and the Daniels would never have put up with you!"

Then she inquired, semi-serious, "What finally happened to this nation of yours...and when did you abandon presidential politics for fossils?"

"I don't have the foggiest notion. I can't even remember who was president when the experiment in revolution expired. But I've always boasted of one thing."

"What's that...or dare I ask?"

"Arikmica lived and faded into obscurity without ever going to war."

"It's a good thing. You might have been hanged for treason!"

They both chuckled at the unlikely scenario.

Josh embellished the legend of the rise and fall of the unknown nation while they sat atop the highest boulder dominating the abandoned relic. The two grew silent, holding hands, awed by the splendor.

Eventually, Josh untangled his lanky frame and rose, simultaneously tugging Traci to her feet. As if on cue, Mother Nature and Father Time extended a helping hand to the fledgling romantic. Directly above the derelict dam, straight up through the canopy of gnarled trees guarding the site, dark clouds parted, revealing an oasis of brilliant blue. Josh interpreted this as an omen.

With showers temporarily in abeyance, the optimistic suitor used both arms to slowly pull his dream woman to her feet and press her along the full length of his body. She looked up at him, suddenly aware that this retired president of the defunct nation of Arikmica was one passionate hombre—and every fiber of his being was focused resolutely in her direction!

Random raindrops still skittered down her cheeks as Josh gently brushed them with his lips. She didn't pull away. As his arms moved across the small of her back, her hands entwined around his neck. His arms tightened, drawing her closer. She reached up, took his head in both her hands, and pulled his face down to meet her own. In a nano second, his lips slid sensuously toward her waiting mouth. Again and again they kissed, bringing to reality long-suppressed dreams.

The two publicly dueling intellectuals of the morning had become, by design and desire, private lovers, driven by irresistible powers. Their faces flushed with the heat of a blast furnace. Gasps of labored breathing burst in concert, driven by devouring hunger. For Josh Ryan, this could be no casual, lust-inspired tryst. No room for Darwinian animal lust here. The event shouted love—very pure, very simple—and very sacred.

Responding as though this pinnacle of passion were doomed to be their last, Josh and Traci clung to the moment with more passionate kissing, caressing, and hugging—first gentle, then more urgent—until coming to rest to regain their bearings, searching each others' eyes for ratification.

Fifteen minutes slipped by before Josh was able to shift gears to the ultimate but as yet unrevealed purpose of the trek. He whispered, inches from her ears, "I love you, Traci. I wish I could find some way to be smooth, but you know me, the blunt scholar...I just plain love you!"

Without saying a word, Traci placed her head over his heart, her eyes looking past his shoulder as though focusing on some distant object. Sensing a now-or-never moment, the amateur Don Juan plunged ahead.

"I've been excommunicated from grad school; I have no serious job prospect; and I'm being sued for a million-plus big ones! I don't even have a car to drive to the courthouse to defend myself. Other than personal effects like heirlooms and fossils, my horse Turbo is my only asset."

The abject confession prompted its own critical assessment.

"Bleak credentials, I admit!"

Pausing to restock his rapidly draining courage, he reached for the golden ring. Traci didn't blink. She waited, without a stir.

"While this may be the worst time to bring it up, I got to thinking if you could love me at this rather awkward juncture in my life...then for sure...your love is real."

Although she didn't so much as raise an eyebrow at the halting words, inexplicably he felt the euphoria of jumping into unexplored space, with or without a parachute.

"Never before in my life have I come close to knowing the meaning of love. You're the most genuine, spectacular woman I've ever met. I love you with all my heart...I always will. More than anything in this world, I want to be the man in your life...forever!"

After one last gulp, he popped the big question in a hoarse whisper: "Will you marry me, Traci?"

For a moment, she didn't respond. Instead, she remained mute, immobile as the fossil Archy. After what could have passed for a millennium, she raised her face to his—tears merging with the raindrops cascading down her face. His heart felt like a hunk of petrified wood. All bets were off; his gambler's luck had flown the coop—the loaded dice had landed snake-eyes.

Eventually, the muffled reply he dreaded most came in a chain of barely audible phrases. "Oh, Josh...I didn't know...It's just that...you need to understand..."

As the unsettling message mingled with sniffles and a sob or two, she put her head on his shoulder, staring off into the woods.

Bereft of energy, Josh agonized, trying to conceal his pain. He hadn't the faintest clue what to say or do. His mind churned in helpless despair.

Come to think of it, why should she choose him? he thought. She could have any guy. He had served his purpose as a *Monkey II* pawn. Given reality's high risk, he had overplayed his cards, miscalculated the odds, gambled all he had, and wound up a loser—big time! But as sure as the stars above, no Ryan would wallow in self-pity!

As her rain hat tilted crazily on the side of her head, he kissed her hair, inhaling the luscious scent as he sighed in resignation. He surprised himself at the stoicism of his next line. "You'll always be in my heart, Traci...Maybe we'd better saunter back to the cottage before the rain comes again."

She nodded assent as he straightened the recalcitrant cap, took her hand, and embarked on the tedious climb back to the real world. Neither chose to prolong the pain. Stripped emotionally, the jilted romantic vainly reached for refuge in a wish list of might-have-beens.

As if on cue, an ugly bank of roiling black clouds enveloped the sky, erasing the patch of blue. It was as though the clouds were shedding tears for torn and aching hearts.

Josh and Traci glumly returned to the mill house through the damp forest. The journey seemed eons longer than the outward jaunt. Labored efforts at conversation were suffocated by the gloom. His plan to surprise Traci with the engagement ring his dad had given his mother had gone awry. Now, there was nothing to do but stand tall and take his heartbreak stoically. While he felt no regret at the all-out effort to land this remarkable lady, it stung that she could have used him like a disposable pawn in an intellectual chess game.

In spite of the vanished vision, he resolved to reassemble the remnants of his dreams and press on—although without the first clue as to where.

At the mill house, Traci made no move to re-enter, nor did he extend the invitation. Instead, he busied himself helping her straighten the dripping yellow slicker, scrape the mud from her boots, and readjust the recklessly out-of-kilter rain hat. Only the mascara smudges betrayed her tears.

Looking down at the love of his life, he wanted to erase the ill-fated journey to the dam from his memory and to hug her with all his might. The best he could do was stand stiffly, announcing in matter-of-fact tones, "See you tonight at the trustees' reception. Drive careful now, y'hear?"

Staring at him with hardly a flicker of emotion, she finally donned a wan smile, caressed his face gently with her fingers, and kissed him on the cheek before retreating quickly to her car. In sharp contrast to her bubbly arrival an hour earlier, storm clouds scrubbed away all semblance of gaiety. She summoned a reply with a demur glance. "I'm looking forward to it, Josh."

She melted away in the mist.

# 39

A forlorn Josh lingered on the porch of the cottage, standing as rigid as the bronze statue of General Chamberlain. Numbness snuffed out every remnant of joy. Not even the winsome tinkle of his porch chimes assuaged the emptiness of failed love. The naïve scholar had experienced the bruising rudiments of a crash course in Love 101.

Today's inaugural lesson introduced the fundamentals: True love tolerates no strings or conditions; it blossoms unselfishly without choreography or manipulation; and regardless of reciprocity, it flames eternal. No matter what Traci said or did, he could never abandon his commitment.

Determining not to succumb to the bitterness of regret, he re-entered the mill house, taking the first step toward the rest of his life. Inside, the phone beeper announced messages competing for attention.

Michael's voice greeted him first. "Seeing you kids traipse past the house reminded me to tell you I can pick you up around 6:30 p.m. The trustees' reception begins at 7:00 p.m. I know you aren't thrilled at the prospect of wearing a tux, but this is a formal shindig. And don't forget, your grandmother insists that Bishop Daniels and I join you and the *Monkey II* cast in her private dining room immediately following the reception. I have no idea what she has up her sleeve, but we should humor her—at least until you pick up your horse."

The cultured voice of the Duchess followed the next beep. "Josh, you towered tall this week—feisty and dashing! I know your mother and all the rest of the Morgans would have shared my pride.

"I'm looking forward to seeing you at the trustees' reception and dinner this evening, but remember you have an excuse to escape the stuffy banquet's formalities. I've invited the *Monkey II* cast key players, together with Bishop Daniels and your father, to join me for a private dinner at the club, immediately after the reception.

"Of course, Traci Kilburn will be there...And if you don't mind my saying so, my woman's intuition sees stars in her eyes for you."

"So much for woman's intuition," muttered Josh..

Next he heard the distinctive inflections of Edward Anthony Stone.

"Sorry I missed speaking to you in person, Mr. Ryan. Block off some serious time for a meeting Saturday evening. We'll meet in the university chancellor's executive boardroom. Both you and your dad need to be there. You'll have to trust me as to the agenda. Except for private conversations with your father, this matter must be kept in strict confidence.

"Oh, incidentally, Mr. Ryan, for what it's worth, had I been a *Monkey II* juror, my vote would have been with the majority. I've always taken Darwin hype with a grain of salt."

Josh admired Judge Stone and trusted his judgment. He felt no apprehension as to the hinted agenda. Already reeling from the afternoon's emotional shortfall, it proved easy for him to put news of the Saturday evening conference on the back burner.

The tape next amplified the welcome voice of JT. "Hey, Dar, old man, you broke away from today's performance before I could shake your hand, congratulating you on your skill as a scientist and star witness. That movie thing caught me off guard...I plain forgot my manners. You did a bang-up job!

"Of course, the way I saw you looking at Traci, my money says she might have had something to do with your blowing the place in such an all-fired hurry. I'll be expecting the truth, the whole truth, and nothing but...Only a blow-by-blow description will do. Mr. Straight Arrow can't con his wily, worldly wise attorney.

"Speaking of which, since I'm serving your nefarious legal intrigues pro bono, I'm hoping you feel beholden enough to grant me a favor. You know Granddad is in town to address an ecumenical confab at the National Cathedral tomorrow morning. The legal assignment to guard the constitutional rights of the wayward Ryan scholar prevents my squiring Granddad to his appointments around town. When I asked him if it would be OK for you to act in my place, he said 'Yes,' with a condition: 'Providing he brings that charming Miss Kilburn—to guarantee his best behavior.'"

Still shaken by the fiery loss of the venerated Mustang, Josh felt impotent, wondering aloud, "What can I do for wheels?" But he had learned never to underestimate JT's skills.

"Just in case you're worried about wheels, Dar, I recall your fondness for luxury. A chauffeured limo is already booked in your name...Expect it to pick you up at the mill house first thing in the morning at 8:30 a.m. sharp. It's yours to command for the day. And don't worry about the tariff. The $500 you contributed to lure me to Tennessee last spring has been channeled to underwrite this worthy cause. Since Granddad's plane departs National shortly after lunch, you and your heartthrob can make good use of the limo until the coach becomes a pumpkin...hint, hint, hint.

"Hail to the chief of Arikmica. See you this evening, my man!"

Who could resist the irrepressible Jonathan Thomas Daniels?!

"If JT only knew I struck out big time this afternoon," Josh fretted. "But for sure, I'll look forward to spending some quality time with the bishop."

A cryptic call from Jessica Saunders followed: "Your line has certainly been busy. Just wanted to tip you off that the arrest of a prime suspect for arson is targeted no later than dawn tomorrow. In the meantime, keep me informed as to anything new."

Josh almost overlooked the sixth message, which contained nothing but silence. The phantom harasser loomed alive and well. Josh yanked the tape to hand-carry to Jessica, uncertain as to whether the vacant message warranted status as evidence.

By the time Michael stopped by to taxi his son to the trustees' bash, Josh was good to go. Immaculately attired in tuxedos and cummerbunds, the handsome father-son duo could have blended with the best at any college prom. Josh's exterior of affability left no hint of the gnawing ache at the pit of his stomach. At least the thunderclouds had lifted, offering respite from intermittent rain. Only a gray overcast remained to remind him of the afternoon's wrenching fiasco.

As the Ryans crossed the threshold of the Hampshire Club, the place buzzed with small talk. A Baltimore-based orchestra provided melodious background and ambiance. The sweet sounds of an occasional brassy solo or a percussion drumbeat amplified the mood.

Most trustees arrived early to meet an exclusive roster of the moneyed and corporate high and mighty, fair game for all institutions of higher learning. The coveted invitation to the university's most prestigious social event required a big-bucks donation—traditionally a minimum of five grand per couple. A cadre of well-heeled alumni mingled with philanthropists, swelling the ranks of the privileged. A sprinkling of Capitol Hill politicians glad-handed with polished aplomb—decorative ornaments, exempt from big-ticket contributions. Pace Terhune's diligent staff scurried inconspicuously, catering to all.

Sophisticated socialites in New York fashions lounged in vivacious effervescence, or chatted in casual clusters. Ostentatious displays of diamonds, rubies, and sapphires, mounted in rich settings of gold and silver, danced and dangled in profusion. Dr. Wood Comstock, university chancellor and host for the event, beamed as he presided over the dazzling scene.

A select few teachers and students had received invitations to the festivities. Traditionally, this honor extended to three faculty nominees for the annual award recognizing meritorious service to the university, and to students being nominated for a yearly award.

Applause and commendation alone could motivate most climbers struggling up the rungs of the social ladder. But the cast gold *Chamberlain Medal of Distinction* fortified by a $10,000 cash award, galvanized academia's attention. The silver *Morgan Medal of Distinction* for stellar student achievement replicated the faculty honor and was accompanied by a check for $5,000.

While any worthy campus personality could be nominated, all names were filtered through the administration, leaving it to the trustees to pare the list to three in each category. Ultimate authority for the selection had been vested in the source of funding for the awards—Lady Carrington.

Competition for the selection was always intense. Less-than-subtle politicking and arm twisting often afflicted the nomination process. But come sweepstakes time, the Duchess alone called the shots. Inscrutable and meticulously analytical, she refused to be hoodwinked by flattery. No one messed with Duchess Iron Pants.

Inside the reception hall, the Ryan scion and son went separate ways, casually working clusters of familiar faces, conferring briefly, greeting male acquaintances with a handshake while saving the traditional hug and gentle kiss-on-the-cheek for the ladies.

By 6:45 p.m., most of the *Monkey II* cast, conspicuous additions to the guest list, had arrived. Appearance on the Chamberlain Playhouse stage had propelled the amateur artists to overnight prominence. Regional public television had picked up the one-hour summary of *Monkey II* that aired daily during the week, introducing previously unknown faces to public scrutiny. And thanks to the timely appearance of Attorney Marcus Brogan, the story had made the national news. All performers were rewarded with hospitable welcomes. Collectively, the cast enjoyed a fifteen-minute burst of fame. Only Attorney Marcus Brogan had evaded the acclaim: The barrister spurned the invite, claiming "prior commitments."

A dapper Karl Striker worked the crowd in grandiose style. The exhibition surprised no one, given the academic's impressive ability to turn on the charm when it served his purposes. Dr. Striker's presence signaled his nomination for the *Chamberlain Medal of Distinction* award. But the sight of the two hovering hangers-on raised eyebrows.

Gus Morley's stuffed-shirt manner clashed with Striker's nonchalant ease. A sympathetic Josh hoped no one noticed the whiff of mothballs from Morley's tight-fitting, out-of-style tuxedo. Since good-ol'-Gus couldn't have made it to the event riding his boss's coattails, it figured that this inept but affable champion of student causes must also be a nominee for the annual merit award.

But the presence of the third member of the trio blew Josh Ryan's mind. Unable to explain the inexplicable, he finally concluded, "There's no way a notorious kiss-up like Slade Lassiter could wangle an invitation unless he was nominated for the student award."

Karl Striker preened, pranced, and strutted. He saw the scene as high drama in an academic *struggle for survival*, confident he alone would emerge triumphant. He reasoned, "What better way to achieve overdue recognition of my inherent ability and to ratify credentials tarnished by that farce of a trial."

As for the bedraggled Morley, his appearance inspired curious double-takes. A glib corporate type quipped, "Looks like the proverbial penguin in a bow tie—

only this overfed specimen may be a Charles Darwin bear trying for a comeback as a whale!"

Undeterred, Gus Morley muddled gamely through the morass, hoping that his nomination for the medal and the prize money represented more than someone's careless mistake.

The more worldly wise Slade Lassiter slithered alongside. Charitable observers characterized him as a coattail jockey. Josh saw him as intellectually parasitic.

The younger Ryan would have been flabbergasted had he been privy to the source of the nominations. Seeing the three basking in the limelight, each mentally rehearsing appropriately humble words of acceptance, it was better he didn't detect the manipulative hand of Grandmother Carrington at work. While his own private world teetered in topsy-turvy confusion, he heartily wished that, just this once, the groveling Morley would outshine Karl Striker.

The reverie ended when, jostled by the crowd, he stumbled awkwardly into the bosom of an alluring woman, accidentally tromping her toes for good measure. Before he could turn to apologize, he recognized the wafting fragrance. Carefully coiffured and immaculately attired in a shimmering splash of sky-blue silk, Traci Kilburn skipped aside nimbly, unfazed either by the bump or the afternoon's romp through the woods.

Recovering, Josh blurted an inane, "Oh, excuse me! Imagine meeting you in a place like this!"

Poised as ever, the young woman reacted with unruffled charm. Feigning a slight bow, she countered, "Mr. Darwin, I presume. Discover any interesting fossils lately?"

Eyes dancing mischievously, her words paraphrased their Tennessee introduction the previous April. A worse-for-wear Josh felt ill-equipped for witty small talk but affected blasé nonchalance, countering, "Nothing new in fossils, but recently I've encountered philosophical truth. I still don't buy Darwin's *progress toward perfection* but if his *Descent of Man* was meant to describe humanity's abysmal failures, he just might have had a point."

Struggling to conceal his embarrassment over the afternoon's dead-end, he started to stammer an apology. "Look, about this afternoon...I'm sorry that I..."

Traci put a restraining hand on his arm, stopping his apology in mid-sentence. "Not now," she whispered.

Just then, two giant arms from behind enveloped them both. Powerful hands pressed at their shoulders, pushing them together like rag dolls clasped in unchoreographed embrace.

"Josh! Traci! Congratulations! You both performed magnificently!"

The words boomed in rich resonance as the spirited Bishop Brock Daniels announced his presence to lifelong friends. The bishop treasured Josh Ryan as he would an adopted grandson and had enjoyed a social relationship with the Kilburns through the Hampshire Country Club since before Traci was born.

Brock Daniels' entrance required no town-crier proclamation. A towering 6 feet, 2 inches in his stocking feet, encased in a frame that carried 225 pounds of muscle, the finely chiseled appearance belied his 75 years. His tawny-bronze features and mane of snow-white hair still turned heads, and his splendorous baritone voice inspired heart and soul.

A ten-year-old Jonathan had once confided to Josh, "Grandpa could tell an audience that Interstate 95 ran all the way from Washington, D.C. to New York City, and they would be impressed that he was sharing some big secret."

Born in Tulsa's Greenwood suburb in 1921, Bishop Daniels had grown up amidst a conflagration sparked by the undercurrents of racial hate. When the inferno flared, his quick-thinking parents escaped to the woods, sparing their family the mindless fury that snuffed out the lives of a hundred innocents. When the smoke cleared, thirty-five blocks of Greenwood smoldered in ashes—an evil testament to man's inhumanity to man.

After three-quarters of a century, Brock Daniels had risen tall above the brimstone and brick bats hurled by petty bigots. Every day of his life epitomized the triumph of good over evil.

Though Josh had never heard the great man preach, Brock Daniels had cast a spell over Josh's pathway. He always greeted the bishop with deference, as he would his own grandfather, but never hesitated to share wisecracks. Michael encouraged his son to address the grandfather figure as "Bishop" or "Mr." This evening, Josh opted for the less-formal salutation.

"It's an honor to have you with us, Mr. Daniels. Glad you enjoyed today's show-and-tell."

"I'm with *Mr. Darwin* on that," opined Traci, pointing at Josh while trying not to make the spurned suitor uncomfortable.

"The honor is mine," rejoined the bishop. "I predict a bright future for the two of you." Partially tongue-in-cheek, he cracked, "Have either of you considered donning the mantle of the cloth? You could be great warriors for good in battles of the spirit...Nowadays the clergy is color-blind and gender-neutral!"

They shared a chuckle before Josh admitted sublime ignorance. "I hardly know what a church looks like inside, much less what preachers preach."

Bishop Daniels brightened. "I'll give you a personal tour tomorrow morning. Jonathan informs me that you two are willing to keep me out of trouble while I make my ecumenical rounds at the National Cathedral."

Traci shot a questioning glance that was studiously ignored by Josh. Instead, he grinned shamelessly, nodding mute concurrence. With a mind running on overload from disappointed love, he had neglected to tip off Traci about JT's invite. But good trooper that she was, she didn't disappoint Josh twice in the same day. Flying blind while exuding gracious enthusiasm, she diplomatically paraphrased the biblical Ruth, "Wherever you go, we will go—gladly!"

"Have to hand it to her," thought Josh, "she's good!"

"Then it's a date! If you don't mind picking me up at the Hay Adams by 9:30 a.m., that should be more than enough time to make the 10 a.m. session at the National Cathedral."

"We'll be there bright and early," Traci assured airily, having not the faintest idea as to the master plan or the means of transportation.

Josh jumped in. "Thanks to the generosity of your grandson, we will pick you up in style—limo wheels, no less!"

"So I understand!" The old man seemed pleased at JT's audacity. Looking at Josh, he volunteered some advice.

"By the way, Jonathan confirmed that this could be your first-ever non-wedding or funeral church appearance. If so, you've picked a winner. The cathedral itself preaches an eloquent sermon. Be sure to look for the moon rock built into the stained-glass window above the south wall."

With professed meekness, Josh acknowledged the accuracy of JT's assessment. "I'll confess to being guilty as charged. Hope you won't be too rough on us sinners!"

"You can relax, Josh! The word is out that *'all have sinned and fall short.'*[189] Besides, we begin each day with a clear screen." Then with a wink directed Traci's way, "Never fear, young man! God hates sin—not the sinner!"

No sooner had he finished offering that insight than the bishop found himself corralled by an opportunistic political type hungry for a photo op. The intruder unceremoniously wheeled the bishop around ninety degrees to face a blinding flash, while babbling something about, "Just one more for the family album...if you don't mind."

Left alone, Traci and Josh marveled at the privilege of growing up with greatness. As the moment for the award presentations approached, they diverted their attention to the curious spectacle of two professors, trailed by a fish-out-of-water grad student, inching center-front. The teacher/student trio shared the conviction that their hour of overdue recognition had arrived. Hadn't the persuasive Lady Carrington implied as much when she followed up the written invitations with confidential telephone calls? Delivered in words of fine-spun silk, her words lingered, titillating their greedy imaginations.

"Recognition of academic sciences at Mid-Atlantic University is overdue. It's only fair to confide that the sciences will be front-and-center when this year's *Medals of Distinction* are awarded. I would be terribly disappointed if you didn't take advantage of my invitation for you to attend this headline-grabbing event."

The cocky Striker calculated he had bagged the award. The buttons on the shirt of the pear-shaped Morley almost popped with pleasure. As for the smugly overconfident Lassiter, he knew the Duchess possessed more savvy than to make an award to her controversial grandson. He imagined his stint as editor of the Science Research Foundation journal, where he doubled as consummate Morley flak, was about to pay off.

Minutes before the grand presentation, Gus Morley stumbled into a tray stand of half-empty glasses. The clattering crystal doused partygoers indiscriminately in swirling sprays of ice and wine. Morley hung his head, apoplectic with consternation. The ruffles of Striker's immaculate white shirt hung in disarray, stained with a spattering of red wine. Lassiter made a futile attempt to whisk away the spreading stain. A mortified Striker could barely contain himself, but with iron discipline stifled his ire, determined that the incident would not mar an academic coronation.

Biting his lip with clenched teeth and fake smile, he shooed away onlookers, assuring, "That's OK; that's OK! No damage done! Could happen to anyone."

Daggers of contempt alerted Morley to his boss's real thinking.

While club staffers acted with calm efficiency to retrieve and restore, university chancellor Wood Comstock mounted center stage and tapped the mike in a ritual standard for such events.

"Is this thing on?" Squeals and howls from the audio equipment erupted, assuring listeners that the electronic system of the computer era could still mimic the static and whistles of pre-World War II vacuum tubes and hand-wired amplifiers.

The old pro was no stranger to the ceremonial litanies that plague academic circles. He relaxed, comfortably in command of the audience, tolerating nothing less than unanimous attention. He made no effort to speak until conversations faded. Eventually, partygoers lost interest in the Morley *faux pas*, the sound system reached equilibrium, and decorum returned.

The chancellor's "few opening remarks" were hardly few—and mostly a rehash of uninspired phrases. The benign ambiguity of the remarks could have passed for closing or introductory commentary. He praised the university's successful centennial celebration, touting the "profound meaning of these *Medals of Distinction*" and the administration's intention to use this year's event to highlight the "auspicious achievements of scientific disciplines that bring prestige to our historic campus."

After an interminable fifteen minutes of overblown prose, Dr. Comstock invited a pair of "distinguished guests" to come forward and officiate in the presentations. Two United States Senators rose in lock-step response. Their mutual claim to fame centered around their survival on the Hill without taint of personal scandal or public controversy. They were better known for their appearances at glad-handing events than for sponsoring earth-shattering legislation.

Midway through the droning monologue, a restless Josh sidled mischievously over to JT. Nudging his friend with an elbow, Josh whispered a provocative quip. "Hey Bro, listen to these political talking heads! Wind blows and they stand and salute, thinking it's voter applause!"

JT reacted with a snicker disguised as a sneeze. Josh tried again. "I bet politicians are so verbally ambidextrous that they could perform with equal incompetence as actors—or even clergymen!"

This time, the boys from Magruders Mill village suppressed coughs in sync, trying not to break up! It took a sage word from JT to restore sobriety.

"You're probably right, Josh—but only in part. While these two windbags would sound like sounding brass or tinkling cymbals as clergymen, you can bet your shirt Granddad would make one fabulous Senator—or U.S. President, for that matter!"

Duly contrite, Josh seconded the nomination. "He's got my vote—any time!"

Senator V. Lester Blott, the epitome of all things mediocre, received a polite smattering of applause when called to the stage. *Landslide Lester's* six terms in the U.S. Senate were the product of razor-thin electoral margins, typically hanging in the balance until absentee ballots were counted. Pundits claimed that Landslide had never met a poll he didn't bow to or a motherhood-and-apple-pie issue he wouldn't embrace. He had managed to live luxuriously off the public till most of his adult life, thanks to his good-old-boy skills and the ability to filibuster for hours on end without so much as a pit stop.

Landslide's luck expired in the 1996 primary when voters elected to put their tax money to better use. The newly christened *Lame Duck Lester* enjoyed this one of a dwindling number of public appearances thanks to the loyalty of the Duchess—as a Mid-Atlantic University student, Blott had earned his room and board, plus tuition, by serving as a part-time Morgan estate stable hand.

Senator Blott held the envelope with the name of the student medallist. Summoning a mediocre imitation of Hollywood's Academy Awards, his fumbling fingers tore at the envelope with slow, deliberate motions. He perused the contents, as if reassuring onlookers of his literacy. Eventually the thunderous filibustering voice feigned enthusiastic surprise.

"And this year's winner is the tireless editor of Mid-Atlantic University's own student newspaper, Mr. Ira Lonergan."

Blott couldn't have identified Ira in a police line-up if offered a million dollars, but this did not deter him from lavishing high praise on the recipient.

"This outstanding young American is recognized for his skillful, unselfish efforts in editing the videotapes of the unsurpassed *Monkey II* production and for successfully placing the series on public television."

Applause erupted, sustained and genuine. Most attendees had spotted media reports on the series, and many had personally witnessed at least one act of the show trial. Communicator Lonergan laid no claim to science credentials, but he had contributed mightily to the university's scientific disciplines, to say nothing of its financial coffers.

Ira reacted with shocked surprise. He had lacked the time and the funds to rent a tuxedo to match the formal garb of other guests. After the senator draped the medal, attached to a red, white, and blue ribbon, around his neck, Dr. Comstock followed with a cashier's check for five grand. All the usually loquacious student could muster was a heartfelt Elvis quote, "Thank you! Thank you very much!"

Few were aware that the young media genius had earned his way through college, but for sure the Duchess knew!

Student editor Lonergan was the most surprised person in the reception hall, except for one. Slade Lassiter bolted, intending to abandon ship. Dr. Striker would have none of it. Locking the recalcitrant student's wrist in an iron grip, Striker hissed, "No way are you going to rain on my parade. I insist you remain here to applaud when I get my reward!"

Striker relaxed his hold, edging a step closer to the podium in time to see Billy Bob Yakley, the senior senator from someplace far away, uncork his contrived affability. Senator Yakley had never met a voter he didn't pretend to love nor a principled position he wouldn't abandon at the first whiff of campaign dollars. Glib spinmeister Billy Bob pandered shamelessly, his speeches dripping with redundant references to his allegiance to the "American people!" He had more than his share of brains but devoted his intellectual resources to self-interest. Challenging the timeworn cliché overrating public intelligence, Billy Bob could fool "most all the people most all the time."

His matchless fund-raising skills, harnessed to underwrite extravagant campaigns, outmatched most Washington politicians. His charismatic chutzpah, diverting public money to pork-barrel projects, surpassed all comers. Reportedly the toe-tapping senator had managed to squander tax dollars in bogus feasibility studies confirming the impossibility of constructing a down-home naval base in a swampy part of a landlocked state. The smiling extrovert used strong-arm tactics behind the scenes to target offenders who stood in his way.

Billy Bob perceived the ultimate national issue of the day to be sustaining his personal career! Admirers and detractors alike surmised that his far-sighted but jaundiced eye coveted the power of the Oval Office.

The Duchess saw through the froth and the slick persona but agreed to Billy Bob's appearance when advised by Pace Terhune that "Senator Yakley authored a congratulatory message, published in the *Congressional Record*, singing the praises of Sam Houston Morgan and commemorating Mid-Atlantic's centennial."

Senator Yakley performed up to snuff. Deftly he ripped open the second sealed envelope and read with eloquence, as though presiding at a federal election ballot count.

"The 1996 faculty *Medal of Distinction* goes to Pace Terhune for his exceptional achievement in focusing national attention on Mid-Atlantic University's academic sciences by showcasing *Monkey II* and by initiating programs promising generous funding for MAU's scientific research."

Thunderous ovation ratified the popular choice. Students, faculty, and community benefactors joined in enthusiastic acclaim. Terhune didn't suffer from false modesty and rarely was rarely at a loss for words. But with the coveted medal dangling beneath his bow tie and a $10,000 check stuffed securely in his pocket, the best he could do was echo a line from a James Cagney movie.

"My mother thanks you, my father thanks you, and I thank you. This medal will be treasured with pride for the rest of my life." Then he brought the house down, in classic Terhune style. "As for the $10,000, this will be my contribution to a student scholarship fund memorializing the memory of this university's all-time champion of science, Dr. Brooke Fielding."

Kibitzing luminaries, already on their feet, added crescendo to the applause.

Morley had suspected all along that his own nomination seemed too good to be true. As for Striker, the humiliating loss to a non-scientist competitor rubbed salt into deep wounds. Livid, he stomped out the door in another ill-tempered tantrum.

This time, not a soul noticed—except Lady Carrington.

Precisely at 8:00 p.m., the row of double doors to the master ballroom swung open, signaling welcome to the elite "400," inviting each to find pre-assigned place settings. The Duchess, gently shepherding her nine handpicked guests, slipped away unobtrusively to a private dinner in the Sam Houston Morgan banquet hall—the most lavish setting available to Hampshire Club members.

Despite having lived almost fourscore years, the grand lady of Morgan Manor had "only just begun" an evening to remember.

# 40

The posh Sam Houston Morgan banquet room could comfortably accommodate two hundred guests. This evening, the ornate setting belonged exclusively to Lady Carrington and her roster of nine invitees.

Red oak paneling embellished the walls. A twenty-foot high trey ceiling looked down on luxurious carpeting. A gilded chain anchored a massive chandelier that reflected a thousand points of flickering candlelight. An Irish linen tablecloth draped a circular table, its embroidered perimeter caressing the floor; while an array of Waterford crystal traded rainbow reflections with the chandelier. Hand-etched silver nameplates marked the place settings of English bone china emblazoned with the Morgan coat-of-arms and surrounded with matching sets of sterling flatware.

The thickest hand-sculpted rug Josh had ever seen ringed the banquet table, spilling out another ten feet behind the diners. Glove-leather armchairs mounted on rolling brass wheels beckoned comfort.

A massive fireplace, built from stone quarried from the Morgan home place, dominated the room, while a bigger-than-life oil portrait of the gambling Gettysburg survivor stared down from above the mantelpiece. Spotlights accented the gold-plated frame around the artist's depiction of the Morgan clan's founding scion.

Ensconced comfortably at the base of the painting, a cadre of elite musicians, conscripted by the Duchess from the ranks of the orchestra entertaining the trustees across the hall, delivered a repertoire of background sounds from the big band era—selections in keeping with the tastes of the hostess.

Michael Ryan occupied the chair immediately to the left of Lady Carrington. Grandson Josh sat in the seat of honor, immediately to her right. Leaving no doubt as to her hostess role, the Duchess' leather chair stood imperceptibly larger and higher than the others, facing directly across the room toward the portrait of Grandfather Sam.

At the invitation of the hostess, the guests settled in. All seemed intrigued and more than a little pleased to be present. Only Michael Ryan clung to suspicions. After a quarter-century of boycott, he wondered what this elderly erstwhile mother-in-law could be up to at this late date? But having never reciprocated her petulant animosity, he accommodated easily to the role of gallant gentleman awaiting her next move. It came in a string of ad-libbed reminiscences of happy times. The grand lady exuded charm, motioning the guests to take the assigned seats while she remained standing, directing traffic.

"I've looked forward to an occasion like this for many a day. This centennial year celebrating the founding of Mid-Atlantic University would have pleased Grandfather Morgan. He never had the privilege of attending college, but as you all know, the charming rascal managed to do pretty well for himself, regardless."

All the guests smiled affirmation, fully aware of Sam's reputation as lucky gambler and seat-of-the-pants entrepreneur.

"The mastermind of the centennial has just been awarded the university's Medal of Distinction. He above all deserves it."

Smiles and polite clapping ensued, while Pace Terhune waved in slight salute to the Duchess, mouthing, "Thank you!"

"A centennial event of ultimate interest to me concluded successfully earlier today. You will have to pardon this old lady's proud bias, but in case you didn't know already, the star witness is none other than my own grandson...who carries the Ryan surname but shares Morgan clan genes!"

Those sitting around the table responded to this recitation of common knowledge with condescending smiles and winks. Michael remained quizzical, wondering what the conniving Duchess had up her sleeve.

"*Monkey II* achieved smash hit status in this community. All of you have contributed. Tonight this is my own way of honoring and thanking each of you on behalf of the Morgans, the Carringtons...and particularly, the Ryans!"

She looked steadily at Michael when she referred to the Ryan surname. Now, offering a clue to what might be coming later, she continued, "I hope you won't think it selfish of me but rather will join in indulging my eccentric whim. Since my grandson is not only the wrong gender, but also much too old to qualify as a debutante, this seems the perfect time to tell the world that I love this young man."

Carefully choreographing her master plan, she announced that commencing with her introductory words, each of the other nine would be invited to offer a toast to anything or anyone special in their respective lives. She then took her seat, after explaining that the dinner would proceed immediately and that all toasts would be saved for the end rather than, as traditional, at the beginning.

Josh rolled his eyes in mock protest, then shrugged, willing to play along with her game.

The banquet proceeded with gusto. Conversation flowed uninhibited while the diners mentally scripted ideas for creative toasts. Only one person present, other than the Duchess, understood the reason for the flourish and fuss.

Bishop Daniels remained as always a staunch friend and loyal supporter of the proud socialite. More than a pastor, he responded with sage counsel whenever she felt overwhelmed by mood swings. He had witnessed firsthand the anguish Regina Ann suffered at the death of her only child and the self-inflicted wound imposed by her pride. He knew the malignancy of her bitterness flourished so intensely that years passed before she would allow herself to touch Leslie Anne's personal effects.

When she belatedly opened her daughter's diary, its faded pages revealed agonizing secrets that left the Duchess racked with remorse and regret. Confidential disclosures, hidden for years, revealed how wrongly she misjudged Michael Ryan. But even with the surprise discovery, her arrogant pique prevailed, preventing her from reaching out to a banished son-in-law and an unacknowledged grandson.

She knew she could never recapture the squandered years. An aching void filled her soul, instead of the exhilarating embrace of warm family memories. She longed to reach out to fill the emptiness before the loss became irretrievable. Although she didn't know how to reach out, Brock Daniels, empathetic confidant, could help!

It was the bishop who drafted the map and blazed the trail to reconciliation. After listening patiently to her excruciating pain, it was he who responded with a grand strategy to bridge the chasm of grief she herself had created.

It was he who purchased and presented her with the *Monkey II* tickets—with his compliments. It was he who had inspired her with the courage to surprise her grandson with an invitation to Morgan Manor the previous Tuesday. He it was who had encouraged her to present Josh with the most cherished keepsakes reminding her of Leslie Anne—the oil portrait, the wood candlesticks, and the precious diary. And it was Bishop Daniels who had suggested this evening's dinner to bring estranged grandson and son-in-law together, with her, in an attempted reconciliation.

The astute clergyman tempered his counsel with caution. "Understand, Regina, you can't rewrite history. There is no guarantee you can heal the rift. But knowing the Ryan men as I do, you have everything to gain by making an honest overture and absolutely nothing to lose.

"Don't ever underestimate the power of love. Remember, the only way to cross a river is to take the first step and start walking—like Moses at the Red Sea!"

This Friday, two years shy of fourscore years, the fabled Lady Regina Ann Morgan-Carrington craved family forgiveness and reunification more than power or possessions. Given the coolly elegant performance of the Duchess this threshold evening, no casual observer could detect the internal churning of her heart. With waiters outnumbering guests, and gourmet courses uncounted, two hours quickly disappeared.

Toast time had arrived! Bishop Daniels chose tonic water. The rest selected from a menu of California champagnes and chilled sodas.

As promised, the hostess rose first.

"Some people, like my grandson Joshua, seem wise beyond their years. Others, myself included, have to spend a lifetime learning from our mistakes, but stumbling hopefully in the direction of what's important—love and family. Blind ignorance and selfish pride dog my tracks. I've made classic errors in judgment, to my own irreplaceable loss.

"Tonight I'm taking my first step, crossing a bridge in a new direction. I treasure this opportunity to open my arms in love to my grandson, Joshua Chamberlain Ryan." With that she spread her arms wide, adding, "And to his incredibly remarkable father, Michael Joseph."

Raising her glass to the "Hear, Hear!" of the listeners, she swallowed her pride, washed down with an enormous swig of the bubbly. She didn't dare look at either Ryan, choosing instead to sink back regally into the comforting confines of her soft leather throne.

The candid overture blindsided Michael Ryan. Carefully composed phrases were now caught and jumbled in the blizzard of emotions swirling in his head. Awkwardly, he stood erect, reciting a mechanical mantra that could mean anything to anybody.

"Here's to family and friends...to all things true and good...fragrant memories of the past...blessings of the present...and bright promise of the future!"

It was the best he could do. The Duchess responded with a smiling nod, grateful for the public civility and the faintest trace of a partial pardon. She moved past the moment by introducing a spry, eighty-eight-year-old guest seated immediately to Michael Ryan's left.

"Wilma Humphreys sat in a Dayton, Tennessee, high-school biology class taught by John Scopes. She witnessed the 1925 *Monkey* trial and saw the Bryan/Darrow confrontation. She even played her clarinet in the school band that entertained visitors on the Rhea County Courthouse lawn during the trial. A retired teacher in her own right, she lives in Johnson City, Tennessee, and made the trip here as my guest to help celebrate the success of *Monkey II*."

The other guests clapped spontaneously, saluting this vivacious eyewitness to legal history. When the Duchess invited her to say a few words, Mrs. Humphreys obliged with bright-eyed clarity.

"All the students liked John Scopes. Don't recall him ever teaching us anything about evolution. Wouldn't have mattered much, anyway…Most kids in town believed in Creation. The case was concocted in Robinson's Drugstore…Dayton wanted to beat Chattanooga to court to test a new Tennessee law.

"I have good memories about Mr. Bryan…I saw him as a splendid man. Can't say the same for Mr. Darrow…He seemed mean…standing there, snapping those red suspenders, waiting for answers to his questions…trying to make Mr. Bryan nervous."

The gracious Southern lady wound up her soliloquy by raising a glass of sparkling water with the words, "Here's to yesterday's happy times and to courageous champions of all things good!"

The softly accented message drew a chorus of "Hear, Hear!" and the slightest trickle of a tear from the eyes of the Duchess.

Jessica Saunders took her turn next, deviating from tradition. Pulling an envelope from her purse addressed to Traci Kilburn, she glanced toward Judge Stone, who eyed the proceedings while retaining his pose of judicial sobriety.

"Judge Stone, it is with some reluctance that I must confess questionable conduct on the part of five of the celebrated *Monkey II* performers. This envelope contains $50, representing a $10 investment placed by each of the errant five with the entire sum to be given to the most successful predictor of the jury's verdict. It is with no small regret that I confess to being one of the five—not because I bet, but because I lost."

Jessica's personality sparkled like a diamond tiara in the crystal chandelier's glow. Easy smiles graced the faces of all the targeted "culprits," each aware of the direction of the playful humor. Even Judge Stone's stern mask of pretended disapproval cracked. Having set the stage, Jessica delivered the envelope with the punch line.

"There is some solace in knowing, however, that the 4-1-1 decision was predicted, on the money, by the other lady in the pool—Traci Kilburn!"

Laughter flowed uproarious and jubilant. Judge Stone could only shake his head, pretending disdain, and quipping, "I won't find you in contempt this time, young lady, but would you please report, for the record, who fell farthest off the mark?"

Impishly faking chagrin, she pointed an accusing finger toward Pace Terhune. "This evening's Medal of Distinction winner, no less! Better stick with Hollywood promotions, Pace…and as for star witness Josh Ryan, I predict a future of poverty if he relies on his legendary mathematics skill to ensure a positive cash flow at the track."

Peals of laughter erupted again as Jessica handed the prize envelope across the table to the wagering winner, who endured the teasing and accepted the prize with her usual aplomb.

The next to stand, Bishop Daniels spoke with clarity of thought, rich tones rolling in rhythmic resonance. His first words endorsed Josh's *Monkey II* testimony. "Random chance parents chaos."

Then while raising his glass of iced tonic water, he pledged, "In the footsteps of Dr. Martin Luther King, Jr., Judge Frank M. Johnson, Jr., and all the other giants who have marched on ahead, I stake my life on the eventual triumph of freedom, justice...and love!"

With misty eyes, the Duchess pledged support. "I'm with you on that, my dear friend."

Pace Terhune's turn to toast arrived. He still wore an ear-to-ear grin at the surprise Medal of Distinction, an expression now enhanced by the good-natured assault upon his deficient gambling instincts.

"Here's to academic freedom, the power of a free press, and the unfettered quest for truth!"

After the cheers subsided, Pace wound up his toast with more celebratory news, offering a vintage promoter's commercial. "This afternoon I learned that big bucks are in the wind for Mid-Atlantic University: TV syndication rights to an excerpted production of *Monkey II* have been licensed at a healthy fee; and a movie based on our stage play will hit the silver screen!

"Furthermore, thanks to tonight's trustee dinner, pledges inspired by *Monkey II* will exceed one million big ones. Here's to great universities everywhere and to higher learning's adventure...Hopefully, Senator Yakley won't insist on a percentage of the take for his favorite PAC!"

Cheers at news of the boost in the university's fortunes degenerated into laughter at the expense of the senator's political image. As the snickers faded, Judge Stone took the floor, his eyes shifting to JT Thomas.

"It's my pleasure to announce Cabot, Calvert, and Stone's success in recruiting the brightest and best law grads as associates. After watching Traci Kilburn and Jonathan Thomas Daniels perform, I'm proud of our demonstrated wisdom in snaring their considerable talents."

The judge's starchy image of judicious restraint had held up throughout *Monkey II*. But now, Ed Stone enjoyed the opportunity to relax, kick back, and let down his hair.

"Grapevine hearsay normally is unreliable. But after a week's practice at bending the rules, it should come as no shock that I'm prepared to add a hearsay report from an unnamed but authoritative source relating to the future career of Mr. Daniels.

"While it may seem like an embarrassment of riches, this talented young man appears to have flummoxed a Hollywood mogul who promises a movie contract based on this week's thespian performance."

The judge showed he was no stranger to humor, suggesting, "Since you signed on with us first, Jonathan, Cabot, Calvert, and Stone can act as your agent...for a small fee, of course.

"Again, my boy, the point is clear—the law firm wants you as a law clerk today, an associate after you clobber the bar, and a partner someday. The triple-threat credentials of distinguished lawyer, stellar athlete, and heartthrob movie actor promises a one-of-a-kind résumé."

The proposal ranked as high drama for Judge Stone, obviously pleased with the flexibility of his firm's recruiting policies. Edward Anthony Stone did not fly blind. He had observed much more than JT's professional, albeit theatrical, courtroom performance. His working behind the scenes for free on the civil action brought against Josh Ryan had impressed the firm's senior partners. JT had demonstrated tireless skill and innovative brilliance.

Judge Stone completed the toast, anticipating a fait accompli. "Welcome aboard, Traci and Jonathan! Cheers to a proud moment!"

Glasses clinked decorously, adding sonorous background to a repeat chorus of "Hear, Hear!" Backslapping and congratulatory high-fives followed.

Brock Daniels glowed as JT's time to toast arrived. The barrister-to-be stood, paused, summoned his patented smile, and declared, "Here's to the jealous mistress!" Recalling his apprentice labor-negotiating skills polished at Center Hill Lake, he jested, "By the way, I hope you pay more than I earned last spring working in Tennessee for this Ryan tyrant."

More laughter of jovial camaraderie as Judge Stone assured, "The going rate, Mr. Daniels...the going rate!"

JT had intended to exploit his toast time to tease Josh and Traci about their allegedly budding romance, but fortunately for all concerned, thought better of it. An alert Traci bounced to her feet before he could change his mind. She delivered a toast of elegant propriety.

"Here's to the United States of America where all are created equal. Here's to the judicial system and law firms like Cabot, Calvert, and Stone, devoted to individual rights, irrespective of gender. And here's to the splendid traditions of our families who thrive in freedom's environment: The Kilburns; the Daniels; the Morgans; the Carringtons, and tonight, in particular, the Ryans."

More cries of "Hear, Hear!" tumbled enthusiastically while Josh tried to detect some subliminal meaning in the lady's suggestive phrase, *in particular, the Ryans*. But there was no time to interpret or to unscramble imaginary messages or subtle innuendo. The round of toasts awaited completion; he was both next and last to speak.

Josh could not remember offering a toast other than at a friend's wedding reception or some slap-happy, undergrad party. He wanted to somehow please his hitherto aloof grandmother. But his priority was to coin phrases aimed at the heart and soul of Michael Joseph Ryan. Traci and the other guests would never know that he had intended to spring news of his engagement in the course of the dinner—except for the disastrous turn of the afternoon's events. With his psyche off-balance, he studiously avoided reference to romance.

Artfully concealing the emotional batterings he had endured during a tumultuous week, Josh emulated the example of Bishop Daniels and stood tall. He lead with a fictitious anecdote.

"This week's adventure reminds me of a guest speaker who was given flowery plaudits for his success in building a reputed $370 million uranium fortune in two short weeks. When he rose to speak, he corrected public perception.

"'It was uranium oxide, not uranium; it took only ten days, rather than two full weeks; the number was $730 million, not $370 million; and to tell you the truth, I didn't make it—I lost it!'[190]

"That tale of twists and turns summarizes my *Monkey II* week."

The humorous remarks drew hearty laughs. Josh continued in a sober vein. "My toast is to family: to a grandmother who has entered my life; to a mother I never knew but love more than I can describe; and to my dad, Michael Joseph Ryan, who's been both a father and a mother to me every day of my life—my all-time best friend, and the finest human edition I've ever met.

"Here's to the Twentieth Maine and the epitome of a twentieth-century human edition!"

Michael's head tilted lower, concealing the trace of a tear. The sapphire-blue eyes of Regina Ann Carrington-Morgan brimmed.

"Thank God for that first step," she thought. "I'm ready for the second."

Instinctively, the Duchess reached out, tenderly caressing Joshua's hand, signaling her gratitude. She wanted to turn back time, so she could hug and kiss her grandson. But Josh missed the furtive stroke. Perhaps he didn't notice; perhaps he carried the burden of emotional overload; or perhaps he had pushed the envelope as far as he dared, still unready for reciprocal love for a grandparent out of touch and sight for twenty-three years.

She understood, slowly relaxing her grip and retrieving the wrinkled hand.

While glasses clinked in syncopated approval, Brock Daniels bounced quickly to his feet, followed instinctively by all but the Duchess. Raising high his crystal glass of tonic water, with closed eyes turned heavenward, the booming baritone intoned a single word of benedictory blessing:

"Amen!"

Nothing more needed to be said. The record for Friday, September 6, 1996, closed, winding down a week that had brought kaleidoscopic change to the life of Joshua Chamberlain Ryan. But compared to the surprises lurking in the next twenty-four hours, the muddy road detours of the *Monkey II* week would seem like incidental blips.

*Warren LeRoi Johns*

# VII
# Day Seven: Saturday

# *Celebration*

"FISHY" GENEALOGY

"IF MY ANCESTOR WAS A FISH,
WHAT HAPPENED TO MY INSIDE BONES?"

# 41

Josh awoke refreshed at 7 a.m., uninterrupted by calls from the mysterious midnight telephone intruder. Stretching and bending, the now-retired expert witness rejected any thought of a show-biz career, even though completing a doctorate in geology/paleontology had seemingly slipped out of reach. Still, despite his harrowing week of personal disasters, his natural optimism caused him to look forward to further adventures. Exhilaration inexplicably filled his soul.

The young scholar's computer mind raced ahead, concocting master plans for his future: Scan the newspaper ads for a reasonably priced replacement Mustang; begin a fast-track quest for some reputable university where he could patch together his doctoral program; find a temporary job paying enough to live on; and trust Cabot, Calvert, and Stone to extricate his body, soul, and bank account from the grasping claws of pending civil litigation. But even a 4.0 GPA mind proved inadequate to cope with priority *numero uno*: how to keep from second-guessing about what might-have-been in a shattered romance.

An hour before limo pick-up time, Josh began to worry about what to wear. Never having attended religious services other than routine marrying and burying rites, he had no idea what was expected attire. The young fossil hunter, most comfortable hiking rugged terrain, lacked a collection of Sunday-go-to-meeting-clothes. Accordingly, it didn't take him long to assemble a stylish mix from his sparse wardrobe: his one-and-only beige suit; freshly buffed cordovan shoes; and a light-blue oxford shirt with a button-down collar. A splashy, cocoa-brown tie accented the ensemble. It was the best he could muster as designated host for the distinguished bishop. Subliminally he hoped to tantalize Traci Kilburn, and inspire her with regret at the catch that got away.

Before he could settle into his routine and pick up September 7's *News Press*, the phone jingled. JT was looking out for his client.

"Howdy, Dar! Have you seen today's TV news?"

"No. I've been struggling to make myself a worthy host for a famous visitor to the U.S. capital."

"Then I suggest you get to channel surfing and catch a blurb of local history. I guarantee an adrenaline rush!

"And by the way, I've arranged for the limo to drop by Morgan Manor to pick up Traci on the way to your place. Wanted you to start the day spurred on by that vision of beauty..."

"But what...?"

"Gotta go, man! Get to the tube. Looks like our paths won't be crossing until 7 p.m. this evening when I'll be seeing you in Chancellor Wood Comstock's inner sanctum! Rare atmosphere and company, Mr. Ryan...better practice bowing.

"Have a good one," JT signed off before the curious Josh could extract more information.

Josh clicked the TV remote to an early-morning news update. He already knew about Hurricane Fran and the celebrated home run feat of Oriole slugger Eddie Murray. But the account of a drug bust galvanized him. The voice of an inside-the-Beltway newscaster described the crime scene.

A before-dawn arrest netted a half-dozen suspects in a cocaine ring under surveillance for the past half-year. At least one faced arson charges stemming from an alleged Mid-Atlantic University firebomb incident.

Manacled prisoners, hanging their heads to avoid being identified, passed single file toward the open doors of police vans. The barest glimpse of one downcast profile induced a double-take. The hunched figure matched the face of the owner of the Rolex he had identified to Jessica Saunders—MAU's student editor of Dr. Striker's science journal!

Josh knew from day one that Slade Lassiter was a certifiable jerk. But why on earth the guy would bother to firebomb the old Mustang still baffled. Intuitively, he assumed that the villainous Lassiter had also punctured the tires on his car Wednesday evening and made the harassing phone calls. But what about the spooky E-mail death threat? Bewildered by the implausible, Josh shook his head. Data still didn't compute. True, no love was lost between him and Slade, but criminal violence?

A toot in the driveway announced the arrival of the limo. Its immaculate pearl-white finish gleamed as though dispatched from a dealer's showroom. Leave it to JT to cover the angles. A visor-capped driver greeted him pleasantly while simultaneously opening the rear door, which led into a white, leather-clad interior.

An apparition of beauty awaited. The limo luxury ranked as extraordinary, but the radiant presence of the love of his life turned Josh's mind and heart to jelly. He was hard-pressed to separate the component parts of the dream, except to recognize that the lady wore a carefully tailored ivory suit and a coffee-colored blouse draped with a three-strand gold chain necklace. The riveting hazel eyes peeked out from a frame of exquisitely styled chestnut hair gleaming in the morning sunshine. And if that wasn't enough to discombobulate, this vision of glory greeted him with legs crossed casually at the knee, one foot beckoning provocatively in a come-hither cadence.

Had she been a wax figure, the awestruck Josh would have been impressed. Living, breathing, vibrant creature that she was, it required all he could do to manage a nonchalant "Good morning" without his voice cracking or dwindling to a whisper.

"Welcome aboard, Mr. Ryan. I hope you're looking forward to this day as much as I am."

"Too much for us mortal types," thought Josh. "I wonder how many proposals she's turned down in the past month?" Truthfully, he looked forward to the day also, determined to find joy wherever he could. At least they could jointly celebrate *Monkey II's* triumphal run.

As they pulled out of Fort Ryan's main drive, Josh took a stab at statesmanship, advising the driver to head for the Hay-Adams just across the park from the White House. Too polite to reveal that he had a detailed written schedule of the day's events tucked securely in his pocket, the driver acknowledged the instruction, assuring, "We're on our way."

Josh envisioned a banner day for small talk. Traci obliged, wasting no time on light chatter. "Now tell me the truth, Mr. Ryan. Is it true that you have never attended an honest-to-goodness regular church service?"

"Not that I recall."

"Then if this is your first time, you picked a winner. The incomparable National Cathedral took most of a century to build. And you know Bishop Daniels' reputation as clergyman orator."

"I must qualify as a despicable sinner for not paying more attention, but I'm not particularly attracted to weekly rituals, having to pony up money to ensure that I get blessed, or holier-than-thou blacklistings that disfellowship dissenters. I don't see how a God of love stands ready to condemn erring souls to eternal hell for resisting coercive religious systems!"

The fervent observation came without apology or rancor...more of a blunt warning shot across the bow! Although Traci had attended church off and on most of her life, she was not the least bit defensive. Resting her hand on his for a moment, she said soothingly, "Maybe it's something like science, Josh. There are all kinds of theories—some possibly right, others wrong. Sometimes God takes a bad rap when self-righteous believers make him sound like an arbitrary tyrant rather than the embodiment of infinite love."

As Josh nodded assent to the logic, she returned to *Monkey II* issues.

"You clearly reject the fact-free science attributed to Charles Darwin, yet declare that you are not a creationist."

"It's true...I've never seriously studied creationism!"

"I don't mean to cross-examine, but if you don't buy the Darwinian line, and have never given creationism the time of day, what's your alternative?"

"Maybe someday I'll check out some rocks or genes in quest of an *alternative.* For now, it's enough to realize the Darwinian circular trail leads to never-never land."

Traci listened with interest. Encouraged, Josh continued, pleased to steer the conversation in any direction other than Friday afternoon's fiasco on the banks of the Hawkings River!

"Gullibility is a social phenomenon, vulnerable to exploitation. That's how a Charles Robert Darwin can be revered as secular religion's icon."

"I'm not sure I understand."

"Independent thought gets blindsided by a three-pronged collective force: recognition of some urgent human need; dogma delivered in a bells-and-whistles package appealing to the senses and emotions, and adaptable to individual need; and coercive peer pressure promising personal security through membership in a community that requires collective obeisance.

"Individual dignity surrenders to arbitrary power—often exploited by a charismatic cult hero. Hitler employed the formula with ruthless cunning, boasting scornfully that government is fortunate when people don't think."

"Surely you don't mean to compare the mild-mannered Darwin to Adolph Hitler?"

"Of course not. My apologies to Charles Robert. I'm only saying that *Homo sapiens* tend to rally in support of a cause perceived as worthy, but are often deceived by a manipulative power-broker touting phony political, religious, or scientific dogma!"

"You sound wise beyond your years, Josh Ryan!"

Josh sensed the playful toying with his ego, but didn't mind. The lady could project starry-eyed rapture when the occasion demanded. She proved equally adept at pushing his buttons and refocusing the exchange.

"Is there a bottom line to this insight of yours? It seems I'm about to be blessed with some heavy stuff from a scientist-turned-philosopher—or even preacher! Have at it, great thinker!"

Undeterred by her teasing, he unloaded one last gem. "Free choice and expression are the key to discovering truth in the marketplace of ideas. Surrender to the whims of a dictator propagates error. Throw in a touch of stardust, and a cult-like power structure is off to the races. Some spot it in fanatical fringe religious groups. I see it in mega-evolution's uncorroborated suppositions."

He took a breath, using the clincher to exonerate his break with Striker. "It takes old-fashioned guts to escape the tyranny of tradition!"

"Sounds reasonable, Josh. You'd better watch out, though! Looks to me as if you are caught in a philosophical no-man's-land. Keep talking like that, and both religious and scientific groups may vie for the right to burn you at the stake to snuff out this dangerous heresy of independence. You may need a lawyer to defend you...for the rest of your life. As you can imagine, that benefit comes with a price."

As Josh quizzically searched her face for some obscure message implied in the last remark, she quickly added: "Men are so blind!"

The words entered one tin ear and came out the other, leaving Josh without a clue to her subliminal meaning. The rest of the trip downtown inspired banter ranging from the mundane to the ridiculous. Preoccupied by mutual wit and wisdom, time flew.

During a brief interval of silence, Josh reviewed the bishop's bio.

*******

Brock Daniels had graduated with honors from Morehouse College in time to join the U.S. Army and to confront a ruthless Nazi scythe slashing across Europe. He emerged from Air Corps flight training with a lieutenant's commission and a place in history as a Tuskegee Airman. The cool-handed pilot proved gutsy in combat, chasing Hitler's fighters across continental skies.

Although he survived the war unscathed physically, he came home with burdensome psychological baggage. Returning stateside as a combat hero, bitter realities stung!

Wearing the proud garb of an officer of the United States Army Air Corps was not enough to gain access to an Army PX—a ban based solely on skin color. Not so coincidentally, German prisoners-of-war routinely enjoyed the courtesy of admission. Bristling at the obscenity, he vowed to fight, only this time his weapons would not be those of steel and gunpowder. Brock Daniels dedicated the remainder of his time on earth to battles of the spirit! Propelled by the stark vision of man's inhumanity to man, he elected to put on *the whole armor of God* and to serve humankind as a man of the cloth!

Brock Daniels' eloquence was forged in the front-line trenches of experience. Blessed by a resonant baritone voice, he held audiences enraptured with his powerful messages extolling goodness, equal justice, and above all, inner peace through unselfish love.

At the beginning of his ministry, it took but a few months before major metropolitan congregations were competing for his services. The day he announced his farewell to the three small, rural congregations comprising his first pastorate, he found himself blessed by an unlikely sermon, compliments of his faithful flock of parishioners. Impromptu tears greeted his departure at the church door following his somewhat flippant announcement of the planned move. The stark lesson hit him head-on: While he had given his intellectual best, the caring parishioners had returned heartfelt love, without strings—the core value of his own words.

Brock Daniels never forgot that lesson. From then on, his heart embraced this as the central theme for a widening ministry.

More than once he marched side-by-side with Martin Luther King. He crossed the Selma bridge with someday Congressman John Lewis, in sync with the non-violent display of the inviolability of the human spirit. The long and dreary journey to justice withstood and overcame police batons, dogs, and fire hoses. Many times, the fearless preacher led the marches and like the dauntless Apostle Paul, spent more than one night in cold, dank jail cells—guilty of nothing except harboring aspirations of equity and goodness.

*******

On schedule, the driver rolled under the covered entry to the fabled Hay-Adams, in close proximity to the White House. The always-punctual Brock Daniels waited patiently with luggage in hand. Tipping the doorman generously for stowing the luggage in the trunk, the bishop hopped nimbly into the limousine, belying his seventy-five years. The jovial cleric greeted his host and hostess with sage wit.

"What a pleasure to be escorted by the finest-looking couple I know."

Unencumbered, except by a New English Version of the Bible cradled in his right hand, the bishop added, with a twinkle in his eye, "You see, Josh, I've brought along my library of sixty-six books in this compact package for your edification."

Knowing that his young friend would recognize the gesture's intended good will, he added, "And since you say this may be your first-ever sermon, I promise to make it easy for you. Rather than discussing the plight of a *confessed sinner*, I plan to share my understanding of *The Truth About God*."

"Whatever you say, Sir. I will listen to every word with total respect despite the woeful limitations of one who is religiously uninitiated."

"Fair enough," came the resonant response. "This promises to be a day of celebration indeed."

The trio approached National Cathedral with fifteen minutes to spare. The passengers exited at the side entrance, with Bishop Daniels lending a hand to Traci as she stepped to the pavement. He next stopped to point up to the window with the moon rock. Then with a wave, he strode off briskly to join his colleagues. Josh escorted Traci into the sanctuary, pretending to be familiar with the passages of the cathedral. After being shown two seats by a helpful usher, they located the moon rock, centerpiece to a riot of colorful stained glass.

Nearly a thousand dignitaries, representing a full spectrum of religions, occupied the reserved front seats. Families and guests like Josh and Traci were seated behind them and off to the side. Josh responded to the muted ambience with a whispered wisecrack.

"Think how confused God must be when he looks down on this array of diverse religious leadership, many claiming exclusive possession of the truth about God."

As though reprimanding a child, Traci slapped his knee with an admonishing "Shhhh..." Duly chastened, he behaved as ordered.

The convocation had been underway since 8 a.m., dealing with agenda items in preparation for Bishop Daniels' keynote address at the 10 a.m. session. Promptly on the hour, the mid-morning segment came to order with a hymn, followed by a prayer recited from a book. The student scientist paid little heed to liturgy's unfamiliar sights and sounds. So far, Josh had heard nothing motivating him to swap his club seat at the Redskins' Sunday home games for a church pew. But he had to admit that the intoxicating scent of Traci's perfume did give the religious ritual an unfair advantage.

Josh sat up attentively when a retired astronaut stood up for the Scripture reading. Catching sight of the cosmic explorer, Josh recalled the recent comment of a young Arizona State University astronomer and co-creator of the computer-enhanced *Pillars of Creation*—a stunning graphic depicting the Eagle Nebula, 7,000 light years away. Taking a cue from Astronaut John Glenn, the scientist had marveled at the creative power of God.

*"My experience is that the more I learn about the universe, the more I believe there has to be a Creator...just because it's so complex."*[191]

The reference sounded faintly similar to the astronaut's thought-provoking reading this morning, straight from the pages of Job, one of the sixty-six books toted reverently by Bishop Daniels.

*"'Can you fathom the mysteries of God? Can you probe the limits of the Almighty? They are higher than the heavens—what can you know? Their measure is longer than the earth and wider than the sea.'"*[192]

The reading left the young scholar with questions of his own, rolling about in his mind like agate marbles in a child's box of keepsakes. "Who was this guy Job, anyway? When did he write about those *mysteries*? Did he ever see God? Speak to him? What evidence did he have? For sure, I don't have the answers!"

The passages from the pulpit continued while Josh pondered.

*"'The heavens declare the glory of God; the skies proclaim the work of his hands.'*[193] *'In the beginning was the Word, and the Word was with God, and the Word was God. He was God in the beginning. Through him all things were made; without him nothing was made that has been made. In him was life, and that life was the light of men.'"*[194]

Josh's meanderings were interrupted by a verbose presenter monotonously describing the keynote speaker's storied reputation—a grand old man of the cloth, a captivating orator, and a champion of all things good. Once the extravagant embellishments had been exhausted and Brock Daniels moved to the microphone, the assembly rose as one person to give him a standing ovation.

Gesturing to the delegates to be seated, the speaker grinned and shared a classic Daniels' line borrowed from Ronald Reagan, "I'm glad to be here today...But of course, at my age, I'm glad to be anywhere!"

His homey touch captured audience rapport, holding the delegates spellbound for the next twenty minutes. Despite himself, the ex-*Monkey II* witness listened intently to the piercing rhetoric. Coming straight from the orator's soul, *The Truth About God* was delivered without a single note.

The bishop's vivid words pictured Paul of Tarsus confronting Athenian philosophers on their home turf, adjacent to a public altar erected *To an Unknown God*. As Bishop Daniels quoted catchy bits of Paul's monologue from memory, the two-thousand-year-old drama unfolded in modern language. The words thundered, fresh and rich in meaning.

" *The God who made the world and everything in it is the Lord of heaven and earth...He is not served by human hands, as if He needed anything, because He Himself gives all men life and breath and everything else.* "[195]

The bishop warned sternly, "Human ranks bulge with the gullible who bow to grotesque caricatures of God. Terrified victims of this fraud cower in fear to a Supreme Being unjustly portrayed as harsh, coercive, arbitrary, and vengeful. The misdirected pay homage to false gods of prestige, wealth, and raw power.

"This counterfeit faith does not approach God's truth as preached by Paul."

Bishop Daniels' vision of God appealed to the ecumenical majority. Still, several hard-liners, cherishing preconceived notions of a tyrannical deity, winced in discomfort.

Disdaining microphones, the bishop's ringing tones echoed throughout the cavernous cathedral, reflecting his commitment to the principles of the message. Brock Daniels painted a compelling picture of a Supreme Lifegiver who is infinite, eternal, all-powerful, forgiving, the essence of unchanging love, and the Author of all science.

The articulate theologian had no use for mega-evolution's mutation explanation of humanity's origins, dismissing the idea contemptuously.

"Evolution's mantra promising progress toward perfection perpetuates the original big lie foisted by the arch-deceiver. Mutations, the alleged engines of evolution, degrade the genome and cause diminished genetic capacity—a feeble foundation for the lofty claims of Darwinism."

Josh listened, intrigued. He took the words as a signal that Bishop Daniels shared his sentiments. The reference to a "big lie" didn't compute, but he felt reassured by the bishop's unimpeachable integrity.

"Truth makes free!" Bishop Daniels continued. "Fraud and fanaticism cringe at God's truth. Fraudulent fables obscure and deface the pure; ritualistic fanaticism complicates the simple—*Love God; love your neighbor!* Anything more is redundant; anything less is inadequate. The formula is key to peace: it is internal to the person; external to nations.

"People say, 'Bishop, you've had your share of hurts and downers, but you always seem happy and upbeat! What's your secret?'"

The courtly old gentleman beamed, basking in the peace that passes understanding.

"I assure them, 'It's hardly a secret! I don't pursue happiness, I pursue goodness and God's truth. Happiness tags along as a dividend!'"

The speaker's sincere logic and passionate rhetoric aroused the curiosity of the theologically uninitiated. But nothing prepared Josh for the next pronouncement.

"While it may surprise and shock some sensibilities, I'm inclined to agree with Charles Darwin's description of conflict, cruelty, and struggle as a part of the natural world. Pervasive evil tramples lives even as we speak."

Enlivened by the rapt anticipation, Bishop Daniels continued.

"But I disagree heartily with his explanation! Darwin's *survival of the fittest* mentality worships at the shrine of the god of power, rewarding so-called supermen for their ruthless brutality.

"To the contrary, unrestrained force destroys, causing decline and deterioration in the quality of both human life and the environment. The only antidote is the power of unselfish love!"

The bishop punctuated the pronouncement with his personal perception. "I'm convinced life exists beyond Planet Earth, some of it superior to our own. Moreover, I believe that a life-and-death cosmic struggle rages in which a cunning adversary, with his rebellious evil forces, battles the forces of good for control! In my judgment, all aspects of life on our planet have been tainted by that spiritual conflict."

The bishop's upbeat sermon touched upon his basic philosophy. No gloom and doom for him, but a people-building mission—destroying enemies by fashioning friends through forgiveness and love, inspired by the command of the Supreme Lifegiver.

A collective chorus of "Amens" floated heavenward, only to be absorbed by the vaulted ceiling. Josh and Traci exchanged smiles, as she unobtrusively rested her hand on his knee.

"Maybe I've been missing something in this church thing," he thought.

By now, Bishop Daniels had raised the emotions of his hearers to a concluding high. Pausing en route to the clincher, he ad-libbed a P.S. that Josh sensed was meant for his ears.

"The two professions that should be the most comfortable with the truth about God include my own—the clergy—and the scientific disciplines. Both science and religion are vulnerable to a superstitious ritualism that results in paralysis of mind and tyranny of spirit. Calcified bias in either discipline builds barriers that are impervious to change."

"Wernher von Braun said it best. *'I find it as difficult to understand a scientist who does not acknowledge the presence of a superior rationality behind the existence of the universe as it is to comprehend a theologian who would deny the advances of science.'"* [196]

Listeners familiar with Bishop Daniels' homiletics sat up straighter, curious at this aside.

"If the disciplines appear contradictory, one or the other, or possibly both misunderstand truth. Neither stands alone. Religion without true science does injustice to the Author of science. Science without true religion fosters a circular fantasy."

The bishop pulled no punches. Using the words of Phillip E. Johnson to chide an *intellectual imperialism* that exalts *materialism* and ignores overwhelming *discontinuance* in the fossil record, he warned of unreliable consequences, unworthy of the scientific method.

*"'Darwinism is based on an a priori commitment to materialism, not on a philosophically neutral assessment of the evidence. Separate the philosophy from the science, and the proud tower collapses.'"*[197]

Reinforcing his *Truth About God* message, Bishop Daniels declared that "true science and true religion stand inextricably commingled, mutually complimentary! It shouldn't be surprising that forty percent of *'American scientists believe in a personal God...a deity to whom they can pray.'"*[198]

His basis for belief galvanized listeners! With gusto, he quoted George Will: *"'The idea of God is slightly more plausible than the alternative proposition that, given enough time, some green slime could write Shakespeare's sonnets.'*[199]

"It is no stretch to believe in the existence of a cosmic intelligence superior to our own. Identify this wisdom with infinite power, combined with unlimited love, and the majesty of God warrants reverent worship. The gateway to human understanding requires acceptance of the threshold *Truth About God*: It was He, the Almighty, who created the Earth and all living things—a long time ago!"

With that, Bishop Daniels left the pulpit, waving "Thank you's" to a sustained ovation. Normally he didn't hit and run, but an evening appointment in another state allowed only enough time for a leisurely lunch and drive to National Airport. Josh and Traci slipped out a side door to join the clergyman at the waiting limo. At his request, they headed to the Hay-Adams and a gourmet lunch in the shadow of 1600 Pennsylvania Avenue.

The combination of proximity to the seat of international power and the heady wisdom of the gregarious ambassador for good proved intoxicating. Chatting amiably, the trio explored the day's events within the context of science, religion, and history while other diners smiled in recognition or sauntered over to shake the hand of the affable Bishop Daniels, friend to all, stranger to none.

Scanning the right side of the menu, Josh reacted with barely suppressed panic, realizing that the cash he had brought fell embarrassingly short of the bare minimum necessary to pick up the tab at the swank Hay-Adams. Catching the consternation on the face of his young host, the bishop put him at ease.

"Don't worry, young fella, this is my treat all the way. This place holds happy memories for me. I just wanted to share them with the most dashing young couple I know."

Traci smiled at the gracious hyperbole. Josh tried not to choke, managing a nonchalant, "I've always respected your judgment, Sir."

The bishop continued with an impish grin, "For what it's worth, despite my tight travel schedules, I still squeeze in time to perform weddings for special people."

His dark eyes flashing, he flipped out a worn, leather-bound date book, pretending to leaf its pages with studious intensity. After his fingers flipped through, he shared his findings in triumph.

"Aha! Just as I thought. Looks like there's some flexibility in the summer of 1997!"

Josh gulped as Traci stayed cucumber cool, coyly fluttering her eyelashes. "Someday, should I marry, I promise to pick a date convenient for you to perform the ceremony."

A properly bedazzled bishop extended his hand to seal the covenant. "That's a deal!"

The trio hadn't yet noticed Attorney Marcus Brogan huddled in a far corner of the restaurant, where he was finishing a strenuous consultation with client Karl Striker. Exiting, Striker bypassed the bishop's party. Trailing at his heels, Brogan wheeled abruptly and headed for Brock Daniels' table, inviting confrontation. The tawdry barrister couldn't resist the opportunity to flaunt his attack-dog style. Without breaking stride, he taunted, "When all else fails, I see you turn to religion, Mr. Ryan! Does the bishop administer last rites?"

The brazen affront drew a blank look from Josh and Traci. No stranger to confrontation, the bishop responded with softly spoken counsel. "The Lord works in mysterious ways, Brother Brogan! Don't you agree?"

The deflated barrister stammered for an answer, then bolted for the door in pursuit of his client, confused at the *brother* business.

Brock Daniels grinned ear-to-ear. "Works every time! Maybe Mr. Brogan will end up another Saul of Tarsus...The Lord works in *mysterious ways,* you know!"

Still speechless, his luncheon guests shook their heads in admiration.

Anxious to avoid further wedding talk, Josh steered the conversation back to the bishop's *Truth About God* remarks.

"Thursday evening, Brogan challenged my belief in the existence of someone I've never seen. In fairness, Mr. Daniels, have you ever seen God?" Speaking sincerely and without guile, Josh had no intention of baiting his distinguished friend.

"Rarely in a face-to-face encounter, if that's what you mean. Ordinary mortals would likely be overcome by the incandescent glory that blinded Saul of Tarsus on the Damascus Road. But all those sixty-six books I showed you contain a written record inspired by God.

"You need to read about them to understand. It's the road to wisdom! The very rich and good man Abraham, living in the city of Ur on the Euphrates River, is one. Another, Moses, had been picked as heir-apparent to the Egyptian throne. Peter was a brash, diamond-in-the-rough fisherman. Saul of Tarsus misunderstood the truth about God until he became Paul the Apostle, a tentmaker by trade, a fearless first-century faith crusader by mission!

"Perhaps it's something like television. You see the image on the screen and you believe that those people exist even though you can't see them face-to-face. The authors of those sixty-six books collectively agree that the Great Lifegiver created the human edition in his own image, each of us empowered with the ability to accomplish good things if we choose."

"So you believe in power you can't see?"

"Just as you, a scientist, believe in the unseen. Evidence crops up everywhere. You can't see television or radio waves, but you cheer for the Redskins and Orioles thanks to the sights and sounds reproduced on screen or by your radio. You cannot see gravity, but its existence prevents our floating into outer space! Ultra-wide band signals pulsating at ten-trillionths of a second can't be seen but are about to change communications technology forever.

"Except in a smoggy city, air is invisible—but we can't live without it!"

Pointing to the ceiling, Bishop Daniels reminded the couple of the soft glow of indirect lighting in the Hay-Adams dining room, available night and day at the touch of a switch, compliments of invisible electric power.

"Human eyes are blind to magnetic force. Yet the compass needle responds to its magic. Ultraviolet and infrared rays work round the clock, silent and out of sight. Photosynthesis powers nature's life-support cycle. Sound waves elude optic detection. Even the wind that sends ripples on a lake, powering a sailboat, signals unseen power."

Appreciative of his guests' courteous interest, the preacher continued his private sermon with a favorite anecdote.

"X-rays pack a wallop we neither see, hear, nor feel, but they mark trails. They can penetrate the secret recesses of the human body or peer into a sealed treasure chest. Harnessed radiation also fights cancer, as mathematically targeted proton beams destroy hard-to-reach carcinomas.

"One old-timer undergoing treatment at a California medical center using a proton beam accelerator, complained to the supervising technician, 'You must be cheating me. I can't see or feel a thing. How do I know I'm getting my money's worth?' The technician responded: 'By the healing results.'"

Reflecting pensively, Bishop Daniels concluded with emphasis, "The same Power Source heals the human soul!"

"You obviously believe in God?"

Joshua's question lingered in the air as Brock Daniels pondered.

"I'm no fool, Josh!"

Josh missed the twinkle in his eye, so tried to backpedal. "Oh, I'm sorry, Mr. Daniels, I didn't mean to imply..."

The clergyman interrupted with a gracious chuckle and a quote from one of his favorites among the sixty-six books. "No, no, of course you didn't. I just borrowed an answer from an ancient sage who didn't mince words: '*The fool says in his heart, there is no God.*'[200]

"Believe? Talk about evidence! Everywhere in nature I detect the fingerprint of God! On my tally sheet, the bottom line score is: God, 100; random chance, zip!"

The bold analysis captured since it came from a distinguished veteran of the cloth who lived in the real world.

"Faith is a core value of religion. Someone suggested that '*for the believer, no proof is necessary, and for the unbeliever, no proof is possible.*' But let's be fair. Doesn't it take a much greater stretch of faith to buy into a secular religion built on random chance chaos?"

"Sounds logical, but does logic alone prove God?" Josh asked.

"Evidence staring us in the face can be sifted through logic! I like the concept you shared with Jonathan last spring during your dig in Tennessee. When scientists retrieve an artifact of stone, brass, bronze, or gold depicting a replica of the human edition, no rational archaeologist would claim that such a delicately designed work of art evolved by random chance without the touch of an artisan. Does it make sense that an even more beautiful work of art, a living model, emerged accidentally from chaos without benefit of an intelligent designer?"

Josh appreciated the bishop's skill at sharing ideas without sounding preachy. Looking at the gorgeous Traci, he felt disinclined to argue with the clerical logic pointing to a Master Artisan!

Believing he had delivered sufficient food for thought, Bishop Daniels signed off with a smile. "You youngsters will have to forgive us preacher types for unloading a lifetime of learning in a single, top-heavy dose." With a twinkle in his voice, he quipped, "Politicos and lawyers are no strangers to verbosity, either!"

Josh realized that JT had it right: When Brock Daniels speaks, the world listens!

Glancing at his watch, the bishop asked for the bill and plunked down a generous tip. Rising to return to the limo, he urged, "Since we have a few extra minutes, if you kids don't mind, I'd like the driver to take us by the Vietnam Memorial. I hope you understand that it's hard for me to go by without taking time to salute the memory of Parker Daniels."

They understood.

# 42

Westward-bound on Constitution Avenue, it took the limo but moments to deliver the trio to the Vietnam Memorial, a stone's throw from the Hay-Adams. Except for Josh's instructions to the driver, few words were exchanged. The mood was somber. Josh thought about the time, five years earlier, when he and his dad had visited the landmark stretch of black marble to search for Parker Daniels' name. The memorial stop would be a first for Traci.

Tourists strolled the west end of the Mall singly and in groups—all ages, all cultures, all respectful. A pilgrimage to the national shrine resurrected records of individual courage and heroism, memories tarnished only by acts of political deceit.

Before descending the gently sloping walkway along the face of the wall, Bishop Daniels beckoned his companions to join him on a park bench that enjoyed a panoramic view. The most-visited monument in the nation's capital testified mutely to the thin line dividing life from death.

The bishop mused aloud, addressing no one in particular. "We all ride life's escalator, eventually disappearing from the view of those behind us, moving upward to take our place. Civil War historian Bruce Catton compared death to boarding the *night train* to sleep in peace—waiting for the conductor's call *in the morning.*[201]

"The longer we live, the shorter the time left for the journey...the stronger the trip's pull to darkness."

At this, the usually ramrod-straight cleric sighed and his shoulders sagged. No longer jovial, the aging giant fell silent. Damp-eyed, he pointed past a patch of splendidly manicured trees.

"Out yonder, I remember standing in the ranks behind Martin when he inspired the nation with his *I Have a Dream* speech. For my money, it lives right up there with Lincoln's *Four score and seven years ago* commemorating the valiant dead at Gettysburg, and the eloquent message of freedom and peace attributed to *Thunder Rolling From the Mountain*, a native American we honor as Chief Joseph: '*From where the sun now stands, I will fight no more forever!*'[202]

"The ultimate war is not between nationalities or races but a struggle between good and evil. Merchants of hate devour ambassadors of peace. Evil agents strike down the good guys—Lincoln, Martin Luther King, Gandhi, Sadat, Rabin—but truth lives forever. Others pick up the banner.

"If Martin were around today, I imagine he'd still be going strong—a human dynamo. That compelling dream of his...I guess we've made a bit of progress...but it looks like a long row to hoe..."

The bishop's voice trailed away. His gaze focused on some remote time and place, oblivious to any other presence. A living legend, he was the embodiment of the biblical adage that the *"truth shall set you free."*[203]

Abruptly, he shifted to the present and fixed gentle eyes on his two young companions, who had been absorbing every word. Returning to his jovial self, he said, "I don't want you two youngsters to get the wrong idea. I didn't come here today to mourn for Parker. I came to celebrate his life. I'd swap his Silver Star a million times over for just one split second to hug on him and let him know how proud I am that he was my son and how proud he should be of his son, Jonathan Thomas Daniels."

Before going down to walk the sad length of the monument, the grand old man summoned the recurring nightmare of his own brush with death long years past.

"Whenever my heart aches for Parker, I think of the pain I caused the parents of a young German Luftwaffe pilot I shot down in World War II combat over Europe. May his soul rest in peace—I never learned his name. He had a mom and dad. I still remember that day.

"All Tuskegee Airmen were hotshot pilots, itching for a fight. Most of the guys were born athletes, blessed with lightning fast reflexes—the Michael Jordans of another time and place. State-of-the-art training for flying combat polished razor-sharp reactions. By contrast, as the war dragged on, the cream of the Luftwaffe melted from the skies. Still, we faced tough, raw talent.

"One day a would-be Red Baron locked on the tail of my best buddy. Instinctively, I zoomed in, found the German ace in my sights, and blasted away, setting his plane ablaze. It seemed too easy. As it burst into flame, I pulled alongside the cockpit, so close I could see his face. The instant inferno prevented bailout. Before his plane disintegrated into a plummeting red-orange ball, he turned in my direction, half-saluted, then disappeared forever."

Brock Daniels could never erase the picture of the flame-shrouded, stricken face, an anonymous casualty to war's voracious appetite. His face rested pensively in the palms of his hands. When he looked up, his expression was melancholy.

"After the war, I tried to learn his identity so I could speak to his parents. But all trails led nowhere. Perhaps it's just as well that I never found out.

"For what it's worth, I'm dredging up this memory so you will understand a couple of things: First, that combat revolutionized my life. It's the reason I switched to a career as a clergyman and abandoned a personal dream to fight for justice as a lawyer."

Looking squarely at Traci, he added, "I'm proud that you and Jonathan have chosen to go into the legal profession. But after the war, I felt compelled to walk in the footsteps of the Man of Peace.

"The other thing you should remember is that no agony compares to the loss of an only son or daughter. Never again could I pull a trigger that would inflict that kind of hurt on the heart of a human parent."

Josh interjected, "Do you believe in a hereafter?"

"You bet I do, my boy. Those sixty-six books are loaded with assurance that the just will live again in a re-created Planet Earth. For those of us who believe original life came from the Supreme Lifegiver, it takes no greater leap of faith to believe in an eternal hereafter than to accept the reality of our here-and-now existence."

"The other day I saw a poster picturing the nineteenth-century German philosopher Friedrich Nietzsche who embraced Darwinian gospel. That rabble rouser lived fifty-six years, passing on to hopeless oblivion at the turn of the century. The poster spoke volumes: *Nietzsche claimed 'God is dead.' Today, Nietzsche is dead!*

"Seems to me, if you refuse belief in a Supreme Being, you may end up like Nietzsche, futilely looking for answers while believing in contrived speculation that leads nowhere."

As the congregation of two digested the thought, Bishop Daniels added a punch line. "Atheism traps you in a belief system that prevents acceptance of Creation and precludes spiritual renewal in your heart. If you reject evolution's dry-bones philosophy, you implicitly acknowledge God's existence—whether or not you've joined a church.

"It's a theological exercise: Either you place your faith in materialism, with its ticket to nowhere, or you embrace a reason-based faith that recognizes the evidence for a Supreme Being. Either way, it's a religious choice—worship the secular trappings of materialism or make a spiritual commitment to the Creator."

Traci listened respectfully, but Josh challenged the bishop. "Do you honestly believe in a future eternal bliss of some sort? Or a hell that condemns evil to a cauldron of never-ending torture?"

*"Yes and no!"* he replied. Josh looked puzzled.

"Let me explain. If God created original organic life by the power of his spoken word, is it unreasonable to conclude that he still has the power to re-create? If it's rational to believe that the universe exists compliments of a Master Designer who loves his creation—then eternal life requires no intellectual stretch!"

"But what about all that stuff like *'the smoke of their torment rises forever and ever?'*[204] That doesn't strike me as justice, much less love," said Josh.

The bishop continued, "For once, you and Charles Darwin sing from the same page. Disillusioned and a touch bitter at the loss of loved ones, an elderly Darwin could not understand *'how anyone ought to wish Christianity to be true,'* complaining that the Bible's *'plain language...seems to show that the men who do not believe, and this would include my Father, Brother and almost all my best friends, will be everlastingly punished. And this is a damnable doctrine.'"*[205]

"So what about everlasting punishment?"

"The biblical language comes loaded with graphic symbolism that is clarified by context analysis. Those who intentionally choose evil, voluntarily inherit eternal death. The *hell* that awaits them is actually death and permanent separation from God—the everlasting consequences of an evil life. God plans to banish evil *forever and ever*!

"By contrast, the godly triumph and are rewarded by the gift of life everlasting, bestowed by the Giver of all life. Good people can make mistakes, take temporary detours on life's highway. It's one's basic direction that counts. Cutting through all the labels, there are only two categories of people—those committed to all things good, and those devoted to evil."

The two listeners silently pondered the bishop's words. To Josh, it sounded as foreign as ancient Greek. As much as he respected and admired this icon of truth, the message seemed strange and baffling. But one thing was certain: Both his mom and dad, Leslie Anne and Michael Joseph Ryan, represented all things good—they would never suffer damnation and banishment to eternal torment!

The bishop had finished sermonizing. Taking a deep breath, he rose, stretching his frame to its full stature, and motioned Traci and Josh to follow. Cradling an elbow of each of them with one of his giant palms, he said, "Here—let me show you."

Steering his companions down the grassy slope to the walkway running the length of the monument, the bishop guided them to the midst of the crowd, most standing silently or with heads bowed. The wall of remembrance exhibited a roster of the fallen engraved artfully in the reflective black stone.

Moving with the sober faces, they examined decorative mementos strewn at the monument's base by stricken loved ones: a Purple Heart; a wrinkled, tear-stained letter; a tiny, faded American flag; a cross cast in pewter; a twisted yellow ribbon; a Star of David; even a child's teddy bear. A collection of these mementos, harvested at intervals from the site, was stockpiled safely for some as-yet-unknown future exhibit.

Bishop Daniels led the way, on past the apex of the monument, where the walk began to rise slowly to the east in the direction of the Washington Monument. Some families could be seen placing paper over a name and rubbing with pencil or chalk to produce a cherished reproduction to carry home. Near the eastern end of the walkway, the bishop stopped, slowly raised his hand, and with his index finger pointed to one of the 58,000 names.

"Here it is! Captain Parker Daniels!"

He caressed each letter with his fingers, one at a time, as if ceremoniously retracing the name on his heart. Finally, he covered the etching with the palm of his brawny hand. Leaning there reverently, he closed his eyes in silent meditation while his two young companions hovered empathetically, arms overlapping the bishop's broad hunched shoulders. As always, he choked up at this spot, but the days of weeping had faded long ago. Captain Parker Daniels, Annapolis grad assigned to duty in Vietnam, would not have wanted his family to suffer unending anguish. Brock honored his son's sentiment by celebrating the miracle of his life.

Three or four minutes passed before the bishop dropped his hand, stepped back between Josh and Traci, entwined his sinewy arms with theirs, and moved toward the waiting limo.

Walking through the inspirational setting, Josh kept thinking, "I used to sit on this guy's knee as a kid listening to his stories, playing catch, and responding to his laughter. All my life I've been in the presence of greatness, and took it for granted." Caught up in the gravity of the experience, he asked:

"You expect to see Parker again someday in that everlasting life setting, don't you?"

"As I live and breathe!" The words came with a triumphant smile and reassuring hugs. No further words were exchanged as the trio silently contemplated the meaning of life.

As they climbed back into the limo, it headed west across Memorial Bridge in the direction of Arlington Cemetery, then down the George Washington Parkway to the redesigned and seemingly always-under-construction National Airport. The conversation returned to the upbeat and the future. Bishop Daniels was not the least bit bashful in exploiting clergy prerogatives to give some unsolicited counsel to his captive audience.

Turning to Josh, he observed, "You're a lucky man to have this young lady as a very dear friend."

Josh blushed his full agreement, fresh with regrets that the relationship held meager promise.

"And as for you, Traci, recognizing the flaws inherent in the male gender, this one is just about as good as they get."

Josh thought he saw the faintest hint of color spread across Traci's cheeks as a "what can I say" grin wreathed her face.

The bishop laughed joyously at their reaction. He added a wry, unconvincing disclaimer. "Of course, far be it from me to try to influence the normal course of nature." And then, looking straight at the squirming Josh, he added, "Remember, some of us don't think much of leaving things to blind, *random chance!*"

Bishop Daniels knew they wouldn't take offense. He surmised wisely that a prod here and there might not hurt the cause of true romance. But events of this day foreshadowed far more than anything even the canny bishop could have imagined.

Good-byes at the airport were hearty. Rib-cracking bear hugs were exchanged all around. Grateful for the good man's company, Traci signed off with an affectionate kiss and a pat to the cheek of the grandfatherly figure.

Returning to the limo, she confided to Josh, "Bishop Brock Daniels exemplifies goodness...and a heavy dose of greatness!"

Josh squeezed her hand in agreement.

# 43

After the last farewell, when the limo had begun weaving through the mad rush of weekend traffic at National Airport, it was nearly 1 p.m. The driver edged back onto the road bordering the Potomac River shoreline—in the southbound lane. In short order they had missed the turnoff to the Maryland-bound George Washington Parkway and were cruising toward Olde Town Alexandria, Virginia.

Josh protested, "Hey, wait..."

Before he could get the attention of the driver, he felt Traci's restraining hand on his arm. Softly she urged, "Relax."

"What do you mean, relax? I'm due to meet Judge Stone at the university in a few hours. Besides, this trip is costing JT an arm and a leg. The sooner we check in this limo, the better."

"Trust me," she replied, smugly.

"That's a powerful tall order when we're headed I know not where to see I know not what and will return I know not when—and to pay the overtime mileage I know not how! Other than that, Your Majesty, I trust you completely."

Enjoying the power trip, she couldn't resist laughing at his plight. "You're the precise scientist, always wanting to understand everything. OK! To put your mind at ease, you're entitled to a private briefing. Understand, though, you are not entitled to further interrogation."

"Go ahead and brief. I can hardly wait," he said with resignation.

"You'll be home no later than 6 p.m., never fear! As to where we're going, I wanted to surprise you with a visit to another hillside manor house with a view rivaling that of Fort Ryan. And don't worry about your impoverished condition. JT and I worked it out so any overtime is my treat. Consider it your just reward for risking your neck as a *Monkey II* witness."

Traci was nothing, if not bold and blunt...with beauty and brains to match! Josh had little choice but to kick back and see what the crazy lady had in mind. He pretended a benign pout. "You two lawyer-types are a dangerous tag team. You'd think my boyhood buddy would have warned me!"

She smiled and patted his hand. Four more hours with the lady hardly denoted eternal torment. His mind reconciled, he decided, "Why not enjoy?"

As they sped southbound along Route 1, Alexandria soon disappeared in the rearview mirror. The moment they swung onto Route 235, Josh surmised that the surprise destination had to be George Washington's Mount Vernon estate.

But, "Why now? Why today?" he wondered. The scenario didn't compute.

Sure enough, the limo pulled up to Mount Vernon's main gate. Moments later, he was whisked through the turnstile by his hostess, who exclaimed, "This is perfect! Exactly what I had in mind." Then, looking pensively at her charge, she said hesitantly, "I hope you'll feel that way, too."

When they reached the assembly point for the mansion's walking tours, she proposed a detour. "I want to show you something special that I bet you've never seen before. You might even like it as much as you do that pond by the old broken-down dam on the Hawkings River."

Mystified, he muttered a muffled, "Whatever." Stoking his curiosity, she steered him away from the crowds, north of the mansion, well past the crest of the hill, where every glimpse of the restless Potomac invited wonder.

Finally she stopped. There was not a tourist to be seen in any direction.

"Here we are! This is my spot. I discovered it when I was a kid. Today seems like a perfect time to share it with you, particularly after yester..." Her voice drifted.

Cloaked in a mantle of shrubs and decorative grasses, the lush setting boasted faint hints of autumn hues, opening to the panorama of the powerful Potomac—rolling relentlessly to the legendary Chesapeake Bay and the Atlantic's briny depths.

But before Josh could surrender his senses, reason intruded with a dash of cold water. "She's bringing me here for a 'Dear Josh' good-bye. Here it comes: 'I'm sure we will always be good friends' and all that garbage. I really don't need this. I'm much stronger than she must think—and life does go on!"

Disdaining the likelihood of getting grass stains on her best suit, Traci picked out a patch of clover, sat down, and tugged him gently to her side.

"This is absurd," he thought. Regardless, the warmth of her body pressed tight to his released a welcome tingle.

"Do you like my hideaway?" The inquiry sounded winsome, almost pleading.

"Of course. It's spectacular. What's not to like?"

Somewhat relieved, she went on. "Then I hope you'll like it as much when we leave. Every time I visit Mount Vernon, I rediscover my private getaway."

Turning her face skyward, eyes closed, she leaned back with arms spread wide and explained, "This is where I blend yesterday's history with tomorrow's dreams. It's kind of hard to put into words...Know what I mean?"

"Yeah, I guess..." he said in confusion. Trying to match her unpredictable strides back in time, he remained awkwardly stuck in the present. Her next comment touched a nerve.

"All my life, Josh, I've dreamed about growing up, meeting, and marrying a perfect someone. But you caught me completely by surprise when you up and asked me to marry you yesterday—it came like a lightning bolt from those gray skies."

"Uh, oh, here it comes," he thought. "The friend thing, the ritual 'Good-bye, nice meeting you,' and all that..."

Fingers locked around his knees, he stoically eyed the Potomac, resigned to the inevitable.

She pressed on. "I'm a realist. In a few days, I return for my final year at Harvard Law. Next summer I plan to sit for the Maryland bar. That's heavy-duty stuff, Josh. And I know how torn up you are with your doctoral candidacy up in the air and your career uncertainty. I was aware that our relationship had grown, but..."

He cut her off. Trying to avoid self-pity, he couldn't help thinking, "Let's get this over with...Bring on the guillotine!"

Aloud, he said, "You might as well come out and say it...My status as an unemployed grad student dropout offers lousy marriage credentials."

"I didn't say that," she fired back. Seizing his face with her hands, she turned it firmly toward her, demanding, "Listen to me, Joshua Chamberlain Ryan. I've just finished a week of listening to you spout scientific blather. Now it's my turn."

He nodded acquiescence, expecting a punch line he preferred not to hear.

"I was trying to explain to your trillions of semi-functioning brain cells that the audacious timing of your proposal caught me off guard...and I needed time."

Josh desperately wished he could make a graceful escape before the *coup de grace.*

"There's this fellow in Boston...Miles Brennan...a stockbroker."

The deliberate words weighed heavy as stone. Josh stifled a groan as they struck his battered heart.

Ignoring his pain, she pressed on. "We met at a party. A nice guy, Josh, so when he asked me to marry him last Christmas, I said 'Yes' because it seemed like the right thing to do."

Josh felt a fist crushing the pit of his stomach. He grimly focused on the lazy currents of the river.

She continued, "There were no bells and whistles between Miles and me, Josh—the chemistry was missing. We drifted aimlessly in the currents of romance. We both knew it wouldn't work. When I fell into your arms last spring I no longer wore his ring, but hadn't bothered to return it."

She paused with moist eyes, as she dropped the second shoe. "Late yesterday, I called Boston and returned the ring!"

"That's a new twist," Josh thought with confusion. But as her right arm slid around his neck and her head came to rest on his shoulder, his brain cells shifted into high gear, racing to keep pace with his hormones.

"Don't you see, Josh? I needed time to grasp the reality of your love.

"You may be surprised that I felt an overwhelming love for you from the time I stumbled into your grimy arms at Center Hill Lake. That jolt set off bells and whistles—for the first time in my life!

"Yesterday, it seemed too good to be true that in spite of the turmoil in your life, you still had the courage to say those three little words—'I love you.'"

Josh was not about to break the spell now. "Lead on!" shouted his resuscitating heart. Unless something was missing in the translation, the dream woman was volunteering that her love matched his.

Suddenly, it was as if he saw Fourth of July fireworks arcing over the Mall, crowning the Washington Monument with a veil of red, white, and blue starbursts. His left arm tightened around Traci's waist. At last she delivered the punch line—a radical shift from his negative imaginings.

She whispered, "You know, Josh, contract law requires an offer, acceptance, and consideration to bind a contract."

She watched for his reaction.

"Also, there is the rule that the offer can be accepted by the offeree anytime before revocation. Since you haven't yet notified me of a revocation, can I assume yesterday's marriage offer is still valid?"

"Irrevocable, now and forevermore...in strict legal parlance!" Josh blurted.

"In that case, Mr. Ryan, I accept! And I'm hoping the good bishop can find time next summer to officiate in the formalizing of the most important contract we'll ever execute." With that, she dropped all pretense of legal formalities and hugged him fiercely.

When the reconciled lovers came up for air, Josh flung out a final reference to contractual formality. "But what about the mutuality of consideration?"

Catching the glint in his eye, she playfully covered his neck and face with kisses, and answered the obvious, "Our love, you fool!"

Stretched out neck-deep in clover, the newly engaged duo confirmed the contract by passionate embraces. They alternately rolled, hugged, and clung, exchanging lusty promises and sharing lofty dreams. Consumed with the rapture of true love, the pair were oblivious to the wandering tourists just over the ridge.

Time evaporated. Exhausted, the duo finally sat up, crimson-faced and bleary-eyed, eyeing each other's grass-stained, hopelessly wrinkled Sunday-go-to-meeting clothes. Neither cared a hoot. Clothes were replaceable, but love lived forever!

They fell back in each other's arms, laughing uproariously.

Hard-pressed to break away from the exhilaration of the moment, Josh bounced to his feet first. Ad-libbing an agenda for the newly formed alliance, he looked ahead to post-*Monkey II* life with all the fuss and flurry of organizing for a dinosaur dig.

"If we're going to make it back to my place by six, we need to cut and run. But we need to make two stops: I want to sell your folks on the idea of your marriage to an unemployed student, and my dad needs to know what a lucky dude I am."

"Don't forget a third stop. Let's surprise your Grandma Carrington."

"Another offer accepted. Let's hit the road to celebration."

Brushing the leaves and smudges from their hair and clothes as best they could, the couple emerged from their seclusion and headed straight for the exit gate and the limo. Despite their disheveled appearance, few passers-by paid them any mind.

The chauffeur opened the limo door with an enigmatic smile, the epitome of paid discretion. Once inside, they gulped cold drinks to slake the urgency of their thirst. The two sat scrunched together in air-conditioned luxury, never again to be quite the same individuals as before this day. Josh and Traci would face their tomorrows as a team.

Sinking into the embrace of the limo's extravagant leather interior, with one of Traci's hands resting on his knee and the other behind his neck, Josh had absolutely no recollection of the disappointments of the past week. Doctoral candidacy, career promise, litigation—nothing so pedestrian cluttered his mind. All problems were swept away by the consuming fires of love. Josh was ready to take any job, fight any battle, meet any challenge, as long as Traci walked at his side.

Lost in daydreams, the pair arrived at the elegant threshold of the Kilburn residence. Their unscheduled appearance did not detract from the welcoming hospitality. It seemed as if Traci's parents were anticipating their visit.

Having had no time to compose a speech, and feeling uncharacteristically giddy, Josh got right to the point. Offering encouragement, Traci sat seductively close, hands clasping his right elbow.

"Well, um, you see," he stammered, "we wanted you to be the first to hear the news. This afternoon, Traci and I got engaged. I sure hope you approve!"

Neither shocked nor surprised, both Kilburns offered their enthusiastic endorsement. Traci's dad, the Johns Hopkins surgeon who had once used his skills to salvage Josh's mangled left leg, spoke for both parents. "Young man, if we hadn't approved, you would have gotten the word long before now."

Hardly taking time to savor the response, Josh and Traci hurriedly described their tight schedule and two remaining stops. The hit-and-run tactic raised no eyebrows. But the Kilburns extracted a solemn pledge that the betrothed would return soon for the preparation and small talk that binds wedding-bound families together.

Dancing out the front door, Traci called back, "It's gonna be an August wedding...after I take the Maryland bar exam...We're going to invite Bishop Daniels to preside."

"We'll start planning!" rejoined the Kilburns.

Before they reached the car, little sister Kristi approached at breakneck speed on her bike, ponytail bouncing. She came to a screeching halt, inches from the toes of her targets. Although not privy to the announcement, she shared the psychic wisdom of little sisters. Hopping off her bicycle, she rushed to Traci's arms, shouting for the benefit of any curious neighbor, "I just knew you were going to marry him! He's cute...a lot better than the Boston bean-counter, if you ask me!"

Traci blushed a rosy pink and chided jokingly, "Shush now...You be careful what yarns you spin. You may scare away this handsome hunk!"

Giggling but undeterred, the precocious youngster turned her attention to Josh, grasping his hands and gushing her blessing. "You'll be a perfect brother, Josh...I just know it. I watched you coming by here all summer and thought you were super-nice. I told Traci if she had a lick of sense she sure should marry you if you asked. Then when I caught you two messin' around at the Hampshire pool Wednesday, I just knew..."

Trapped and speechless, Traci's blush deepened. Josh roared with delight, patting Kristi on the head and offering reassurance. "You and I are going to get along just fine, Punkin...Thanks for being on my side and rooting for Magruders Mill over Boston. I need all the help I can get...And thanks for persuading your big sis to say 'Yes.'"

Now re-composed, Traci dangled enticements. "Looks like I'm going to have to invite you to stand up with me at my wedding, Kristi, seeing as how you are so all-fired wise and influential in making good things happen—providing there's no more bean-counter talk!"

At the news, Kristi skipped for joy and the sisters wrapped their arms around each other. Josh couldn't recall a happier afternoon!

The Fort Ryan reaction to the engagement news was a repeat of the Kilburns'. The news elicited broad grins, congratulatory handshakes, and three-way hugs.

"You two are going to be an awesome couple. I'm delighted for you both," enthused Ryan senior. The vote of confidence was framed in gentle teasing. "It's a relief you finally found someone who will put up with your eccentricities, Josh. I always suspected Traci's character matched her beauty and brilliance."

Touched by the reaction, Traci hugged Michael Ryan and kissed him on the cheek. Her next statement articulately summarized their shared emotions. Wrapping each word in love, she said, "You must have been the ultimate father and the best substitute mother in the world to raise the likes of this remarkable guy. Thanks for your gift to the world...and to me!"

The sentiment triggered a barely suppressed tear and a husky order to "Get a move on, you two. Remember, Josh, I'll be picking you up early this evening for the conference at the chancellor's office." With tongue-in-cheek, he advised, "By the way, Son, you might want to consider a change of clothes. Dr. Wood Comstock is no fan of the rumpled look."

The blessing of Lady Carrington would be a bonus, but of considerably less importance to Josh than to Traci. They found the Duchess patrolling the Morgan Manor stables in high spirits. Wearing faded blue jeans, riding boots, and a blue denim shirt embellished by several strings of pearls haphazardly askew, the wiry old lady spied them from across the immaculately kept grounds.

She welcomed the approaching couple with arms extended, first to Traci, then to Josh. Unused to this spontaneous show of affection from someone who was both stranger and grandmother, Josh stood awkwardly on one foot then the other, unsure how to reciprocate.

Finally, he patted her shoulder as she exclaimed, "Now this is some kind of pleasant surprise. The best-looking young couple I know coming by to cheer up a little old lady stranded on this lonesome hill. I just finished feeding the horses and made my daily visit to the tack room to admire your mother's trophies."

Josh wasn't sure how to respond. He figured that Traci knew his grandmother better than he did, so waited for her lead. Traci read his mind and took the cue. She flushed imperceptibly, hazel eyes aflame. "Josh and I wanted you to be privy to the inside scoop of some late-breaking news..."

An animated Duchess interrupted, not letting her finish. "Congratulations, you two! You have my total blessing—even though I know you don't need it. The Kilburns and the Ryans represent a merger of impressive family dynasties...like the Morgans and the Carringtons. When can we start planning the smash hit of the social season?"

Josh reacted with barely concealed panic. Even Traci seemed a bit taken back at the perceptive insight.

"Now don't you two look so startled," exclaimed the Duchess. "Don't forget, I've had several turns around the arena in my lifetime and I know a thing or two about people. All the pretend games can't hide facts of life from a wizened old-timer like me. The two of you have been nuts about each other from day one. It's about time you acted with the brains God dished out to you and become partners in life's madcap adventure."

Amused by the good-natured scolding, Josh tried to appear suave and in control. "It will be next summer...and we want you to be there!"

"Well, I wish I may never...Of course I'll be there," she snapped. Then looking at Traci, she added with presumption, "And if you like, I'll arrange for the finest reception spread the Hampshire Country Club has ever seen. My treat, if you will allow me.

"But of course, decisions like that belong to the Kilburns...and to you, my dear...I just wanted you to know I'll be available at your beck and call to contribute in any way you might choose."

"Of course you will," assured Traci, kissing the weather-beaten socialite's bony cheek. "We'll want you involved every step of the way, won't we, Josh?"

Josh sidestepped diplomatically, saying with a shrug, "For sure, I don't know much about these things." He wasn't sure how his dad might feel, despite the week's overtures of good will.

"And the officiating clergyman?"

"Bishop Brock Daniels, providing he's willing to hitch my future to a heathen like Josh," teased Traci.

Grandma Carrington cackled. "You, my child, are wise beyond your years," she exclaimed, heartily endorsing the choice. In fact, she was thrilled to be a prospective insider at a family wedding. Every year that passed, she ached at the irretrievable opportunity of presiding at the wedding of her daughter, Leslie Anne—an opportunity she had disdained in a fit of pique. This event could help assuage that void. She desperately hoped that the Ryans would opt for her inclusion.

"Josh has a VIP confab this evening with Judge Stone and Dr. Comstock, and I promised to get him home by 6:00. Hope you don't mind our running."

The Duchess beamed her most radiant smile. "Not at all. You kids go on now...Do your thing." Then came another outpouring of hugs and kisses, with Josh stoically submitting to his grandmother's overtures.

Returning to the limo and the last leg of the day's celebration, they were sidetracked by a glimpse of the thundering hoofs of Turbo rushing headlong across the meadow to add his greetings. Having missed the irascible animal, Josh headed over to the fence, where he instinctively reached out to cradle the glistening neck and pat the bobbing nose.

Turning to Traci, he warned, "This spoiled critter comes with the package, you know."

"I should hope so," she replied. "Don't forget, I owe him a zillion sugar cubes. Turbo saved the day, delivering my star witness to Wednesday evening's *Monkey II* performance."

As the couple slipped away from the pasture, Turbo snorted happily. A few steps from the waiting limo, Josh paused to call a promise to the stallion. "I'll pick you up first thing tomorrow and take you home, ol' fellow. And by the way, you're going to be seeing a lot of this beautiful lady."

Turbo whinnied a horsy good-bye, anticipating the grand reunion.

Following Traci's crisp instruction, the limo reached the mill house precisely at 5:55 p.m. Awaiting delivery to her Columbia residence, Traci remained seated while exercising her newly acquired fiancée prerogatives.

"Be sure and call first thing in the morning, Josh. You can tell me about your meeting with the chancellor...I'll be at my folks' place."

"You'll hear from me. That's a binding contract: offer—acceptance, consideration, and delivery. But first I want you to come inside with me. It will only take another minute." He offered his hand as she stepped out of the car. With inquiring eyes, she followed as he led her into the cottage, straight to the bookcase. Unceremoniously he reached behind the diary, took out the blue velvet box, and popped it open before her admiring eyes.

"This is for you."

The diamond sparkled with rainbow hues. While not the biggest, or the brightest, it ranked as the best—hands down! Michael Ryan had given the gem to his beloved, and Josh knew his dad expected him to share the treasure with his own bride-to-be.

The impromptu presentation caught Traci by surprise. She stood riveted in place, speechless.

Josh repeated the sentiments. "This ring is for you, Traci. It once belonged to my mother. There is nothing I own or could buy that would better symbolize my love for you. I love you now more than I can describe, but I will love you more every day of our lives."

Tenderly he reached for her left hand, slid the jeweled band onto her finger, and kissed the delicate emblem in unrehearsed jubilation. The diamond fit perfectly!

Her eyes welled. At first, she couldn't speak. She held her hand before her face, shielding the glare of the afternoon sunlight, trying to focus on the heirloom. At last she looked up.

"It's the most beautiful ring I've ever seen...and you're the most perfect man I've ever known. I'm the luckiest woman on Planet Earth. You will be loved by me more than you will ever know for as long as we live...I just plain love you, Joshua Chamberlain Ryan!"

And with that, the two capped the day's celebration with a kiss so passionate, so unyielding, and so reciprocal that another twenty minutes flew by before they could break away. Feeling emotionally drained, the gallant ex-grad student finally walked her to the limo, helped her in, and stood admiring the precious cargo. As the limo pulled onto Highway 97, Josh Ryan for the first time in days looked ahead without dread to the future.

# 44

The ever-punctual Ryans arrived early for the meeting at the white-columned administration building, nucleus of Mid-Atlantic University. After they showed their IDs and signed in, the guard waved them onto a private elevator programmed to go straight to the waiting room of the chancellor's top-floor executive suite. Greeted by a smiling receptionist who seemed unfazed by the unusual Saturday night meeting, the visitors were ushered courteously into the inner sanctum of the boardroom. Neither father nor son had ever been in the penthouse before. They took their assigned seats minutes before 8 p.m.

The polished mahogany conference table could accommodate eighteen. This evening, only ten chairs were filled—four on each side and a judge's chair at either end. Pace Terhune, Augustus Morley, Karl Striker, and Marcus Brogan sat on one side of the table, across from the Ryans. On the other side, with Michael and Josh, were Jessica Saunders and Jonathan Thomas Daniels, who sat to the right of Josh.

As the Westminster clock chimed the hour, Chancellor Wood Comstock and Judge Edward Anthony Stone emerged from the chancellor's private domain, both wearing inscrutable expressions. Chancellor Comstock took his place at one end of the table, while Judge Stone sat at the other.

Judge Stone had prepared the agenda that called together the unusual assemblage. Neither Ryan had the remotest idea what to expect. But they sat attentively, like good soldiers summoned to the front lines of battle.

By the time the chimes finished tolling, the judge had arranged a sizable stack of paper with military precision and had called the impromptu session to order. Looking up, he broke into a disarming smile.

"It was gracious for each of you to come at such an inconvenient hour and on such short notice. Several urgent matters demand immediate attention. For those wondering what this is all about, I promise an evening of enlightenment."

Marcus Brogan reacted with a yawn, flipping through his ever-present legal pad. Karl Striker stared ahead sternly, while Gus Morley struggled to appear nonchalant. Pace Terhune brightened at the word *enlightenment*. Chancellor Comstock rocked back and forth on his chair, arms folded and eyes closed—but alert as a poised panther.

Jessica Saunders opened a file crammed with mathematical calculations, confirming her evening's assignment. JT sported a stylish three-piece suit, implicitly pledging allegiance to the Cabot, Calvert, and Stone protocol imposed on new hires. The Ryans waited patiently, wondering what surprises Ed Stone had in store to round out the tempestuous week.

The case pitting *Mid-Atlantic University Science Research Foundation v. Joshua Chamberlain Ryan* headed the agenda. Looking over the assembled group, the judge spoke with the authority of the bench rather than the bar's negotiating stance of legal advocacy. He didn't mince words.

"It's time to terminate this travesty!"

Responding on cue, JT distributed a single-page legal document to Brogan, Striker, and the Ryans that summarized the judge's hardball message. Judge Stone went on: "In order to put this atrocious mutation of the legal process behind us, I have prepared this 'Stipulation for Dismissal With Prejudice,' which requires the signatures of all attorneys of record representing the parties—before we adjourn this evening!"

Striker bristled while mouthpiece Brogan spoke for his client. "Now hold on, Ed. You're not sitting on that judicial throne this evening. The Science Research Foundation has been conned out of a bundle of money. This case isn't going anywhere except to a jury!"

"I should think a jury would be the last place you would want this case to go, Mr. Brogan. After this evening, you may feel like suing someone, but I can assure you it won't be Mr. Ryan!"

Resisting Brogan's restraining grip, a belligerent Karl Striker sputtered, then exploded. The self-appointed co-counsel for the plaintiff barged ahead, his words breathing fire. "Really, Judge Stone, the facts appear quite clear. Whether by intent or mere negligence, young Ryan cost the Science Research Foundation a good half-million bucks plus, and we intend to collect!"

"Now just a darn minute...," Michael Ryan sputtered.

Before the elder Ryan could join the fray, Judge Stone cut him off with the assurance, "That's all right, Mr. Ryan. I understand your reaction, but first let's allow the facts to speak for themselves."

Without missing a beat, Ed Stone signaled Jessica Saunders to distribute the documents stored in a bulky file stamped "Confidential." Moving deliberately, she unwrapped a pile of financial data as though revealing a Magna Charta original. Spread before ten pairs of eyes, an independent audit exposed the foundation's financial practices to the scrutiny of daylight.

"Gentlemen, my office had been reviewing the financial records of the Science Research Foundation since long before Josh Ryan's fossil Archy disappeared. Frankly, I've uncovered more questions than answers. But the answer to the question at issue in this litigation is clear as crystal."

As Jessica Saunders ticked off numbers with devastating precision, apprehension creased the foreheads of three of the conferees.

"Aggregate compensation paid Joshua Ryan for his work on the Tennessee research project totals $18,792.19—and that includes a standard $12,000 student stipend—$2000 a month for six months. He has received nothing since April

1996, despite an outstanding cost reimbursement balance of $2,722.13 owed him by the foundation."

"That's absurd!" Dr. Striker shouted, turning first to Attorney Brogan and then to Dr. Morley for support. Busy on a note-taking jag, Brogan ignored his client's outburst while Morley grunted skeptically, "That can't be true," and shook his head.

"That's totally off the wall," Striker said contemptuously, glaring at Judge Stone. "As much as I'd hate to see young Ryan hung out to dry, the books show a batch of checks totaling several hundred thousand dollars made out to him."

Striker and Brogan settled back, but an unrelenting Jessica paid no heed. She continued her report without a flicker of emotion. "You're on the button in one regard, Dr. Striker. There are a large number of canceled checks that say 'Payable to Joshua C. Ryan.' For some reason, billings don't correlate with funds disbursed."

Morley shrugged, explaining, "Shows the risk of using student book-keepers."

"Just a minute, Dr. Morley—there's more. Happily, the Science Research Foundation checks deposited to Mr. Ryan's account at Silver Spring National Bank correspond with the total of his reimbursement requests prior to April 1, 1996: $18,792.19. Unhappily, the other checks allegedly paid him don't appear as deposits to that account."

Striker scoffed, "Everyone knows that a con-artist could readily open another account in another bank anywhere in the country."

"Brilliant conclusion, Sherlock," said the petite CPA, who didn't shy from sarcasm to underscore a point. "That's exactly what the con-artist did. Only the crook wasn't Mr. Ryan. Some third-party bumbler produced a mediocre forgery of Josh's name, then went all the way to a branch of the Silver Spring National Bank in Baltimore to open a fictitious account using a Baltimore Post Office box as an address."

The Striker team looked at each other uneasily.

"Now doesn't that strike you as a strange coincidence? Why would the real Josh Ryan concoct a poor forgery of his signature and travel out of his way to Baltimore County to fool his own Magruders Mill bank?"

"Still..."

"Oh yes, before I forget, the make-believe 'Joshua C. Ryan' managed to mess up the genuine Ryan Social Security number by a single digit. The Social Security Administration says the number with the one-digit error was assigned long ago to someone of the female gender, a person uninvolved and innocent of this fabrication...someone who, in fact, died in infancy.

"Now that's pretty remarkable, don't you agree?"

Brogan leaned forward, listening intently. His clients fidgeted. Jessica Saunders continued, "It occurred to me, gentlemen, that any number of university employees could have had access to Josh Ryan's Social Security number. It's listed in virtually all his records, including his original application for admission. All references are correct except one. That erroneous Social Security number listed in the Science Research Foundation files matches the one-digit mistake used to open the phony bank account—the figure '7' in place of a '1.'"

Striker and Morley stubbornly resisted the implication. "This is ridiculous," Striker shouted. "The next thing we know, you'll be blaming us." Jaw clenched, Striker glared belligerently at Jessica Saunders and Judge Stone.

Ignoring the outburst, Judge Stone returned the focus to the matter of the *Dismissal*. "You will notice, Dr. Striker, that Ms. Saunders has not fingered a culprit. Rather, she has revealed the tip of the evidence iceberg, but more than enough to exonerate the defendant in the litigation.

"Gentlemen, it's an exercise in futility to try to pump oil from a dry hole."

Although Brogan sensed that the judge had something up his sleeve, he stalled, not yet willing to fly a white flag. "I have a fiduciary duty to my clients and see no reason to act precipitously. We haven't even begun discovery. There's nothing like old-fashioned depositions and interrogatories to clear the smoke," Brogan barked pompously.

"Your key problem, Counsel, is that discovery can be time-consuming and costly...particularly when discovery may reveal facts embarrassing to the plaintiff."

Knowing full well the driving force that fueled Brogan's quest for "justice," the judge looked over the top of his half-rims to pointedly advise the lawyer, "Of course, an insolvent entity would be hard-pressed to afford legal fees of $500 an hour."

Somewhat unnerved by the innuendo, Brogan sounded more tentative. "What's that supposed to mean?"

"In dollar talk, it means the Science Research Foundation has fallen on hard times and has dipped into principal to fund its two-million-dollar yearly budget. The brutal truth, Counsel, is that profligate ways have left your client stranded and fresh out of cash."

"I don't understand! Something sounds out-of-kilter...Don't play games!"

Judge Stone obliged in three-dimensional candor. "Last week, the Silver Spring National Bank called the chancellor's office to inquire whether the university would make good on the foundation's $172,000 overdraft. When the answer came back negative, the bank froze the account. Word is that a $55,000 retainer check to the Brogan law firm may have been stamped 'Insufficient Funds' as a result."

"You told me..." Brogan began, glowering menacingly at Karl Striker. The best the befuddled Striker could do was shrug and spread both arms wide imploringly, saying, "I don't understand."

The increasingly distraught Morley whacked the arm of his chair with his fist. Seething self-righteously, he exclaimed, "That despicable Lassiter!"

Judge Stone knew exactly how to get Brogan to embrace the dismissal. He began to reel in the greedy fish. "The good news, Counselor, is that a generous donor, who wants to underwrite a new, solvent foundation, waits in the wings as we speak. Although there would be no legal obligation, this could be a source of settlement for the defunct entity's creditors, such as yourself!"

The barrister's ears perked up, as Judge Stone had predicted. The wily jurist continued: "However, the bad news hits your pocketbook! I doubt any negotiated legal fee would exceed 50 cents on the dollar—at the very most!"

Shocked dismay swept Brogan's face as Judge Stone finished his pitch. "You understand how it is, Mr. Brogan...new donor, new foundation, dead-horse enterprise.

"That potential new foundation is next on my agenda, once we can clear up this matter of your client's litigation against Mr. Ryan. Why don't we take five so you can discuss it privately with Drs. Striker and Morley."

Marcus Brogan conversed eloquently in money talk. Reflexively, the litigator leaped to his feet and ushered his two quite confused clients to the hallway for an impromptu *tête-à-tête*. The opportunistic barrister covered his abrupt exit with a hollow but high-sounding line. "Reason is the mother of justice!"

Judge Stone shook his head in wonderment at the resourceful spin, musing, "That rascal can concoct verbiage worthy of enshrinement over the lintel of a courthouse."

An observant Michael Ryan brightened visibly. "Will they buy it?"

"Does the sun rise in the east and set in the west?" The confidence of the weathered champion of justice put the father's mind at ease.

He turned to Josh, volunteering grandfatherly sentiments. "I predict you're off the hook, young man, but be sure and stay tuned for the next episode. There's a good deal more to come. And don't forget to thank this heartthrob movie actor friend of yours, who's worked like a Trojan. In fact, it was his cunning that concocted tonight's legal blitzkrieg."

JT sat impassively, ignoring the plaudits.

Barely three minutes passed before the door swung open. Shedding the role of battling barrister for the guise of benevolent statesman, Brogan shepherded the subdued scientists to the conference table. Grabbing a chair, he bellowed, "Reason triumphs! My clients are men of stature. They're reluctant to do anything to embarrass the university...or risk harm to Mr. Ryan for that matter.

Therefore, they agree I should sign the 'Stipulation to Dismiss With Prejudice' should be signed."

With that, he scribbled with a flourish and officiously took his seat. Josh Ryan's real-life litigation defense in the federal court had started and stopped in less time than the six-day run of *Monkey II*!

Judge Stone signed for the defendant, without comment.

Hard-bitten and canny, the impatient Brogan cut to the chase, trying not to sound venal. The week's billable hours for time spent with Striker could buy a membership in more than one country club—even at fifty cents on the dollar.

"Now about this donor and a new foundation...?" he hinted greedily.

The metronomic rocking of Chancellor Comstock's chair ceased as the chancellor's eyes opened to narrow slits. The agenda was Judge Stone's, but the chancellor knew in advance everything that would transpire. He sensed that a propitious time to speak would come soon enough. Pace Terhune and Jessica Saunders knew only that a well-to-do benefactor planned to make a substantial contribution to Mid-Atlantic University.

Chancellor Comstock explained, "The gift is something more than a single contribution of a few hundred thousand dollars. Rather, it represents allocation of a multi-million-dollar charitable trust. The portfolio reflects post-World War II corporate prosperity. The donor expressly directs that the resources be earmarked to underwrite scientific research at Mid-Atlantic University.

"The gift also means the derelict Science Research Foundation can be put out of its misery without embarrassment to the university."

Karl Striker knew that public exposure of the foundation's financial debacle would destroy his carefully cultivated image. And since financing for his ego-trip research appeared as unlikely as a downpour in the Sahara, he was attracted to this new revenue stream like metal shavings to a magnet. Personal aggrandizement pre-empted ill-advised litigation in Striker's lexicon.

"What kind of dollars are we talking about, Judge?" Striker asked bluntly.

"In the six million or seven million dollar range," came the answer.

Striker breathed easier, suggesting, "That would be a big boost, enough to see us through the next two or three years!"

"You don't understand, Dr. Striker. That figure represents *annual income*, not principal. At current market, the fund portfolio reaches nine figures."

There was stunned silence, followed by deep breaths and muffled whistles.

The Striker contingent sat bolt upright, looking thunderstruck. A single gift in the $100 million-plus range was unprecedented in the history of Mid-Atlantic University. Even the flamboyant Pace Terhune and the unflappable Jessica Saunders exchanged glances. Questions flowed thick and fast.

Brogan rushed in, first in line. "When will this generous gift be available?" Legalese translation: "When do I get my $25,000?"

371

"The new foundation was incorporated on September 1, 1996. The university's executive board accepted the offer yesterday afternoon."

Chancellor Comstock continued, "All terms and conditions required by the new foundation fund are to the overwhelming advantage of the university and will be implemented faithfully. MAU welcomes the windfall!"

Jessica Saunders asked if any portion of the gift could be allocated to the building fund for the long-planned science center.

Energized, Judge Stone could hardly suppress his exuberance. "You'll be pleased to know that this is one of the conditions. The Silver Spring National Bank has agreed to provide the cost of construction provided the loan will be amortized from trust income Construction begins the moment bids are in and contracts signed.

"The building will be named *The Parker Daniels Science Center.* A full-length portrait of the local hero has been commissioned for the lobby. Large bronze letters mounted above the entry will trumpet the maxim: *Truth Frees.*"

Striker recovered sufficiently from shock at the magnitude of the gift to revert to cantankerous self-interest. "I would think the building deserves the name of some renowned scientist who has served this university with distinction."

After surviving a week on the *Monkey II* hot seat, Josh Ryan welcomed the role of bystander. But, faced with Dr. Striker's arrogance, he reverted to irreverent form. Whispering a wisecrack intended solely for JT's ears, he joked, "I can see it now: *The Striker Center for Mutant Monstrosities!*"

The two friends snickered. Karl Striker overheard just enough to scowl disapprovingly. But even Brogan and Morley appeared taken back by the brassy chutzpah of their colleague.

Suspicious that Dr. Striker cherished the delusion that a title such as *Striker Building* might be preferable, Chancellor Comstock cut him off at the knees. "Perhaps you weren't listening earlier when I indicated that the wishes of the donor will be followed precisely."

Duly chastened, Striker sat back while Gus Morley asked for more information about the terms and conditions. The judge summarized succinctly: Cabot, Calvert, and Stone had created the trust fund in 1948 for a client celebrating the birth of an only child. The now-inflated portfolio had converted to a tax-exempt charitable foundation shortly after the premature death of the original beneficiary, and, targeting education, had built a proud history of scholarship grants to worthy students, as well as generous anonymous contributions to a wide range of charitable enterprises.

"Effective September 1, 1996, the foundation earmarked all resources exclusively for the academic sciences under the tax-exempt educational umbrella of Mid-Atlantic University. This unanimous action by the foundation's five-member board reflects the expressed wish of the donor.

"After income is set aside for the new science center, the annual revenue will underwrite field and laboratory research world wide."

Striker and company nodded in concert, liking most of what they heard.

"The foundation has promised to allocate up to $1 million per year to fund a new, full-color, quarterly publication dedicated to objective reporting of scientific breakthroughs. Unlike its insolvent predecessor, this magazine will be professionally edited, richly illustrated, and intellectually honest.

"Lest there be any misunderstanding, gentlemen, this means that neither faculty member nor student will take credit for another's original work. Opposing viewpoints will be aired. Academic freedom will thrive."

The between-the-lines message struck Striker squarely between the eyes, but the consummate politico didn't flinch. "That's a big order for any division chairman, but I can live with that guideline," he acknowledged with forced enthusiasm.

"You won't have to," countered the judge. "As stipulated by the bequest, the president and CEO of the foundation must be a polished writer, someone experienced in public relations and development."

Striker frowned. "I'm afraid I don't understand," he said.

Chancellor Comstock explained: "The university administration endorses wholeheartedly the foundation board's nomination for this office...our own Pace Terhune."

Striker blanched, visibly shaken. Terhune's jaw dropped.

"Don't be alarmed, Pace. The job carries a quantum leap in salary and offers you command of a global empire of sorts. Projects you supervise will extend internationally. You'd better buy some walking shoes and renew your passport!"

Rarely at a loss for words, Terhune stammered, "I never dreamed...I'm...it's just that...!"

The chancellor bolstered reassuringly, "You embody the consummate skills of a professional communicator. And besides, you'll become famous overnight, thanks to being known as the *Founding Editor* of a splashy new scientific journal. You're welcome to take the weekend to think it over and discuss it with your wife, if you choose."

"Uh, 'yes' and 'no'," came the instant reaction. Trying not to stammer, Pace clarified, "I mean, 'No,' I don't need the weekend, and 'Yes,' of course, I accept."

The chancellor beamed with satisfaction. It was the response he expected from the decisive Terhune. Striker and Morley stared at the table in glum chagrin, neither pleased nor appeased. Brogan didn't much care. His services would be available to the highest bidder, regardless of who called the shots at the foundation.

But the chancellor had not finished. Looking mischievously at Jessica Saunders, he could scarcely conceal his glee. "Ms. Saunders, it has come to my attention that the university has just lost one of its valued vice-presidents, leaving a vacancy to be filled. While the role of Vice President for Public Relations and Development at Mid-Atlantic University may not offer compensation equal to the CEO of a well-endowed foundation, it is a prestigious assignment, nonetheless. And it does carry a hefty salary boost above and beyond your current compensation. Yesterday afternoon the executive board anticipated this possibility and voted unanimously to offer you a well-earned promotion. Now, if you would like to take the weekend to..."

Jessica Saunders burst out laughing, interrupting the chancellor's speech. A lightning-fast response demonstrated her skill at making checkmate moves. "Consider this an immediate 'Yes' so that in the event Mr. Terhune gets cold feet and changes his mind, you can advise him that his old job is filled!"

The conferees chuckled while the irrepressible Terhune promised, "It's all yours, Jessica!" Even the Striker trio shrugged in token approval.

The science division chairman optimistically assumed that he was next in line for a promotion. He led with his chin. "What about the foundation board?"

"There will be five members. Naturally, Pace Terhune will function as a trustee. So will Jessica Saunders. Not only will she be liaison for the university, but her job description also includes riding shotgun on the foundation's revenue stream."

"And the other three?"

"Well, the chairman designated by the foundation happens to be a distinguished entrepreneur from international industry with demonstrated investment skills. A fourth trustee serves as a Pan Oceanic Petroleum vice-president. When he retires sometime next summer, his position is slated to be filled by an already-named Generation X successor.

"And, of course, the fifth trustee is the donor. Whenever an individual resigns or retires, the remaining four board members are empowered to name a successor. The board is self-perpetuating."

At this point, Judge Stone interjected a timely disclosure intended to defuse any back-door criticism charging conflict-of-interest.

"Don't forget, Cabot, Calvert, and Stone will continue as general counsel to the foundation, thanks to the generous long-term contract bestowed by a presumably satisfied client—assuming the university will stipulate by board resolution that the arrangement is satisfactory."

The chancellor said with a smile, "How could I forget it with you to remind me? Seriously, Ed, the university wouldn't have it any other way. That reminds me, you're always threatening retirement. What's the back-up plan?"

"Glad you asked, Wood! Frankly, I have no idea when, or even if, I will retire. It's no coincidence that sitting to my left this evening is an associate scheduled to come aboard next summer, after he graduates from law school. His portfolio will include riding shotgun on the foundation's legal needs. Believe me, he knows more about the new foundation than anyone on earth. Most of you have had occasion already to meet Jonathan Thomas Daniels."

Pace Terhune reacted enthusiastically. Jumping to his feet, he swung around the table, and vigorously pumped the hand of a nonplused JT. The future lawyer had no previous inkling of the cushy assignment, nor had he envisioned the degree of confidence placed in him by Judge Stone.

"Looks like I'm going to have to study extra hard to make sure I pass the bar the first time around," he said with a grin. Reaching across the corner of the table, JT took the congratulating hand of Judge Stone, saying, "Thank you, Your Honor. I guarantee you will never be disappointed."

"I know that, my boy. Cabot, Calvert, and Stone rarely stumbles in its assessment of talent and character."

As the commotion subsided and Pace Terhune returned to his seat, Josh thumped his friend on the back, adding his own congratulations. "JT, you never cease to amaze me. Someday I'll probably be marking a ballot for Senator Jonathan T. Daniels."

The Magruders Mill boys exchanged high fives. Striker and Morley looked grudgingly mollified. As for Attorney Brogan, grateful for the promised retainer, diminished though it was, but recognizing the future likelihood of business with the university as slim to none, he began shuffling papers, ready to abandon ship.

"Not yet," ordered Judge Stone. "Your client still needs your services."

The question that had been nagging Striker's mind spilled out. "I may be missing something, but I'm not sure I understand just how the university's Science Division and its key faculty leaders are to relate to the foundation. We are indebted to Dr. Morley for his long and faithful service...and of course, as chairman, my own line of authority is well defined."

Mustering as much courtesy as he could muster from a sparse reserve, and with horn-rims sliding down to the tip of his nose, Striker's words dripped with feigned humility.

"Just how can I be of greatest service?"

"Don't fret, you're not forgotten, Karl. The foundation needs a director of research to coordinate projects. Yesterday afternoon, MAU's executive board endorsed the foundation's earlier action to invite you to fill this prestigious position. Your commitment to research is legendary."

Although Striker showed no inclination to jump for joy, Morley's ears picked up. He rationalized, "With Striker out of the way, who better to step in as top man in the division than the senior associate?"

"While I appreciate the confidence of the executive board, Dr. Comstock, my responsibilities as chairman of the Science Division are very time-consuming. The added burden you're suggesting could be an overload that..."

"Nonsense. You misunderstood the board's intent. This honor would be an entirely new, full-time assignment."

Everyone in the room got the message, loud and clear. Augustus Morley soared high emotionally, while Striker found himself flirting with crash mode.

"I fear it will be hard for me to relinquish..." The reluctant Striker dug in his heels.

Comstock offered a blunt summation. "Let me clarify! Your distinguished academic credentials exceed your people skills. The university administration is unanimous in its belief that this challenging new opportunity constitutes the best utilization of your professional talents. The appointment guarantees international exposure."

Striker's face brightened at the hint of fame. Still, he stalled. "I just don't know..."

"Oh, I almost forgot to mention that the compensation package promises a starting salary that substantially exceeds your present income...with automatic incremental increase...It even includes a not-so-modest, up-front 'signing bonus,' in the vernacular of professional sports."

That did it. Striker snapped to attention. Fame *and* fortune knocked at his door. "Where do I sign?"

JT stood ready with multiple copies of a simple but tightly drafted agreement that he passed to Judge Stone, who in turn handed the document to Striker and Brogan.

"Take your time, Dr. Striker, and confer with Mr. Brogan, because it's essential for him to countersign approval."

Obediently, Brogan and Striker read. Gus Morley's spirits scaled the stratosphere. Short and to the point, the document covered all bases. When the two looked up, Judge Stone asked, "Any questions?"

"Looks OK to me," came the predictable, shoot-from-the-hip legal opinion of the restless Brogan, long since ready to call it an evening.

But one nagging clause annoyed his client. "I understand the detailed job description; the generous compensation; the definitive line of authority directly to the foundation's chief executive officer, and, of course, the voluntary resignation as chairman of the university's Science Division. I also understand I will be forfeiting classroom time unless my appearance is requested as guest lecturer.

"But I'm not sure I understand this provision." He cited the potential fly in the ointment. "Something about my *rescinding all discretionary terminations of doctoral candidacies retroactive to January 1, 1996, unless such actions are ratified by subsequent formal action of Mid-Atlantic University's academic administration.*"

The legal consequences of the clause galvanized the attention of the Ryans. The chancellor's curt response inflamed Striker's emotions.

"It means exactly what it says, Dr. Striker. The administration intends to look at each case and ratify terminations where justified." Striker reacted with alarm at the chancellor's confirmation that administration intended to override the chairman's academic authority.

"There's evidence that in recent months you've acted arbitrarily in the termination of otherwise qualified doctoral candidates who for reasons of their own didn't march to your drumbeat," the chancellor said pointedly. "One example sits across from you as we speak. You should know that the case of Joshua C. Ryan has already been re-evaluated and this agreement ensures his immediate reinstatement in the doctoral candidacy program. Your signature will preserve a modicum of your own professional integrity, untarnished by any hint of academic abuse."

Chancellor Comstock had dealt with recalcitrant academics in the past and knew how to tactfully turn up the heat. "You know, Karl, I've heard rumors of a petition now circulating among students calling for your censure, demotion, or dismissal. It would be unfortunate if..."

The chancellor left the not-so-subtle innuendo hanging for effect, as Striker saw his career twisting slowly in the wind. Brogan's trepidation showed. He tapped his pen nervously, waiting for the end of this embarrassing spectacle, all the while speculating on his monetary rewards.

"Don't be a fool, Karl...Go ahead, sign, and get it over with," exhorted the impatient legal counsel.

Eager for his own shot at an about-to-be-open chairmanship, Morley chortled, "I've warned you repeatedly, Karl...You should display more sensitivity and even-handedness."

Refusal to sign suggested pettiness. Return to the status quo? Unthinkable! The fame *and* fortune of the new frontier beckoned. Striker hesitated briefly, then capitulated. As the ex-division chairman signed duplicate originals, the room broke into subdued bedlam. Striker at first looked granite-solemn, then relented, showing the faintest hint of a grin.

Accepting congratulations from Michael Ryan, Josh muttered in relieved jubilation, "I wish I may never!" JT slapped him on the back, predicting joyously, "It's about to be *Dr. Darwin*, I presume!"

Good-ol'-Gus thrust a beefy palm across the table, pompously asserting, "I promised long ago, Mr. Ryan, that everything would turn out to your eminent satisfaction."

Brogan broke into a rare grin, adding his slant to the curious turn of events. He tried to paint himself as pivotal to the evening's success. "You see, young man, great legal counsel is key to survival in this cruel world."

Judge Stone prepared to spring more surprises. "Before you get away, there's something you all must see." He pointed to three tables lining the wall, each one draped in red velvet.

Reverting to his instinctive trial-attorney style, Ed Stone strode to the first table as though marching to fife and drum. Dropping the drape to the floor, he revealed an architect's model of a brick building featuring a clock tower in classic Victorian tradition. Josh recognized a redesigned, miniature replica of the century-old Dayton, Tennessee courthouse, scene of the 1925 *Scopes* trial. The brass nameplate identifying the structure grabbed attention: *The Parker Daniels Science Center.*

The conferees applauded spontaneously as Judge Stone rested a protective hand atop the mini-tower, explaining, "Thanks to the Leslie Anne Ryan Foundation and the generosity of its founder, Regina Ann Morgan-Carrington, this trophy will grace the campus in time for the fall quarter, 1998."

Most present had guessed already the donor's identity. In a classic gesture for the Duchess, Lady Carrington's philanthropy would fund the new science foundation. Michael whispered to Josh, "Never underestimate your Grandmother Carrington, my boy."

Everyone acted pleased to be let in on the secret before the next day's news release. But a miffed Augustus Morley sat slumped in his chair, showing his disappointment. He now sensed that the evening's agenda would not award him the dream he so desperately craved—appointment to the Science Division chair.

At the second table, Judge Stone uncovered a single volume, bound in crimson leather and embossed with gold. The blank pages of the mock-up book simulating gaudy, turn-of-the-century literature. Picking up a touch of *Monkey II: It's All in the Genes* terminology, the subtitle and author's identify jumped out: *Darwinism Reviewed in the Context of Mendel's Law.* Brooke Fielding's scholarship lived.

Judge Stone explained. "Never again will this pioneering research be hidden from the public. The day Mrs. Carrington discovered the single remaining copy of the manuscript in her attic, she contracted with a publisher to produce 5,000 leather bound copies to be distributed gratis to colleges and universities around the world. Fifty copies will be retained for permanent reference in MAU's new science library—which not coincidentally will be named the *Brooke Fielding Memorial Library*. Dr. Fielding's original dissertation, together with his now-famous letter to Mrs. Carrington, will be displayed prominently in the library's lobby in a sealed glass case."

Looking at JT and Joshua, he added a footnote. "The library's core reading room will replicate the size and decor of the *Scopes* trial courtroom...to implicitly commemorate the recent, stellar performance of these two!"

Nothing further needed to be said. No question or comment arose. It was a stunning victory for innovative thought and academic freedom.

Continuing his guided tour, Judge Stone shepherded his wide-eyed observers to the final exhibit. "Now, over here, we have a treasure destined to be a centerpiece in the entry lobby to the *Parker Daniels Science Center.* The exhibit embodies the essence of one man's honest quest for truth."

He flipped the concealing velvet mantle, letting it slip to the floor. Morley gasped audibly as the color drained from his face. He blinked rapidly, trying vainly to conceal his shock. There lay the fragments of the long-missing fossil, Archy! A bronze plaque described its discovery, theft, and eventual recovery.

His revelations now complete, Judge Stone returned to the judge's chair and present reality. His demeanor shifted to a steely calm as he prepared to spring his trap on the unsuspecting prey.

"Before we adjourn, it seemed appropriate to share inside information concerning one-time student Slade Lassiter and Archy's mysterious disappearance. Ms. Saunders...excuse me, *Vice-President* Saunders, perhaps you had better enlighten us."

An uneasy Morley hoped the news related exclusively to the fate of the accused felon Lassiter. A touch of color returned to his cheeks. Appointment to the top post in the university's Science Division now seemed within his grasp. He looked to Chancellor Comstock, hoping for an encouraging word. Instead, the newly appointed Vice-President began a matter-of-fact commentary on the travel adventures of that troublesome fossil, Archy.

"Kingpins in a drug ring operating on the East Coast have been under surveillance by federal and state agencies since early this year. Cocaine shipments originating in South America used a variety of overland routes to supply their markets in the District of Columbia, Philadelphia, and New York City. The ring experimented with a wide range of innovative means of transport. Mr. Ryan's Tennessee project inspired the devious idea of moving a trailer full of cocaine from Center Hill Lake to Montgomery County.

"The thief who highjacked the camper and its fossils got a bit greedy, double-crossing his own gang. Before reaching Maryland, he dumped the contraband on the street for a quick buck and he delivered Archy to the Smithsonian, hoping to make a sale. Then he torched the camper and abandoned it in the Maryland woods. Once he got into business for himself, he lied to the syndicate, claiming the load had been highjacked by a band of rival thugs."

Josh Ryan listened, incredulous, then asked JT point blank, "How long have you known about this?"

JT sidestepped. "You'd better listen to Jessica."

Ms. Saunders proceeded with more revelations. "Inadvertently, the Tennessee highjacker laid a trap for himself. His gang bosses held him personally accountable for the loss, demanding reimbursement for the estimated two million dollars they had paid the drug cartel for the shipment. Knowing his life was in danger, he coughed up the cash proceeds left from the street sales...and embezzled the balance from the Science Research Foundation."

She paused, eyeing Striker and Morley. "You see, Dr. Striker, it was someone on your own team who set up the phony reimbursements to Josh Ryan and several other grad students."

Numb with embarrassment, Striker struggled for composure. For a drug-dealing, thieving felon to be operating under his nose was humiliating enough. But for such a person to willfully undermine the financial integrity of his foundation—the financial backbone for his ego-trip research—rated as a capital crime in his book. But all he could do at the moment was grind his teeth and be grateful for the face-saving appearance of the Leslie Anne Ryan Foundation.

His peers uniformly acknowledged the abrasive Striker to be woefully deficient in personal charm, but despite his notorious arrogance, no one had suspected that the unpopular professor or his staff were embezzlers and drug dealers!

Alarmed by the innuendo, Gus Morley venomously accused an obvious target. "It's that damned Lassiter, isn't it?"

Although Morley's anger surprised no one, the outburst contradicted his traditionally benign image. Lassiter had basked in the confidence of Dr. Morley, functioning as a minion under the professor's supervision. Now, the anxious teacher welcomed the opportunity to focus attention on a culprit already indicted and behind bars.

Appearing grievously wronged, Morley shook his head in sad disbelief. His words conveyed the angst of his soul. "Sometimes you just can't ever tell about a person—even someone you trust! Believe me, this news comes as a terrible disappointment!"

"That's certainly understandable, Dr. Morley. Lassiter operated right under your nose. For what it's worth, everyone in this room is entitled to know that this scum-bag sold his soul to his drug addiction. He copped a plea shortly after his arrest, confessing to a laundry list of felonies."

Striker and Brogan sat speechless, betraying no emotion.

Morley began to fidget, growing more and more agitated. "It's not only disheartening to hear this about one of our own students, it's downright disgusting! What, pray tell, was the nature of the confession?"

"He confessed the works, Dr. Morley. Pick a crime: transporting, using, and selling illegal substances; torching Mr. Ryan's Mustang as well as the science foundation's camper; attacking Mr. Ryan with a deadly weapon; harassing phone calls; and of course, some major-league grand theft!"

The revealing litany induced anguished exclamations from Morley's contorted face. "Shocking!"

Jessica Saunders wasn't through. "The independent audit extended beyond this recent incident. The trail of evidence suggests that the siphoning of Science Research Center accounts predates Lassiter's arrival on campus."

"Just what are you trying to suggest, Ms. Saunders?" Morley demanded angrily. "Those funds have been collected and dispersed with meticulous care. I've seen to that. I hope you're not insinuating..."

"No, I'm not insinuating, Dr. Morley. I'm accusing! The trail of fraud leads directly to this room!"

Striker was mortified by the bare-knuckled onslaught. Brogan's vacant stare effectively cloaked whatever reaction he might have had. Gus Morley's mouth went dry, and his face turned chalky white. Sitting motionless, he repeatedly clenched and unclenched both fists. The strained voice wheezed and whistled, reedy thin.

"Now look here, young lady, don't you dare play games with me. There's no way you're going to tie me to the criminal conduct of a conniving maverick student."

"I don't have to, Dr. Morley. Your protégé, Slade Lassiter, has done more than cool his heels in the brig these past few hours. After a night in the clink, facing prolonged hard time up the lazy river, the $100,000 you promised him just wasn't enough to buy his silence. All day long he's been singing like an off-key mockingbird—pointing fingers, naming names, confessing that he was just a low-level lackey."

Morley's breathing became labored.

"To be quite blunt, Dr. Morley, Mr. Lassiter alleges that you introduced him to the use of cocaine; that you made him earn his way as a runner for drugs and money; and that for several years your *modus operandi* has been to recruit and hook students to cocaine—and all to line your own pocket!

"Word is, you've been looting the treasury for years. Lassiter confessed that he took lessons from your devious example, got greedy, and after the Archy heist, did some siphoning of his own to cover his tracks.

"In short, Sir, you've been fingered a narcotics kingpin, and are charged with embezzlement. There's also the suspicion that you were the gunman in the woods who attempted to murder Mr. Ryan Thursday night.

"As they say in the trade, the jig is up!"

Jessica Saunders stared at the crumbling visage of the "good-ol'-boy" campus fraud. Emotionally drained, the deflated hulk slowly pushed back from the conference table, turning crimson and finally purple with rage. The roundtable of witnesses watched in horror as demonic rage twisted his features. He started to shout at his accusers, but his voice trailed off in a strangled curse. Rising unsteadily to his feet, eyes glazed with hate and hands trembling, good ol' Gus rushed out the door.

Pious pretender, fake friend and mentor of students, drug addict, and narcotics kingpin, Doc Morley exited Mid-Atlantic University's administration building for the last time. The instant the door closed, Jessica Saunders moved like a cat, responding to the instinct to pursue and outmaneuver evil. Looking first to the judge and then to the chancellor, she announced, "I'm on my way, Gentlemen. I have some unfinished business to attend to."

She shot out the door, hot on the heels of the fleeing villain.

Exasperated and undone, Striker sat with his face in his hands. The worldly wise Brogan admitted candidly, "Could have fooled me." Pace Terhune thought, "Great plot for a movie."

Only Michael Ryan betrayed no surprise. For twenty-five years he had been exposed to a side of the Morley facade carefully concealed from the public. Josh felt squeamish, astounded at his exploited gullibility. The chancellor, JT, and of course the judge had been privy to every aspect of the investigation.

Judge Stone broke the silence. "Since our business has been completed, this ad hoc meeting stands adjourned. However, I would appreciate it if the two Ryans would join Dr. Comstock and me in the chancellor's private office. Shouldn't take more than a couple of minutes."

As the remaining eight conferees rose, Chancellor Comstock volunteered an ironic twist. "Gentlemen, there is a small footnote to history that Ms. Saunders neglected to share. The Smithsonian's assessment of the fossil Archy confirms that these are the bones of a very old bird—but not an *Archaeopteryx*." Then looking at Josh with a twinkle, he added, "Hopefully, you won't feel offended that the Smithsonian didn't seem impressed with Archy, seeing as how it looks to be just another Pleistocene era fossil prairie hen. The market value of the bird specimen rates not a blip on the financial Richter scale."

The Ryans led the lighthearted chuckles that greeted the disclosure. Even the newly dethroned Science Division chairman cracked a condescending semi-smile.

# 45

Wood Comstock's private office looked nothing like what Josh had imagined. Having never passed through the imposing double doors before, he had no idea what to expect. Unpretentious and relatively small, it seemed far less spacious than Karl Striker's office. A triangular brick fireplace, squeezed artistically into a corner of the room, beckoned visitors to occupy reading chairs arrayed in a cozy half-circle in front of the hearth.

Rather than traditional trappings of power and prestige, the walls bulged with shelf after shelf of leather-bound books—the chancellor's private library. Book titles shouted the preferences of a history buff—including everything from an original set of Gibbon's *Decline and Fall of the Roman Empire* to eyewitness accounts of the American Civil War. First editions of Darwin's *Origin* and *Descent* shared shelf space with a dozen Bible translations, including a collector's-edition replica of a Guttenberg original. Each book represented thoughtful choice, some mastered, most read, and all at least scanned by the chancellor.

Wood Comstock insisted that his visitors make themselves at home on his turf.

Once seated, Judge Stone basked in the academic hospitality and waxed expansive. "Well, gentlemen, I trust you've enjoyed this unusual *tête-à-tête*. Striker is stiff-necked and opinionated. His nice-guy genes must have mutated. But the man is brilliant...and no way is Karl a crook!

"As for Morley, he's more than crooked! He's a fumbling felon, a con-artist disguised as a do-gooder. He fooled most of us."

Michael Ryan offered a cryptic assessment. "He never fooled all the people, Judge!"

Josh pondered his father's insight. The judge ignored the comment, anxious to call it an evening. "Gentlemen, the chancellor and I invited you to share this private moment to top off what I consider a slam-dunk event.

"First of all, Michael, you've earned your spurs as a professional investor in Wall Street markets as well as a computer industry consultant. Knowing the distance that has existed between you and your mother-in-law all these years, it may come as a shock to learn she has asked me to recruit your services as chairman of the board of the Leslie Anne Ryan Foundation."

Michael Ryan could not have been caught more off guard had the invitation summoned him to replace the elected occupant of the White House.

Ed Stone reviewed Lady Carrington's relationship with Michael Ryan as he explained the nomination. "She recognizes the risk of your misinterpreting this as a hollow gesture to erase the years of exclusion and personal pique. Mrs.

Carrington was emphatic. She is making the offer on merit alone, hoping that this responsibility will not interfere with your other business interests. Above all, it would be a wonderful way for you to honor the memory of Leslie Anne!"

Michael struggled, speechless and undone. "Wow!...I don't know...it's just that..."

Josh stepped to the plate. "Dad, you don't owe the Duchess anything, but I think you owe it to Mom."

Misty eyed and vulnerable, the computer master acquiesced. "Yeah, sure, of course I'll do it...certainly for Leslie Anne! Maybe it's also time for me to reach out to the old lady on the hill. For sure, she's changed her tune. I'll try to sing harmony."

Chancellor Comstock chimed in, confirming closure. "That settles it! The *Leslie Anne Ryan Foundation* trusteeship rests in good hands. With Pace Terhune, Jennifer Saunders, and the Duchess already aboard, it looks to me like Dr. Striker will be locked securely in financial checkmate."

Thought of the power move had not yet occurred to Michael. Shaking his head in bewilderment from the unlikely turn of events, he tried to be objective. "The professor is a bright guy who means well...I'll try to be fair."

The Ryans made motions to leave but Judge Stone resisted. "Not so fast, you two! The night is young, Gentlemen.

"You may recall that the Leslie Anne Ryan Foundation's fifth trustee is a Pan Oceanic Petroleum vice-president, a geologist. Although worth his weight in gold to the corporation, the seventy-five-year-old scientist plans to hang it up for golf and fishing in the Ozarks. He's postponed retirement repeatedly, waiting for the corporation to line up a top-notch replacement.

"Ready or not, he plans to call it quits, effective July 1, 1997. The search for a successor with blockbuster credentials is more formidable than you might think. But I'm pleased to report that Pan Oceanic identified their man in recent days. I've been commissioned to see if I can lure him to the oil fields."

Michael Ryan expressed curiosity, given his board chairman assignment. "Sounds good to me, Judge. I own a chunk of Pan Oceanic stock. Do I know the candidate?"

"As It happens, you do, Mr. Ryan. I've been empowered to do what it takes to recruit a young Ph.D. candidate named Joshua Chamberlain Ryan." After pausing for effect, he elaborated, with a shake of his head, "Word is that the poor fellow longs to be married despite his current unemployed status!"

The senior Ryan looked at his astonished son, while the wide-eyed son stared back. Both were too astonished to speak, so the judge jumped in: "I should warn you, young man, not to expect compensation at the level of a senior executive. Despite the considerable responsibility, you'll be the youngest in seniority. That means your initial compensation approximates what my law firm pays its new associates, a bit short of six figures—enough to cover the rent. But your remuneration will rise with time."

Josh listened impassively, trying to get his emotions under control. The judge delivered the clincher with Dutch-uncle pragmatism. "Forget the financial aspects, young man. It seems to me this assignment puts you in a position to be paid handsomely for your lifelong hobby—chasing fossils and scouting rock strata. The world is your playground; now you can play for pay!"

The senior Ryan returned his son to reality with a burst of proud humor. "Go for it, Son. Just think! It'll let me off the hook in paying you that exorbitant two hundred bucks a month for managing Fort Ryan!"

The quartet chuckled as the still-incredulous Josh managed some dry humor of his own. "By a strange coincidence, Judge Stone, I expect to return to the job market next July. If I'm gainfully employed by the time I marry one of your new law associates, you may be spared from having her pressure you for a raise early on! Where do I sign?"

Instantly, Josh grasped the weird paradox of power. "If this means I'm to be the fifth trustee for the Leslie Anne Ryan Foundation, my old pal Striker may end up having to answer to me. He may not call for a celebration when the truth dawns." He could scarcely conceal a mischievous grin. Then he replayed his dad's words, guaranteeing to be fair.

With a sly wink, Chancellor Comstock descended from his throne of academia to supplement the irony. "Just think, Mr. Ryan. Every future foundation project will require the initialed clearance of Joshua Chamberlain Ryan, the only trustee credentialed as a geologist/paleontologist.

"Research Director Striker will have to learn to cope with your looking over his shoulder. Once the good doctor realizes his career prospers at your whim, he's liable to snap his heels and salute you at the hint of an 'Achtung.'"

"Justice triumphs," exulted Judge Stone.

Handshakes sealed commitments. Laughter capped the evening. But more surprises awaited the Ryans. Suddenly, the wail of sirens pierced the air. The orange glow of raging flames could be seen just over the ridge—in the direction of Morgan Manor.

# 46

The direct route to Morgan Manor crossed a confusing quagmire of unmarked dirt service roads. Michael drove with reckless abandon in the general direction. Despite erratically gouged ruts and the slosh and splatter of mud puddles, he managed to knock at least three or four minutes off the time it took to traverse the usual route. Maneuvering with the skill of a race-car driver, he finally crested the last hill. Father and son were relieved to see that the Manor House had escaped the conflagration.

"Looks like the barn, Josh. There's so much smoke, it's hard to tell for sure. I'm going to try to get closer."

They had forgotten that some one hundred yards from the fabled Victorian Morgan Manor Stables, a padlocked steel gate formed a forbidding barricade to all vehicles other than an Abrams tank. The Ryans careened to a stop, inches short of the barrier.

"Let's see what we can do," Michael shouted. Vaulting the gate, he took off running toward the conflagration's red-amber glow.

Josh followed in his wake, but was hopelessly out of his league. The crippled left leg hobbled his best efforts. In a flash, his dad dashed yards ahead.

But the frustratingly slow pace positioned Josh to capture the scene's big picture at a glance. A stone's throw to his left, he spotted a shadow stumbling frantically to escape the licking flames, arms flailing. Rubbing his eyes, Josh thought he recognized the staggering stumbler, desperately clawing his way out of the smoke. Next he saw another, lithe figure streaking like a deer, intent on a collision course. It was Jessica Saunders in pursuit.

Josh recognized Jessica's voice barking at the fleeing arsonist to "Stop," declaring the suspect to be "under arrest." The frantic figure ignored the order, muttering gibberish. Suddenly, Josh recognized that the pitiful, whining figure was the overweight Morley, phony friend of students.

A struggle between the trim, athletic Jessica and the out-of-shape Morley was no contest. The black belt agent of law enforcement took Morley to the ground with two or three strategically placed blows and handcuffed his flabby arms behind his back. The surreal struggle took only a few seconds at most.

A stream of incomprehensible blather, salted with invectives, spewed out of Gus at no one in particular. His career of deception was at an end; his masquerade was exposed. Abject and prone, he was broken in spirit and wallowing in shame.

"How the mighty have fallen," Josh marveled.

Obviously Morley carried a burr under his saddle for the Morgans, but why? For whatever cause, good ol' Gus despised the Duchess and her heirs.

Having seen the artful Ms. Saunders yank the suspect to his feet and march him off in the direction of her car, Josh took off again in the footsteps of his father, trying not to think about what awaited him at the stable.

By then, the first fire trucks had arrived, backed by a convoy of emergency vehicles. Clearly, the oak-paneled Victorian barn could not be saved. Almost retching, Josh watched it collapse and crumble in smoke and ashes, erasing forever the remains of precious memories. His mother's cherished tack room memorabilia had been snatched away by the hand of hate. Cold fingers twisted his heart.

As he scanned the raging inferno, looking for signs of his dad, remorse and hurt turned to alarm. Josh had never said a prayer in his life and had no idea of the appropriate formalities. The best he could do came spontaneously from his innermost soul, simple yet profound.

"God help us!"

It proved a masterpiece of theology!

The instant the plea escaped his lips, he spied a figure battling close to the ground, coughing, choking, crawling away from the consuming torrent of heat. Forgetting his own safety, Michael was gallantly doing his best to drag something or someone away from the flames.

Galvanized, a relieved Josh looked heavenward, uttering an impassioned "Thank you." Then he rushed toward the inferno where he joined his father, tugging and pulling with every fiber of his being.

Finally, just beyond the searing ferocity of the flames, father and son stopped to gulp fresh oxygen. Josh recognized the fragile bundle in his father's arms as Lady Carrington. Drawn to the stable at sight of the first flicker of flame, she had abandoned the security of Morgan Manor and put her life on the line to save a horse.

Firefighters had moved in to attack the billows of smoke and glowing charcoal embers. Hissing steam mingled with smoke, weaving eerie tongues of fire befitting Dante's Inferno. Two firemen lifted Michael to his feet, asking if he was OK, and offering to carry the limp frame of the little old lady who had inhaled more than her share of acrid smoke.

Michael refused, saying, "Thanks, but I've got her. Let's get her some oxygen."

The firemen led the way to the wide-open doors of a waiting ambulance where Michael gently delivered his burden to a stretcher. The Duchess' deep-blue, bloodshot eyes popped open at the commotion. Taking Michael's hand in hers, she coughed her gratitude. "Thank you, dear heart. After all these years, now I owe you my life."

"You don't owe me anything, Duchess. You'd better lie still and sniff this oxygen."

With that, one of the medics placed an oxygen mask across the wan face. But Lady Carrington would have none of it. After several deep breaths, she pushed the mask aside, asserting defiantly, "Don't think for a minute that I'm some fragile old woman who needs to be pampered. I appreciate the attention from you handsome young men. But as any fool can plainly see, I'm as healthy as a horse."

At the word *horse*, her defiance melted, and she became somber. Seeing her grandson, she reached for his hand, and pulled him close.

"Oh, Josh, I'm so sorry...I tried my best...I tried desperately...If only I'd been a tad younger..." Her sad eyes misted as her voice drifted off in the darkness.

Anguish crushed the soul of Josh Ryan. Given the stress of the moment, he had completely forgotten that Turbo vacationed within the luxurious spread of the Morgan Manor Stables. He grasped the picture reluctantly.

"That awful man did it, Josh. I saw him lighting matches and throwing them into the straw. If I could have put my hand on a shotgun, I probably would have pulled the trigger...but all I could do was scream. He paid no attention until the blaze exploded up the stable walls. Then he ran like the coward he is.

"He hates us all, Josh...He hates me in particular. But he hates your father, the memory of your mother...and you!"

"But why does he hate us, Grandma?"

His words startled but pleased the old lady. For the first time, he had called her, "Grandma."

"Many reasons, my son. You'll understand more after you read your mother's diary. He never forgave me for blocking his appointment to head up the university's Science Division. Thanks to me, he's had to play second fiddle to Karl Striker all these years. The diary reveals much more...

"It's time for you to do that outside reading I assigned!"

By the time her personal physician arrived, the Duchess had propped herself up, determined to prove she was ready to hike back to the manor house unaided, as sprightly as any 20-year-old. Spying her doctor pushing through the crowd, she issued a proclamation even he dared not ignore.

"Just in case you're wondering, Doctor Pence, I'm not going to allow you to spirit me away to the hospital tonight—or anytime soon, for that matter."

Everyone admired her feisty performance. The attentive physician had no inclination to twist the tail of the tiger.

"OK. You look like a lucky lady this evening, and I'll respect your wishes. However, there are conditions: To spare exertion, I insist you relax on the gurney so these guys can wheel you safely inside for a good night's sleep. Pretend this is your throne, Duchess, and these paramedics are your private courtiers. I'll hang around to make sure you don't get too rambunctious."

She complied meekly, without protest. The caregiver knew how to play her games. She could still spit fire, spout vinegar, or tell jokes—whatever the occasion demanded. This evening she settled for the role of gracious hostess.

"Would you like to join me in a dose of Chardonnay, Doctor? Or do you claim some less tasty prescription more appropriate?"

The physician ignored her impertinence. Plainly, the strong-willed Duchess probably didn't need the stretcher transport, but as the old saying goes, she "went along for the ride."

Throughout the exchange, the Duchess clung tenaciously to Josh's extended hand. As they rolled toward the manor house, she tightened her grip, pulling her grandson close to her face to share words meant solely for his ears.

"Josh, my dear...I'm devastated about your horse. I know what Turbo meant to you. Losing a favorite Arabian champion hurts almost as much as losing a member of the family. Believe me, I know. I've lost some fabulous thoroughbreds, and still ache something awful..." After a labored pause, she poured out her ultimate hurt.

"Then I lost your mother, Josh! I don't want to lose you, too!"

Her cheek close to his, he could smell the smoke that saturated her disheveled hair. He had no idea how to respond. Finally, he pulled back, unnerved at sight of her brimming sapphire eyes. He mustered a thin smile, trying to conceal this excruciating moment of grief—the irreplaceable loss of his gallant mount, Turbo.

As the stretcher moved closer to the mansion, the Duchess again attempted to negotiate peace with her grandson. Tugging on his hand hard enough to stretch heartstrings, and with tears cascading down both cheeks, the haughty Lady Carrington threw pride and caution to the wind.

"I love you, Josh! You are my darling grandson! I will love you forever!"

Waving farewell as she was whisked away, she beamed valiantly through her tears. "I've always been a lucky lady. I hope someday I can beat those random chance odds so you can find it in your heart to forgive...and to love me, too."

Always in command, she managed one last order, disguised as grandmotherly concern. "Now you best run along home, and get some rest, you hear me? You want to look your best next time your fiancée sees the love of her life!"

With that she was gone.

At the age of five, Josh Ryan had mastered the art of choking back tears. Consequently, as the now fragile wisp of a woman was carted off, no reciprocal tears or words from Josh Ryan jeopardized this well-practiced stoicism. But he noticed mist in his dad's eyes. Excusing himself, he sauntered off to the barn. When Michael started to follow, Josh admonished, "Thanks, Dad, but I've got to deal with this alone."

Josh tentatively approached the smoldering ashes, dreading what he would find. He could make out the profile of a massive, motionless form stretched out within a few feet of the blackened ruin. The Duchess had almost succeeded in leading the stouthearted Turbo to safety. But when it came to matters of life and death, almost was never enough.

Despite the surging waves of blistering heat, Josh never slackened his pace until he knelt to stroke Turbo's terribly singed mane. Pain sliced at his heart, throbbing deeper than the physical trauma of the crippling injury to his leg.

Clumsily pulling the noble head into his lap, he tenderly cradled the lifeless form. More than any time since that day early in his life when he swore never to cry again, he wanted to shed tears. But he couldn't. Instead, he sorted through his memories, conversing in low tones.

"I'm sorry, old friend...to lose you like this. There's no way to excuse evil in the lives of humans. It's terribly unfair and unjust that you had to suffer. For what it's worth, I'll remember all the happy times...You were a turbocharged con-artist, but you were always good at it...You kept me guessing most of the time.

"This week's *Ride to Glory* will outlive us both...

"We sure fooled them, didn't we...you and me! How I wish I had found the time to bring you home to Fort Ryan before this...I just didn't suspect..."

For a long time, Josh sat silent, senses tuned to the angry hiss of steam and showers of sparks. Strange irony: two arsons; two days; two lost mustangs—one a replaceable, inanimate package of steel, the other a living friend. Considerate firemen milled about the grieving horseman, dousing hot spots and mopping up.

As he rose from the ground with a whispered, "So long, old friend," a new commotion at the main entrance to the manor caught his attention. Surely the Duchess had not suffered a setback! The jumble of emergency vehicles with their flashing lights illumined small clusters of spectators mingling with officialdom. Michael and Jessica Saunders huddled to one side in sober silence.

"What's going on?" Josh asked, trying to maintain his equilibrium.

"You might say it's a final footnote to what has been an unusual day," Michael replied, without a hint of expression. Motioning to Jessica, he suggested, "Why don't you brief him."

She hesitated but didn't mince words. "Justice sometimes takes slow, unpredictable routes. Tonight's turn of events proves that the wheels of justice can also grind at lightning speed!"

Methodically she documented Augustus Morley's abortive escape attempt. "Handcuffed in the back seat of my van, Morley was awaiting the arrival of the Montgomery County police to take him into custody. In a careless moment of urgency, I forgot to take the keys from the ignition. Somehow, he either crawled between the bucket seats or managed to exit the back door and enter the front seat. Trying for one last dash to freedom, he engaged the ignition, wrestled the car into gear, and with his foot on the accelerator did his best to steer with his knees. His hands were shackled behind his back."

She paused, looking first at the senior Ryan before eyeing the son.

"Did he get away?" Josh asked.

"In a manner of speaking, yes. Dr. Augustus Morley will never stand trial!" After hesitating, she added, "Nor set fires...nor deal drugs...nor steal...nor kill..."

Momentarily puzzled, Josh waited for details.

"The car accelerated dangerously as it headed downhill. Morley must've hit seventy or eighty miles an hour. But the guy couldn't steer. Careening full tilt, he could neither turn nor stop. At the bottom of the incline, where the road intersects the main highway, he managed to miss the traffic, but crashed headlong into the giant oak at the base of the slope. He never had a chance. He died instantly."

Stunned, Josh ventured a spiritual appraisal of human life. "I sure hate to see anyone die...even a gutless felon!"

Jessica Saunders eyed the Ryans cautiously, waiting for Josh to recover his bearings before continuing the briefing.

"Before you mourn prematurely, Josh, remember that Morley was wicked and rotten to the core. He lived a lie, pretending to offer a sympathetic ear to students while badmouthing all of you to Dr. Striker whenever it served his selfish purposes.

"A ruthless narcotics kingpin, he dragged more than one susceptible student into the evil web of drug dependency."

Still sorting things out, Josh wondered, "Was it intentional?"

"Accident or suicide—we'll never know! Most of us think he killed himself.

"One other thing—you know that someone with a bad aim tried to murder you after the show Thursday evening. Investigators found two empty shells in the woods next to the Chamberlain Playhouse parking lot—the fingerprints belonged to the late Augustus Morley!"

The shocker shook both Ryans. The son groped for answers. "But why? It doesn't compute. Lassiter tried to slit my throat. Was he in cahoots with Morley? I can understand Lassiter, but it doesn't make sense that two operated in tandem to try to rub me out!"

"You thought Lassiter was Striker's sycophant. In reality, he was Morley's drug-dealing puppet. As far as wanting you rubbed out, the mystery doesn't correlate outside of one overpowering fact: For reasons unknown to the police, Morley hated your grandmother with an obsessive passion. His vitriolic malevolence overflowed to encompass the Ryans—or anyone related to Lady Carrington!"

Deep in silent reflection, neither Josh nor Michael ventured a response.

Jessica Saunders stepped in to banish the shadows. "It's time to celebrate life, young man! As we speak, the two of us are in the market for new cars...And from the looks of things, the perks that go with our new job assignments should pre-qualify each of us to pick any model we choose!"

The incomparable Jessica Saunders always performed her sworn duty with professional cool. Flashing a scintillating smile to her friends, the Ryans, she bid them good evening and turned to complete official business.

# 47

"Son, most people live a lifetime without the drama you've managed to jam-pack into these past seven days. Can you handle two fatherly observations?"

Josh nodded receptively, as Michael smiled affectionately. "I'm about the proudest dad that ever walked this earth! "You are something else! Gutsy, brilliant, brave...and, thank God, alive! You've even managed to convince that little Traci you are one handsome young dude—as good lookin' as you are smart."

"Whatever it takes, Dad."

They both chuckled.

"The other thing is purely a fatherly suggestion. I can't tell you what to do, but it's obvious your Grandma Carrington loves you and is suffering a major guilt trip for the shabby treatment and short-shrift she's given us for most of your life. She's done her best to apologize, telling you she's sorry. And she's canny enough to know she'll never make peace with me unless she first makes peace with my son.

"Love can't be commanded or coerced. Your grandma is finally getting that message. She's a late-bloomer, for sure, and slow on the uptake, but better late than never.

"She had a close call tonight, risking her life because of love for you. Just maybe you owe her one, Son. I'd like to think you can make peace with the Duchess. And once you've crossed that bridge, love might sneak up and surprise you. It can grab you anytime! But when it strikes, love lasts forever!"

Josh contemplated in silence. He didn't react until the car rolled up to the Fort Ryan entry. "I appreciate the advice, Dad. You've never given me a bum steer yet."

As Michael dropped Josh off at the doorstep of the mill house, he shifted subjects, offering moral support. "If you're up to it in the morning, I thought we'd head back to Morgan Manor to oversee Turbo's burial...and maybe pop in to check on your Grandma Carrington."

Strolling pensively toward the cottage, Josh kept mulling his private thoughts. Finally, unlocking the front door, he turned around to endorse his dad's message of forgiveness.

"Count me in, Dad! Just one more thing troubles me."

"Sure, Son. Whatever!"

"Could we talk about why Morley despised the Duchess and her clan? Although Jessica told what she knew, I got the feeling that you could fill in some blanks."

Bearing a pained, faraway expression, Michael nodded acquiescence and mumbled again vaguely, "Sure, Son...whatever."

With that, he disappeared. Emotionally drained by the day's kaleidoscopic ups and downs, Josh concentrated on the highs, feeling exonerated and optimistic about the future.

Josh reawakened to reality as the clock chimed eleven strokes. The upside-down, topsy-turvy week would close with all the flourish that had trailed the final curtain of *Monkey II.*

Suddenly Josh remembered the candlesticks fashioned from the wood of the fallen Morgan Manor tree. Fishing through a kitchen drawer, he found a match, struck it, and for the first time sat alone in candlelight. He remembered the Duchess' words urging him to reach back in time and to read his mother's diary. He stepped over to the bookcase, lured by the secrets hiding in the gold-embossed *Diary of Leslie Anne Carrington-Ryan.*

Josh sensed correctly that the remaining family mysteries were about to be unveiled. By the flickering light of the candles, he eagerly scanned the exquisite penmanship.

News that his name would have been Regina Morgan Ryan had he been born a girl drew a smile. He laughed out loud at his doting mother's biased description, picturing him as *"the handsomest baby boy ever born."* He felt trepidation in the remark, *"I can't wait to bring Joshua home to introduce him to Mom Carrington. I'm sure she won't be able to resist his charm and all will be peace between the Ryans and the Carringtons."*

Leafing back, he found the entry memorializing his parents' wedding day.

*"Michael is the most magnificent man I've ever met. Although I know Mother wanted an elaborate church wedding in the worst way, she's given us no choice but to elope without her knowledge or permission. Surely someday she will understand some things too painful to share with her yet. When she does, I'm sure she won't be able to resist her handsome son-in-law.*

*"Michael is an all-time prize."*

Several weeks after the elopement, she wrote with bitter disappointment:

*"I thought I knew my mother, but I never realized the height of her pride nor her capacity for anger. Except for her blunt note advising that I have been disinherited, she refuses our calls and returns all letters unopened. Michael didn't marry me for money and I'm sure we will get along fine, regardless.*

*"We can handle being cut out of her estate. Being cut out of her life is something else...I'm heartsick at this tragic state of affairs."*

The poignant words described the harsh edges of a grandmother he had known little about during his short lifetime.

In contrast, the guileless sentiments that spilled from the pen of the mother he never met offered catharsis and vindication. Obviously, his mom lived a three-dimensional life. She was every bit as articulate and intelligent as beautiful. Feeling as he did about Traci, he had an enhanced perception of the unbreakable link of love.

The pages of the diary related a number of quaint anecdotes about family history, exuberantly recited by the vivacious great-granddaughter of Sam Houston Morgan. Lady Carrington's transfer of the diary to his custody came at an auspicious time—whether the gesture was coincidental or calculated.

Transfixed by the treasure trove of stories about previously unknown events, he raced through the pages. The diary offered more than routine, day-by-day accounts of a young woman's life. It reached into her soul, arranging colorful scraps of childhood memories into the tapestry of a vibrant and interesting life. Hungrily he devoured each handwritten syllable.

At a very young age, Leslie Anne had displayed rare musical talent. At one time she considered a career as a concert pianist.

Friends engulfed her life, most genuine, some hangers-on, grasping for the golden ring of life. Her all-American wholesomeness leaped from every photograph, but until now, Josh had not realized that his mother's beauty had earned her recognition as runner-up in a Miss Maryland contest. It came as no surprise that she cherished horses, perpetuating the Morgan family's reputation for reckless riding.

Regina Ann's legendary socialite life left her daughter singularly unimpressed. Still, Leslie Anne adored the Duchess. Although she enjoyed kicking up her heels occasionally, she shied away from phoniness and make-believe. She attended the Duchess' church of choice, but never joined. She believed in God, but rejected mechanical ritual and self-righteous posturing.

Michael Joseph Ryan was not aware of the diary's existence. The day Leslie Anne died, the distraught Duchess stowed all her daughter's childhood personal effects, including the diary. Although Michael had been wracked with agony by the unexpected loss, the Duchess left him dangling, deliberately denying him access to the refuge of his wife's reminiscences.

The chimes of the grandfather clock announced that midnight was but thirty minutes away. Josh read on, paying no heed.

As the candles flickered, Josh stopped reading long enough to return to the present, imagining how much his mother would have liked to share the momentous events of the past week: His engagement to Traci; the dramatic dismissal of the civil litigation; a derailed doctoral program back on track; the arsonist's destruction of his Mustang and Lady Carrington's barn, and his loss of Turbo; the unsought career placement with Pan Oceanic Petroleum—his mind raced over the jumble of events that had occurred since the start of his performance on the *Monkey II* stage. Josh felt certain his missing mom would have blanched at the pain and been overjoyed at his successes.

Returning to the diary, he jumped to the last page, reaching eagerly for the bottom line. The penmanship lost its exuberant style; the writing was shaky, the sentences abbreviated.

*"Don't feel as strong today...Doctors say blood loss...wish Michael had not been called to New York business emergency...bad timing, just after Joshua's birth...trying to reach him by telephone...glad Mom coming soon...I'm sure I'll be OK...Joshua, such an awesome baby, needs me...dead ringer for his dad...feeling tired..."*

That was it! The end!

Pausing again in reverie, Josh realized he had yet to encounter any clues about the alienation that had banished the Ryans from Morgan Manor. He thumbed through the diary, scanning the pages.

More anecdotes surfaced. Clearly, Leslie Anne enjoyed social and romantic attachments typical of a maturing teenager. She complained of a few back-biting peers, accusing her falsely as *"fickle and stuck-up...I think I like most everybody. 'Particular': guilty as charged! 'Stuck-up': never, I hope."* She dated quite a lot but admitted, *"I'm never going to fall for just anyone, even if I end up locked in single bliss."*

Most of her social encounters described the bubbling enthusiasm of carefree youth. But one contact struck a sour note, in large part because *"Mom thinks he is a real catch."*

The grad-student *catch* posed as a big man on campus. Average in appearance and academic performance, he made up for the shortfall by artfully exaggerating his mediocre credentials. He caught the attention of the Duchess, who remained stubbornly blinded to his inadequacies.

*"Mom thinks this guy deserves my attention. I wish she hadn't set up the party designed for us to meet 'accidentally.' The guy comes across as shallow and boring. Usually she can see right through people, but in this case she seems blissfully unaware that he's a cunning, self-serving opportunist. Any interest he shows in me would be motivated by the Morgan/Carrington connection. Of this I am positive."*

On another page, the mother-daughter dilemma emerged full-blown.

*"At Mom's insistence, I went out with the jerk. If she knew what I know, she'd have his scalp. He cheats on tests; he drinks excessively; and while I can't prove it, I suspect he does recreational drugs. For my money, he's a low life con-artist. As for romance, this guy has a one and only love—himself."*

There was a two-week gap in the diary's chronology. In the next entry both elation and concern spilled across several pages.

*"The tea dance in Annapolis proved more than I bargained for. I met the cutest guy. 'Love at first sight' may be a cliché, but I don't think it's simply the Navy uniform...although I'll gladly admit Michael is one handsome young plebe.*

*When Mom asked me if I enjoyed the trip to Annapolis, I made the mistake of mentioning the meeting with Michael. In return, she provided an impromptu ten-minute lecture warning about 'sailors who travel the world and keep a girl in every port.' She seems to have her mind set on my finding some academic Einstein wanna-be."*

Subsequent pages read like a Gothic novel. The romance with Michael Joseph Ryan flourished. The grad student never stood a chance. To say she was "distraught" was gross understatement of the Duchess' ballistic reaction.

*"Michael is fully aware of Mother's hostility. To his credit, he never reciprocates. He is the kindest, most loving person I've ever met. Yesterday, while Mom ritualized in her 'Church of the High and Mighty,' Michael and I rendezvoused at the Jefferson Memorial. In front of God and all those tourists, he knelt on one knee then came right out and asked me to marry him. I exclaimed 'Yes,' almost before he finished. We even set our wedding date. I'm looking forward to that traditional Annapolis ceremony where the happy couple leaves the church under the canopy of crossed swords pointed skyward. We agreed to keep our plan secret—for obvious reasons."*

Josh mused, "Dad told me it was true romance, but he sure skipped a lot of the juicy details." A few pages further, the language of the entries sounded more ominous.

*"The campus 'admirer' is nothing if not persistent and demanding. He refuses to take 'No' for an answer. Sometimes I get the creepy feeling I'm being stalked. He pops up at the most inopportune times in unexpected places. When I told Michael, he said maybe I had no choice but to tell the guy to bug off because of another commitment. Or in the words* of Michael: *'Better yet, I'll deliver the news directly in a manner he won't forget!'"*

She tried the strategy. A few days later she recorded her frustration.

*"This afternoon I was accosted in the parking lot. This time I shoved him away, announcing bluntly, 'I'm taken.' This 'sterling,' so-called rep of the elite academic community blew his stack. 'You'll be sorry. I guarantee it. It's probably that damned sailor.' I never used Michael's name but somehow he knew. Apparently, we have been followed without our knowledge."*

Barely a day passed before the next entry.

*"I'm heartsick! Mom confronted me with 'news' she had gotten from an informant and insisted on an explanation. I had no choice but to be up-front, hoping she would understand and give Michael a chance. Sadly, no such luck. Instead, with an angry, determined face that must have rivaled that of great-grandfather Sam Houston Morgan riding to the Battle of Gettysburg, she outlined a battle plan. Michael Joseph Ryan will never graduate from the United States Naval Academy. Mom doesn't issue idle threats."*

Michael had never disclosed to Josh the reason why he resigned from the Naval Academy. Whatever, it marked a turning point that eventually led the budding mathematician to divert his skills from service in United States Navy to the emerging computer industry.

His interest aroused, Josh leafed through the ivory pages, brushing away history's cobwebs. He stood at the brink of discovery. Barely three pages later, he reached a tawdry disclosure—the part the Duchess urged him to read the day she delivered the heirloom to his keeping.

*"The smirking grad student caught me unawares in the tack room of the Morgan stable. His mouth reeked as foul as his mind. I screamed for help, but all hands were working in the far pasture, out of range. I threatened, 'You'll be sorry.' I pleaded for him to go, but he moved menacingly closer. He sneered, 'Now I'm going to finish the job.'*

*"I reached for a pitchfork, but it was too late. The loathsome brute grabbed my wrist with demonic determination, fingernails digging deep. Spinning me around, he yanked me to the floor. At first I was afraid he was going to kill me. When the nightmare ended, I wished he had. I kicked, clawed, bit, and tore handfuls of greasy hair from his filthy scalp. It was no use. He was twice my size. I can't bring myself to give a blow-by-blow description here. The monster assaulted and then raped me like the vermin he is. He's the most devious, despicable excuse for a human being I've ever met!"*

Shocked, Josh Ryan found himself shaking in fury a quarter-of-a-century after the fact. The next few pages finished the heartrending account.

*"Michael rushed over immediately after I called. He wanted to hunt down my assailant like a hound dog and thrash him within an inch of his life. When I begged him not to because of the shame I feared would tarnish the Morgan/Carrington name, he insisted on calling the police. Again, I begged him not to, for the same reason. It seemed to me it would be more than Mom could bear. Her world of pretense and prominence could tumble. I worried that her social friendships might evaporate. Michael disagreed, but was cool-headed enough to recognize I was the victim and it was my call.*

*"However right or wrong to conceal the crime, the Duchess must not know nor be embarrassed. Mom has suffered more than her share of hurts!"*

The next entry appeared three days later.

*"Michael continues awesome! He's my rock of Gibraltar!*

*"He worries I might be pregnant as a result of the attack. He reasons that the best way to deal with the ugly incident is for us to arrange an immediate marriage in some out-of-the-way place at the earliest moment possible. This astounding love of my life assures me he could love any child of mine, irrespective of the father. I love him so much, of course I said 'Yes' to the elopement proposal. But I also told him I would never carry a child of a rapist.*

*More than anything, I want a family, conceived in love, not in an act of hate. The two of us must be the parents!"*

When the couple returned, married, the Duchess determined to sever forever the relationship between daughter Leslie Anne and her sailor husband. Her volcanic irrationality crushed the congenitally high-spirited Leslie Anne. Stalwartly resolute, Michael anchored her hopes.

*"Secretly I hoped against hope that once Michael and I were husband and wife, Mom would melt. Instead, the rapist widened the gulf. Since he alone had knowledge of the attack, he had to be the instigating source of the vicious gossip swirling through the Mid-Atlantic University campus that 'Poor Leslie Anne Carrington had to get married...The sailor must have gotten her pregnant.'*

*"By the time we returned to Magruders Mill, Mom wouldn't take my calls. Today a curt note arrived, hand-delivered to the door by one of her hired hands. Her blind accusations cut my heart! 'You have blemished the proud reputation of the Morgans and the Carringtons. Your reckless conduct is inconceivable. Make no mistake, you will never receive financial resources from the Carringtons. Neither will I tolerate contacts between us any time in the future.*

*"'You've chosen your destiny with the sailor.'*

*"When Michael returned home, I showed him Mom's note. I made him promise to never tell the real story to her or to anyone else. Gentleman that he is, he promised. Hugging me, he offered reassurance that as long as we had each other, true love would always bloom. He said that if we always stand tall, perhaps someday the Duchess will understand what really happened. He's right, of course, but she will be missed in our lives."*

The identity of the villain remained the only mystery in the diary. By this time, Josh had suspicions. The answer jumped from a page penned more than a year after the elopement.

*"Thank God I wasn't pregnant and didn't have to think about an abortion. Because of campus rumors, Michael shared my feeling that we should wait a few months before starting a family. By then, I would be out of school and his career as a computer consultant should be zooming through the stratosphere. Michael says he'll be happy with a boy and a girl; I say he deserves two of each.*

*"Hopefully, the first will be a boy. I want a clone of his dad.*

*"I've never been happier in my life. I do miss Mom terribly, but I keep hoping that someday she'll come around. Last I heard, though, she thinks that despicable Gus M. hung the moon and expects him to have a great future at Mid-Atlantic University."*

There it was!

The linchpin to the mystery of the family's bitter estrangement and the *Monkey II* harassments beginning with the Tennessee fossil dig had been exposed in bold relief—Augustus Morley, destroyer of lives and ultimate fraud!

Now Josh knew exactly what the Duchess discovered when she belatedly thumbed through Leslie Anne's diary. Michael Joseph Ryan had known everything, always. But forever true to his promise to Leslie Anne, he never hinted at the core of the problem, patiently taking the heat without complaint or explanation.

"Don't I wish I could be what Mom wanted—a duplicate of my old man."

This was the only thought Josh could muster. He had been denied the luxury of despising a living villain. But musing in the soft darkness, engrossed in previously unknown events, he luxuriated in the comfort of a spiritual cleansing of the soul.

Watershed events of the past seven days had shaped a psychological mirror that attracted critical evaluation. At first glance, he didn't much care for the profile that returned his stare. So what if Striker clung cantankerously to disputed views, yielding no quarter? In all honesty, Josh knew he had to plead guilty to the same malady—unconscionable arrogance. Suddenly he understood. As he was pridefully patting himself on the back for defending truth, Striker, the hard-liner, also fought stubbornly for his own, ingrained perceptions.

"I'll reach out and make peace with you, Herr Doctor. That's the least I can do," he said aloud. "At least you have the integrity to be your own arbitrary self. There's no reason two arrogant, poles-apart thinkers can't coexist! Who knows, maybe someday I can persuade you that mega-evolution is based on fantasy...and maybe I can find answers to my own bazillion 'I don't knows!'" He grinned. "I guess it's only a fat random chance that old dogs will welcome new tricks...But again, stranger things have happened!"

Josh thought of Bishop Daniels, and firmly committed himself to the pathway to personal peace exemplified by his life and teachings.

Newly created memories of a mother he had never met occupied a pedestal in his head and his heart. As for Michael Joseph Ryan, he qualified as a poster father-figure if such ever existed. Sharing with others the hurtful secret of Leslie Anne's cruel ordeal would have been easier, or even wiser, but for Michael Ryan, a promise was a promise. He let the true story of his courtship and marriage rest, allowing time and good sense to heal festering wounds.

Grandmother, son-in-law, and grandson—all had suffered from the deprivation of family ties and quality time together. But Lady Carrington came charging back, as best she knew how, to make up for squandered years. A quarter-century of alienation had dissipated in a single, momentous week. The ugly scars of estrangement had begun to fade, erased by love's gentle embrace.

Josh made a move to blow out the low-burning candles. The time had come to close the diary and to seize the future. Before heading for the sack, he rocked contentedly in Grandfather Magruder's decrepit wooden rocking chair. His thoughts skipped about in a flurry of merry circles, bouncing from his father

Michael to his mother Leslie Anne, then back again to images of his Grandmother Carrington.

Moments before midnight and the dawning of the rest of his life, a parade of names raced through his head: *Iron Pants*; the *Duchess*; *Lady Carrington*—all names from the past. At last, the time had come to discard what used to be and meet his honest-to-goodness living grandmother.

Stubborn pride, unyielding arrogance, and a childhood spoiled with material wealth had produced a domineering matriarch who was in reality an emotional cripple. The fragile stick of a woman he had seen breathing fire and smoke earlier in the evening was full of contradictions. After going out of her way during the week to reach deep into his heart, she put her life on the line Saturday night to try to save Turbo. In one respect, she failed miserably—the horse fell to the flames. But in the grand scheme of things, she had succeeded brilliantly, lighting a spark of love that had burst into a festival of reconciliation that drew together her fragmented family.

"Enough," he declared.

For the first time in twenty-three years, Joshua Chamberlain Ryan grasped the big picture. Lady Carrington had gone from a two-dimensional caricature to his own "Grandma Carrington." In his mind's eye, he visualized the frail dowager grasping his hand as the medics whisked her away to the comfort and safety of her manor house. In his head he heard her fervent declaration, "I love you, Josh!"

It was never too late to love!

The memory of her touch struck home, piercing his heart! The raging conflagration melted through his being, devouring the residual bitterness of rejection.

"I love you, too, Grandma! Do I ever...!"

As if on cue, the grandfather clock struck midnight!

For the first time since he was five years old, Josh's pent-up hurts and resentments crumbled. Alone in the blackness, he put his face in his hands and succumbed to a torrent of emotions.

Overwhelmed, Josh Ryan cried.

*******

Still sitting upright in the rocker, Josh dozed, then drifted into the deep sleep of the emotionally reconciled, inhaling the sweet *"peace that passes understanding."* An hour later, he awoke, feeling chilled, and retired to the bedroom for his most refreshing rest in a week. He had no way of knowing that events beyond his control conspired to smother his plans for future memories.

Mercifully, he was shielded from the grim headlines that prefaced the first rays of morning sunshine. Even as he slept, editorial offices at the *News Press* buzzed with late-breaking news as it geared up for Sunday's final edition. A routine, front-page filler had been pulled and replaced by an urgent Montgomery County news story.

Somber excerpts from the story revealed the tragic event:

### *SOCIALITE SUCCUMBS IN SLEEP*
### *DUCHESS DEAD AT 78*

*"Lady Regina Ann Morgan-Carrington, Montgomery County philanthropist and socialite, passed away early today from cardiac arrest. Saturday evening, she had suffered severe smoke inhalation trying to rescue a prized family horse trapped in a stable fire set by an arsonist.*

*" 'The exertion proved too much for a congenital heart defect she had concealed valiantly from closest friends,' according to her personal physician, who was present at her passing...*

*"Funeral arrangements have not been announced, but informed observers predict that long-time friend and confident, civil-rights champion Bishop Brock Daniels, will likely preside...*

*"Lady Carrington, affectionately dubbed 'the Duchess,' is survived by grandson Joshua Chamberlain Ryan, 23, Mid-Atlantic University grad student who attracted recent attention by his appearance in a mock campus trial and an audacious cross-country horseback ride...*

*"Mr. Ryan is believed to be the sole heir to the fabled Morgan-Carrington fortune which reportedly includes a controlling interest in Pan Oceanic Petroleum..."*

# ENDNOTES

*Ride to Glory's* fictional storyline incorporates authentic literary quotations in the text by highlighting the quoted words in italics and identifying the source in endnote citations. Some citations postdate the time frame chronology of the theme to maximize scientific accuracy and assure contemporary relevance. Although authors quoted may not necessarily agree with the interpretations offered, an effort has been made to avoid out-of-context references.

## Preface

\* Adrian Desmond and James Moore, *Darwin* (New York: Warner Books, Inc., 1991), p. 475.

\*\* Norman Macbeth, *Darwin Retried: an Appeal to Reason*, (Boston: The Harvard Common Press, 1978), p. 5.

## Acknowledgements

\*\*\* Isaiah 45:12, *The Holy Bible New International Version*, p. 745.

## I
## Day One: Sunday

1 Charles Darwin, *The Descent of Man and Selection in Relation to Sex* (Princeton, N.J.: Princeton University Press, 1981), vol. II, p. 389. Copyright 1981 by Princeton University Press. Reprinted by permission of Princeton University Press.

2 Ibid., pp. 389, 390.

3 Pat Shipman, *Taking Wing* (New York: Simon & Schuster, 1998), p. 27. See also Adrian Desmond and James Moore, *Darwin* (New York: Warner Books, Inc., 1991), p. 461; and Ian T. Taylor, *In the Minds of Men* (Minneapolis: TFE Publishing, 1996), p. 133. Estimates of Darwin's annual income range from 2,000 to 8,000 pounds sterling with an estate valued at 250,000 British pounds at

death. Dr. Charlie Kramer computes Darwin's 1861 annual income of 8,000 British pounds sterling to be the equivalent of $526,928 U.S. dollars in 1996—or $43,910 per month.

4 Adrian Desmond and James Moore, *Darwin* (New York: Warner Books, Inc., 1991), p. 456; citing Charles Darwin letter to Asa Gray.

5 Loren Eiseley, *Darwin's Century* (Garden City, N.Y.: Anchor Books/Doubleday & Company, Inc., 1961), p. 295.

6 Phillip E. Johnson, *Darwin on Trial* (Washington, D.C.: Regnery Publishing, 1995), p. 179; citing Charles Darwin 1879 letter to Joseph Hooker as quoted by Kenneth Sporne's 1971 monograph, "The Mysterious Origin of Flower Plants." All quotes from this book are Copyright 1995 by Regnery Publishing. All rights reserved. Reprinted by special permission of Regnery Publishing, Inc., Washington, D. C.

7 Johnson, *Darwin on Trial*, p. 180; citing Corner's essay "Evolution," *Contemporary Biological Thought* (McLeod & Colby, 1961 ed.), pp. 95, 97.

8 Michael J. Behe, *Darwin's Black Box* (New York: The Free Press, 1996), p. 156, citing John Maynard Smith, "Life at the Edge of Chaos?" *New York Review* (March 2, 1995), pp. 28-30.

9 Michael K. Richardson, "Haeckel's Embryos, Continued," *Science* 281 (August 28, 1998), p. 1289. Copyright 1998 American Association for the Advancement of Science.

10 See Duane T. Gish, *Evolution: The Fossils Still Say No!* (El Cajon, Calif.: Institute for Creation Research, 1995), pp. 19, 20; Michael Denton, *Evolution: A Theory in Crisis* (Bethesda, Md.: Adler & Adler, 1986), pp. 330, 331; and Harold Coffin with Robert H. Brown, *Origin by Design* (Hagerstown, Md.: Review and Herald Publ. Assn., 1983), p. 382.

11 Denton, *Evolution: A Theory in Crisis*, p. 330.

12 Lane P. Lester and Raymond G. Bohlin, *The Natural Limits to Biological Change* (Dallas: Probe Books, 1989), p. 85; citing Fred Hoyle and N. A. Wickramasinghe, *Evolution From Space* (London: Dent, 1981).

13 Duane T. Gish, *Creation Scientists Answer Their Critics* (El Cajon, Calif.: Institute for Creation Research, 1993), p. 375, citing Sir Fred Hoyle, *New Scientist* (November 19, 1981), pp. 526, 527; see also Ian T. Taylor, *In the Minds of Men* (Minneapolis: TFE Publishing, October 1996), p. 202.

## II
## Day Two: Monday

14 Richard Milton, *Shattering the Myths of Darwinism* (Rochester, Vt.: Park Street Press, 1997), p. 101.

15 Elizabeth Pennisi, "Genome Data Shake Tree of Life," excerpted with permission from *Science* 280 (May 1, 1998), p. 673. Copyright 1998 American Association for the Advancement of Science.

16 Denton, *Evolution: A Theory in Crisis,* pp. 328, 329.

17 See Michael J. Behe, *Darwin's Black Box* (New York: The Free Press, 1996).

18 Hugh Ross, *The Creator and the Cosmos* (Colorado Springs, Colo.: NavPress, 1993), p. 134. All quotes from this book are used by permission of NavPress/Pinon Press. All rights reserved. For copies call 1-800-366-7788.

19 Denton, *Evolution: A Theory in Crisis,* p. 323.

20 Ross, *The Creator and the Cosmos*, pp. 139, 140; citing Harold Morowitz's calculations and referencing Robert Shapiro, *Origins: A Skeptic's Guide to the Creation of Life on Earth* (New York: Summit Books, 1986), p. 128.

21 Ross, *The Creator and the Cosmos*, p. 109.

22 Ibid.

23 Coffin, *Origin by Design,* p. 376; citing James F. Coppedge, *Evolution: Possible or Impossible?* (Grand Rapids, Mich.: Zondervan Publishing House, 1973), p. 109.

24 Coffin, *Origin by Design,* p. 376; citing Harold T. Morowitz, *Energy Flow in Biology* (New York: Academic Press, 1968).

25 Ross, *The Creator and the Cosmos,* p. 139; citing Charles B. Thaxton, Walter L. Bradley, and Roger L. Olsen, The *Mystery of Life's Origin: Reassessing Current Theories* (Philosophical Library), pp. 69-98.

26 Ralph O. Muncaster, *Creation Versus Evolution* (Mission Viejo, Calif.: Strong Basis to Believe, 1997), p. 17.

27 Ross, *The Creator and the Cosmos,* p. 139; citing Gordon Schlesinger and Stanley L. Miller, "Prebiotic Synthesis in Atmospheres Containing CH, CO, CO2," *Journal of Molecular Evolution* 19 (1983), pp. 376-382.

28 Walt Brown, *In the Beginning: Compelling Evidence for Creation and the Flood* (Phoenix, Ariz.: Center for Scientific Creation, 1996), pp. 11, 12.

29 Denton, *Evolution: A Theory in Crisis,* p. 250.

30 Ross, *The Creator and the Cosmos*, p. 141.

31 Robert H. Brown letter to Warren L. Johns, October 22, 1995, citing Gunter Faure, *Principles of Isotope Geology* (Somerset, N.J.: John Wiley and Sons, Inc., 1986), pp. 120, 121, 291; Copyright 1986, John Wiley & Sons, Inc. Reprinted by permission of John Wiley & Sons, Inc.

32 Milton, *Shattering the Myths of Darwinism*, pp. 38, 39.

33 Ibid., pp. 53-55.

34 Steven A. Austin, "Excess Argon with Mineral Concentrations From the New Dacite Lava Dome at Mount St. Helens Volcano," *Creation Ex Nihilo Technical Journal* 10 (1996), part 3; cited by *Acts and Facts*, Institute for Creation Research 26:5 (May 1997).

35 Milton, *Shattering the Myths of Darwinism,* p. 46.

36 Ibid., p. 33.

37 Taylor, *In the Minds of Men*, p. 333.

38 Ibid., pp. 311, 312; see also Robert V. Gentry, *Creation's Tiny Mystery* (Knoxville, Tenn.: Earth Science Associates, 1992).

39 Michael A. Cremo and Richard L. Thompson, *Forbidden Archeology* (Los Angeles: Bhaktivedanta Book Publishing, Inc., 1996), p. 694.

40 Andrew A. Snelling, "The Cause of Anomalous Potassium-Argon 'Ages' for Recent Andesite Flows at Mt. Ngauruhoe, New Zealand, and the Implications for Potassium-Argon Dating," *ICC Symposium Sessions* (Pittsburgh: Creation Science Fellowship, Inc., 1998), p. 510.

41 This summary is based upon the eyewitness account of Dr. Harold G. Coffin, paleontologist, who walked Surtsey in July 1967.

42 Cremo and Thompson, *Forbidden Archeology,* Table A1.6, p. 780.

43 Taylor, *In the Minds of Men*, p. 316.

44 Robert S. Root-Bernstein, "Darwin's Rib," *Discover* (September 1995), p. 38.

45 Gish, *Creation Scientists Answer Their Critics*, p. 321; citing Soren Løvtrup, *Darwinism: The Refutation of a Myth* (New York: Croom Helm, 1987), p. 422.

## III
## Day Three: Tuesday

46 See Sharon Begley, with Thomas Hayden, "How Low Can You Go?" *Newsweek* (February 22, 1999), p. 50. Copyright 1999 Newsweek, Inc. All rights reserved. Reprinted by permission.

47 Darwin, *Descent,* vol. I, p. 203.

48 Ibid.

49 See ibid.

50 Ibid., p. 213.

51 Ibid., pp., 206, 207.

52 Ibid., p. 207; see also ibid., vol. II, pp. 389, 390.

53 Byron C. Nelson, *After Its Kind* (Minneapolis: Augsburg Publishing House, 1927), pp. 99, 101.

54 Darwin, *Descent,* vol. I, p. 280. Darwin labeled Pangenesis a *hypothesis* given the reservations expressed by Huxley and Hooker. Still, *"More and more he fell back on this, making pangenesis essential...Nothing would induce him to emulate Huxley and strangle his baby god."* See Desmond and Moore, *Darwin*, pp. 531; 550, 551; 617.

55 Darwin, *Origin*, p. 175.

56 Ibid., p. 246.

57 See ibid., pp. 276, 277.

58 Lester and Bohlin, *The Natural Limits to Biological Change* (Dallas: Probe Books, 1989), p. 54.

59 Denton, *Evolution: A Theory in Crisis*, pp. 62, 358.

60 Darwin, *Origin,* p. 175.

61 Nina Federoff and David Botstein, *The Dynamic Genome* (Plainview, N.Y.: Cold Spring Harbor Laboratory Press, 1992), p. 335, citing reprint of Barbara McClintock, "Mechanisms That Rapidly Reorganize the Genome," Stadler Symp. vol. 10 (1978), University of Missouri, Columbia, p. 25.

62 Henry M. Morris, "What They Say," *Back to Genesis* (March 1999), p. a.

63 Tim Friend, "Gene Defect is Linked to Parkinson's," *USA Today* (June 27, 1997).

64 *The Star*, Ventura, California (June 24, 1997).

65 Josie Glausiusz, "Fast Forward Aging," *Discover* (November 1996), p. 44; copyright 1996. Reprinted with permission of *Discover* Magazine.

66 Elizabeth Pennsi, "New Gene Found for Inherited Macular Degeneration," excerpted with permission from *Science* 281 (July 3, 1998), p. 31. Copyright 1998 American Association for the Advancement of Science.

67 Marcia Barinaga, "Tracking Down Mutations That Can Stop the Heart," Excerpted with permission from *Science* 281 (July 3, 1998), p. 32. Copyright 1998 American Association for the Advancement of Science.

68 Rick Weiss, "Defect Tied to Doubling of Risk for Colon Cancer," *The Washington Post* (August 26, 1997); Copyright 1997, *The Washington Post*. Reprinted with permission.

69 Daniel C. Weaver, "The River of Life," *Discover* (November 1997), p. 55; copyright 1997. Reprinted with permission of *Discover* Magazine.

70 Karen P. Steel and Steve D. M. Brown, "More Deafness Genes." Excerpted with permission from *Science* 280 (May 29, 1998), p. 1403. Copyright 1998 American Association for the Advancement of Science.

71 Josie Glausiusz, "The Genes of 1996," *Discover* (January 1997), p. 36; copyright 1997. Reprinted with permission of *Discover* Magazine.

72 David A. Demick, "The Blind Gunman," *Impact* (El Cajon, Calif.: Institute for Creation Research, February, 1999), p. iv.

73 Peter Radetsky, "Immune to a Plague," *Discover* (June 1997), pp. 60-67; copyright 1997. Reprinted with permission of *Discover* Magazine.

74 Rob Stein, "Sex May Rid Us of DNA Flaws," *The Washington Post* (February 1, 1999), p. A9; Copyright 1999, *The Washington Post*. Reprinted with permission.

75 Lee M. Spetner, *Not by Chance* (Brooklyn: The Judaica Press, Inc., 1997), pp. 131, 141, 143.

76 Carl Wieland, "Superbugs Not Super After All," *Creation* 20:1 (December 1997/February 1998), pp. 10-13; citing R. McGuire, "Eerie: Human Arctic Fossils Yield Resistant Bacteria," *Medical Tribune* (December 29, 1988), pp. 1, 23. Date of death reported to be 1845, approximately a century prior to introduction of penicillin.

## IV
## Day Four: Wednesday

77 Darwin, *Descent,* vol. I, pp. 37, 38.

78 Darwin, *Origin*, pp. 251-253.

79 David Brown, "Ornithology: The Hard Work of Hovering in the Air," *The Washington Post* (November 10, 1997); Copyright 1997, *The Washington Post.* Reprinted with permission.

80 Darwin, *Origin,* p. 247.

81 Kathy Sawyer, "New Light on a Mysterious Epoch," *The Washington Post* (February 5, 1998); Copyright 1998, The *Washington Post.* Reprinted with permission.

82 James Gibson, letter to Warren L. Johns (August 28, 1997); citing David Raup, *Zoologic Record* published in *Paleobiology* 2 (1976), pp. 279-288.

83 Darwin, *Origin*, p. 232.

84 Milton, *Shattering the Myths of Darwinism,* p. 170.

85 Ibid., p. 169.

86 Ibid., pp. 134, 135.

87 Ibid., p. 153; see also front dust jacket flap.

88 Ibid., p. 52.

89 Milton, *Shattering the Myths of Darwinism,* p. 33; citing Melvin A. Cook, "Do Radiological Clocks Need Repair?" *Creation Research Society Quarterly* 5 (October 1968), p. 70.

90 Denton, *Evolution: A Theory in Crisis*, p. 249.

91 Darwin, *Origin*, p. 647.

92 Leonard Brand, *Faith, Reason, and Earth History* (Berrien Springs, Mich.: Andrews University Press, 1997), p. 145.

93 Alan Hayward, *Creation and Evolution* (Minneapolis: Bethany House Publishers, 1995), p. 42; citing Francis Hitching, *The Neck of the Giraffe* (London: Pan, 1982), p. 19.

94 Brand, *Faith, Reason, and Earth History,* p. 145.

95 Milton, *Shattering the Myths of Darwinism*, p. 184; quoting Michael Denton, *Evolution: A Theory in Crises*.

96 Milton, *Shattering the Myths of Darwinism*, p. 179.


# V
## Day Five: Thursday

97 Ida Thompson, *National Audubon Society Field Guide to North American Fossils* (New York: Alfred A. Knopf, Inc., 1994), p. 765.

98 Gretel Schueller, "Earth News: Death in the Dunes," *Earth* (June 1998), p. 11.

99 Darwin, *Origin*, p. 213.

100 Brand, *Faith, Reason, and Earth History*, pp. 254, 255.

101 Darwin, *Origin*, p. 453.

102 Hayward, *Creation and Evolution*, p. 35; referencing F. Hoyle, *The Universe: Past and Present Reflections* (University College, Cardiff, 1981).

103 Darwin, *Origin*, p. 212.

104 Norman Macbeth, *Darwin Retried: An Appeal to Reason* (Boston: The Harvard Common Press, 1978), p. 60.

105 Ibid., citing George Gaylord Simpson, *The Meaning of Evolution* (Yale University Press, 1967).

106 Steven A. Austin, ed., *Grand Canyon: Monument to Catastrophe* (Santee, Calif.: Institute for Creation Research, 1994), p. 149.

107 Darwin, *Origin*, p. 223.

108 Peter D. Ward, "Coils of Time," *Discover* (March 1998), p. 106; copyright 1998. Reprinted with permission of *Discover* Magazine.

109 Darwin, *Origin*, pp. 248, 249.

110 Ibid., p. 573.

111 Timothy F. Flannery, "Debating Extinction," *Science* 283 (January 8, 1999), p. 182; excerpted with permission from *Science*, copyright 1999, American Association for the Advancement of Science; quoting from Alfred Russell Wallace, *The Geographical Distribution of Animals, With a Study of the*

*Relations of Living and Extinct Faunas as Elucidating Past Changes of the Earth's Surface* (New York: Harper, 1876), p. 150.

112 D. S. Allen and J. B. Delair, *Cataclysm* (Santa Fe, N. M.: Bear & Co., 1997), p. 107.

113 Darwin, *Origin*, p. 648.

114 Gretel Schueller, "Australia's Ups and Downs," *Earth* (August 1998), p. 16.

115 *The Scopes Trial* (Birmingham, Ala.: The Legal Classics Library, 1984; a reprint of *The World's Most Famous Trial* [Cincinnati: National Book Co., 1925]), pp. 238-241.

116 Milton, *Shattering the Myths of Darwinism*, pp. 77, 78.

117 Ibid., pp. 92, 93.

118 Ariel A. Roth, *Origins* (Hagerstown, Md.: Review and Herald Publ. Assn., 1998), p. 216; citing J. S. Shelton, *Geology Illustrated* (San Francisco and London: W. H. Freeman and Co.), p. 28.

119 See Allan and Delair, *Cataclysm*. Major *Homo sapiens* fossil discoveries in North and South America have been dated within a time frame that correlates during or after the 11,500 years before the present when glaciers retreated from mid-America and residual fossil graveyards from cataclysm were formed: Lapa Vermelha Woman, 11,500; Arlington Springs Woman, 10,960; Buhl Woman, 10,600; Spirit Caveman, 9,400; Prince of Wales Man, 9,200; Wizards Beach Man, 9,200; Browns Valley Man, 8,700; Kennewick Man, 8,000; and Pelican Rapids Woman, 7,840. See Sharon Begley and Andrew Murr, "The First Americans," *Newsweek* (April 26, 1999), pp. 50-57. Copyright 1999 Newsweek, Inc. All rights reserved. Excerpted by permission.

120 Kathy Sawyer, "New Light on a Mysterious Epoch," *The Washington Post* (February 5, 1998); Copyright 1998, *The Washington Post*. Reprinted with permission.

121 Luis Chiappe, "Dinosaur Embryos," *National Geographic* (December 1998), p. 38.

122 Michael E. Soule and L. Scott Mills, "No Need to Isolate Genetics," *Science* (November 27, 1998), p. 1659; citing M. E. Gilpin and M. E. Soule, in *The Science of Scarcity and Diversity* (Sunderland, Mass.: A, Sinauer, 1986), pp. 35-56; P. L. Leberg, *J. Fish Biol.* 37 (1990), p. 193; D. Newman and D. Pilson, *Evolution* 51 (1997), p. 354; and L. Saccheri, et al., *Nature* 392 (1998), p. 491. Excerpted with permission from *Science* (November 27, 1998). Copyright 1998 American Association for the Advancement of Science.

123 See W. Shotyk, D. Weiss, P. G. Appleby, A.K. Cheburkin, R. Frei, M. Gloor, J. D. Kramers, S. Reese, and W. O. Van Der Knaap, "History of Atmospheric Lead Deposition Since 12,370 14C yr BP From a Peat Bog, Jura Mountains, Switzerland." Excerpted with permission from *Science* 281 (11 September 1998), pp. 1635-1640. Copyright 1998 American Association for the Advancement of Science.

124 See Heather Pringle, "Traces of Ancient Mariners Found in Peru." Excerpted with permission from *Science* 281 (September 18, 1998), pp. 1775, 1776. Copyright 1998 American Association for the Advancement of Science.

125 Roth, *Origins*, pp. 262-269.

126 Taylor, *In the Minds of Men*, pp. 322-339; citing *The New York Times* (July 21, 1969), p. 1.

127 John Casti, "What Scientists Don't Know—And Why They Don't Know It," *The Washington Post* (November 30, 1997); Copyright 1997, *The Washington Post*. Reprinted with permission.

128 Adams, *The Scopes Trial,* p. 237.

129 Taylor, "The Ultimate Hoax: Archaeoptyrx Lithographica," *ICC Symposium Sessions* (Pittsburgh: Creation Science Fellowship, Inc., 1990) vol. II, pp. 279-291.

130 Gish, *Evolution: The Fossils Still Say No!*, p. 137; referencing Tim Beardsley, *Nature* 322 (1986), p. 677; Richard Monastersky, *Science News* 140 (1991), pp. 104, 105; and Alan Anderson, *Science* 253 (1991), p. 35.

131 R. A. Thulborn, "The Avian Relationships of *Archaeopteryx* and the Origin of Birds," *Zoological Journal of the Linnean Society* 82 (1984), p. 119, as cited by Walt Brown, *In the Beginning: Compelling Evidence for Creation and the Flood* (Phoenix: Center for Scientific Creationism, 1996).

132 Virginia Morell, "A Cold, Hard Look at Dinosaurs," *Discover* (December 1996), p. 102; copyright 1996. Reprinted with permission of *Discover* Magazine.

133 John Schwartz, "Paleontology: Another Aspect in the Bird Debate," *The Washington Post* (November 17, 1997); citing a report in the November 14, 1997 issue of *Science*; Copyright 1997, *The Washington Post*. Reprinted with permission.

134 Bernice Wuethrich, "Stunning Fossil Shows Breath of a Dinosaur," *Science* 283 (January 22, 1999), p. 468. Excerpted with permission from *Science*. Copyright 1999 American Association for the Advancement of Science.

# VI
## Day Six: Friday

135 Darwin, *Origin*, p. 643.

136 Macbeth, *Darwin Retried*, p. 36; citing Wilbur Hall, *Partner of Nature* (Appleton-Century, 1939).

137 See Tim Appenzeller, "Test Tube Evolution Catches Time in a Bottle," *Science* (25 June 1999) vol. 284, p. 2108-2110.

138 Macbeth, *Darwin Retried*, p. 30; citing Charles Darwin, *Origin* (First Edition), p. 184.

139 Macbeth, *Darwin Retried*, p. 57; citing George Bernard Shaw, *Back to Methuselah* (Penguin paperback), p. 1921.

140 Ruth Hubbard and Elijah Wald, *Exploding the Gene Myth* (Boston: Beacon Press, 1997), p. 14; citing Francis Galton, *Inquiries Into Human Faculty* (London: Macmillan, 1883), pp. 24, 25.

141 Darwin, *Descent*, vol. I, p. 169.

142 Ibid., p. 201.

143 Ibid., p. 146.

144 Ibid.., p. 169.

145 Ibid., p. 168.

146 Ibid., vol. II, pp. 389, 390.

147 Adams, *The Scopes Trial*, pp. 336, 337; citing Benjamin Kidd, *The Science of Power* (1918), pp. 46, 47, 67.

148 Marvin L. Lubenow, *Bones of Contention* (Grand Rapids: Baker Books, 1992), p. 58.

149 Ibid., p. 73.

150 Robert Kunzig, "The Face of an Ancestral Child," *Discover* (December 1997), pp. 96, 100; copyright 1997. All quotes from this article are reprinted with permission of *Discover* Magazine.

151 Kathy Sawyer, "Ancient Footprints Discovered," *The Washington Post* (August 15, 1997); Copyright 1997, *The Washington Post.* Reprinted with permission.

152 Shanti Menon, *"Neanderthal* Noses," *Discover* (March 1997), p. 30. Reprinted with permission of *Discover* Magazine. See also Curt Suplee, "DNA Suggests *Neanderthal* Not a Direct Human Ancestor," *The Washington Post* (July 11, 1997); Copyright 1997, *The Washington Post.* Reprinted with permission. Svante Paabo tested mitochondrial DNA taken from the right upper arm bone of the prototype *Neanderthal* discovered in 1856. Paabo's group reported that the *"*Neanderthal *specimen had 27 differences from modern humans."*

153 Paul G. Bahn, *"Neanderthals* Emancipated," Reprinted with permission from *Nature* 394 (August 20, 1998), p. 721; copyright 1998, Macmillan Magazines Limited.

154 Lubenow, *Bones of Contention*, p. 120.

155 Curt Suplee, "Modern Humans May Have Coexisted With Ancestor," *The Washington Post* (December 13, 1996); Copyright 1997, *The Washington Post.* Reprinted with permission.

156 Matt Cartmill, "The Third Man," *Discover* (September 1997), pp. 56-62. Reprinted with permission of *Discover* Magazine.

157 Kunzig, "The Face of an Ancestral Child," p. 98.

158 Bernard Wood and Mark Collard, "The Human Genus," *Science* 284 (April 2, 1999), p. 70. Excerpted with permission from *Science* 284 (April 2, 1999), p. 70. Copyright 1999 American Association for the Advancement of Science.

159 See John Schwartz, "An Endangered Primate's Paradigm: 'Make Love, Not War," *The Washington Post* (July 14, 1997); Copyright 1997, *The Washington Post*. Reprinted with permission.

160 See Dita Smith and Laura Stanton, "Population Momentum," *The Washington Post* (1996); citing Population Reference Bureau, World Bank, "World Population Projections." Copyright 1996, *The Washington Post.* Reprinted with permission.

161 See Carl E. Baugh with Clifford A. Wilson, *Dinosaur* (Orange, Calif.: Promise Publishing Company, 1991).

162 Geoffrey Barraclough, ed., *The Times Atlas of World History* (London: Times Books Limited; Maplewood, N.J.: Hammond Inc., 1979), p. 36. Reprinted with permission of Harper Collins Publishers LTD, London.

163 Kunzig, "The Face of an Ancestral Child," p. 93.

164 Ibid., p. 100.

165 Ibid., pp. 90, 97.

166 Lubenow, *Bones of Contention*, p. 165.

167 Mary D. Leakey, "Footprints in the Ashes of Time," *National Geographic* (April 1979), p. 446.

168 Ibid., p. 452.

169 Ibid., p. 453.

170 Lubenow, *Bones of Contention,* p. 175.

171 Leakey, "Footprints in the Ashes of Time," pp. 454, 456.

172 Lubenow, *Bones of Contention,* p. 175.

173 Gish, *Evolution: The Fossils Still Say No!,* p. 276.

174 Lubenow, *Bones of Contention*, pp. 175, 176; citing William Howells, "*Homo erectus* in Human Descent: Ideas and Problems," Homo erectus: *Papers in Honor of Davidson Black*, Becky A. Sigmon and Jerome S. Cybulski, eds. (Toronto: University of Toronto Press, 1981).

175 Lubenow, *Bones of Contention*, p. 53; citing Henry M. McHenry, "Fossils and the Mosaic Nature of Human Evolution," *Science* 190 (October 31, 1975), p. 428.

176 Lubenow, *Bones of Contention*, pp. 52, 53, 178, 179; see also Michael A. Cremo and Richard L. Thompson, "Summary of Anomalous Evidence Related to Human Antiquity," Appendix 3, *Forbidden Archaeology* (Los Angeles: Bhaktivedanta Book Publishing, Inc., 1996), pp. 815-828.

177 Glenn C. Conroy, Gerhard W. Weber, Horst Seidler, Phillip V. Tobias, Alex Kane, and Barry Brunsden, "Endocranial Capacity in an Early Hominid Cranium From Sterkfontein, South Africa," excerpted with permission from *Science* (June 12, 1998), p. 1731. Copyright 1998 American Association for the Advancement of Science.

178 See Elizabeth Pennisi, "Genetic Study Shakes Up Out-of-Africa Theory," *Science* 283 (March 19, 1999), p. 1828.

179 Adams, *The Scopes Trial,* pp. 75, 84.

180 Darwin, *Descent,* vol. II, p. 327.

181 Ibid., p. 389.

182 Darwin, *Origin* (First Edition), p. 184.

183 See Sarah Richardson, "Tarzan's Little Brain," *Discover* (November 1996), pp. 100, 102; copyright 1996. Reprinted with permission of *Discover* Magazine.

184 Darwin, *Descent*, vol. II, p. 385.

185 Ibid., p. 389.

186 Ibid., vol. I, p. 169.

187 Ibid., p. 201.

188 Ibid., vol. II, pp. 327, 328.

189 Romans 3:23, *The Holy Bible New International Version* (Grand Rapids: Zondervan Bible Publishers, 1979), p. 1490.

190 Spetner, *Not by Chance*, p. 132.

# VII
## Day Seven: Saturday

191 Joel Achenbach, "A Beautiful Illusion," *The Washington Post Magazine* (July 13, 1997), p. 16; Copyright 1997, *The Washington Post*. Reprinted with permission.

192 Job 11:7-9, *The Holy Bible New International Version*, pp. 628, 629.

193 Psalm 19:1, *The Holy Bible New International Version*, p. 683.

194 John 1:1-4, The Holy Bible New International Version, p. 1409.

195 Acts 17:24, 25, *The Holy Bible New International Version*, p. 1134.

196 James Perloff, *Tornado in a Junkyard* (Arlington, Massachusetts: Refuge Books) p. 253.

197 Phillip E. Johnson, *Objections Sustained* (Downers Grove, Illinois: InterVarsity Press, 1998), pp. 72, 76.

198 Sharon Begley, "Science Finds God," *Newsweek* (July 20, 1998), p. 50. Copyright 1998 Newsweek, Inc. All rights reserved. Reprinted by permission.

199 George F. Will, "The Gospel From Science," *Newsweek* (November 9, 1998), p. 88, quoting from playwright Tom Stoppard. Copyright 1998 Newsweek, Inc. All rights reserved. Reprinted by permission.

200 Psalm 14:1, *The Holy Bible New International Version*, p. 676.

201 Bruce Catton, "Night Train," as presented in *Bruce Catton's America,* Olver Jensen, Editor (American Heritage Publishing Co., Inc., 1979), p. 222.

202 Betty Ballantine and Ian Ballantine, eds, *The Native Americans* (Atlanta: Turner Publishing, Inc., 1993), p. 174. (Note: views differ as to the accuracy of some words traditionally attributed to Chief Joseph. The quote used appears in this reference.)

203 John 8:32, *The Holy Bible New International Version*, p. 1423.

204 Revelation 14:11, *The Holy Bible New International Version*, p. 1626.

205 Taylor, *In the Minds of Men*, p. 122, citing Nora Barlow, ed., *The Autobiography of Charles Darwin, 1809-1882* (London: Collins, 1958); see also Adrian Desmond and James Moore, *Darwin* (New York: Warner Books, Inc., 1991), p. 623.

*Warren LeRoi Johns*

# BIBLIOGRAPHY

**Achenbach, Joel.** "A Beautiful Illusion." *The Washington Post Magazine* (July 13, 1997).

**Adams, Leslie B., Jr.,** Publisher. *The Scopes Trial.* Birmingham, Alabama: The Legal Classics Library, 1984; a reprint of *The World's Most Famous Trial* (Third Edition). Cincinnati: National Book Company, 1925.

"Agriculture: Fertile Crescent as Early Breadbasket." *The Washington Post* (November 17, 1997).

**Allan, D. S.,** and **J. B. Delair.** *Cataclysm.* Santa Fe, N.M.: Bear & Company, 1997.

**Anderson, Alan.** *Science,* 253, 1991.

**Appenzeller, Tim.** "The Genes of 1996." *Discover* (January 1997).

_____. "Test Tube Evolution Catches Time in a Bottle." *Science* 284 (June 25, 1999).

**Austin, Steven A.** "Excess Argon With Mineral Concentrates From the New Dacite Lava Dome at Mount St. Helens Volcano." *Creation Ex Nihilo Technical Journal,* vol. 10, part 3 (1996); as reported in "Acts and Facts," *Institute for Creation Research* 26:5 (May 1997).

_____, Editor. *Grand Canyon Monument to Catastrophe.* Santee, Calif.: Institute for Creation Research, 1994.

**Bahn, Paul G.** "*Neanderthals* Emancipated." *Nature* 394 (August 20, 1998).

**Ballantine, Betty,** and **Ian Ballantine,** Editors. *The Native Americans.* Atlanta: Turner Publishing, Inc., 1993.

**Barinaga, Marcia.** "New Gene Found for Inherited Macular Degeneration." *Science* 281 (July 3, 1998).

**Barlow, Nora, ed.** *The Autobiography of Charles Darwin.* London: Collins, 1958.

**Barraclough, Geoffrey,** Editor. *The Times Atlas of World History.* London: Times Books Limited; Maplewood, N.J.: Hammond Inc., 1979.

**Baugh, Carl E.** with **Clifford A. Wilson.** *Dinosaur.* Orange, Calif.: Promise Publishing Co., 1991.

**Beardsley, Tim.** *Nature* 322, 1986.

**Begley, Sharon.** "Science Finds God." *Newsweek* (July 20, 1998).

_____, with **Thomas Hayden.** "How Low Can You Go?" *Newsweek* (February 22, 1999).

_____, with **Andrew Murr.** "The First Americans." *Newsweek* (April 26, 1999).

**Behe, Michael J.** *Darwin's Black Box.* New York: The Free Press, 1996.

**Bird, W. R.** *The Origin of Species Revisited.* Nashville: Thomas Nelson, Inc., 1991.

**Brand, Leonard.** *Faith, Reason, and Earth History.* Berrien Springs, Mich.: Andrews University Press, 1997.

**Brown, David.** "Ornithology: The Hard Work of Hovering in the Air," *The Washington Post* (November 10, 1997).

**Brown, Walt.** *In the Beginning: Compelling Evidence for Creation and the Flood..* Phoenix: Center for Scientific Creationism, 1995.

**Camp, Ashby L.** *The Myth of Natural Origins.* Tempe, Ariz.: Ktisis Publishing, 1994.

**Cartmill, Matt.** "The Third Man." *Discover* (September 1997).

**Casti, John.** "What Scientists Don't Know—And Why They Don't Know It," *The Washington Post* (November 30, 1997).

**Catton, Bruce.** "Night Train." *Bruce Catton's America*, Oliver Jensen, Editor. American Heritage Publishing Co., Inc., 1979.

**Chiappe, Luis.** "Dinosaur Embryos." *National Geographic* (December 1998).

**Conroy, Glenn C., Gerhard W. Weber, Horst Seidler, Phillip V. Tobias, Alex Kane,** and **Barry Brunsden.** "Endocranial Capacity in an Early Hominid Cranium From Sterkfontein, South Africa." *Science* (June 12, 1998).

**Cook, Melvin A.** "Do Radiological Clocks Need Repair?" *Creation Research Society Quarterly* 5 (October, 1968).

**Coppedge, James F.** *Evolution: Possible or Impossible?* Grand Rapids, Mich.: Zondervan Publishing House, 1973.

**Coffin, Harold,** with **Robert H. Brown.** *Origin by Design.* Hagerstown, Md.: Review and Herald Publishing Association, 1983.

**Cremo, Michael A.**, and **Richard L. Thompson**. *Forbidden Archaeology.* Los Angeles: Bhaktivedanta Book Publishing, Inc., 1996.

**Darwin, Charles**. *The Origin of Species* (Sixth Edition). New York: Random House, Inc., 1993.

_____. *The Descent of Man, and Selection in Relation to Sex.* Princeton, N.J.: Princeton University Press, 1981. Photo reproduction of the 1871 edition published by J. Murray, London.

**Demick, David A.** "The Blind Gunman." *Impact* (February 1999). N. Santee, Calif.: Institute for Creation Research.

**Denton, Michael**. *Evolution: A Theory in Crisis.* Bethesda, Md.: Adler & Adler, 1986.

**Desmond, Adrian**, and **James Moore**. *Darwin.* New York: Warner Books, Inc., 1991.

**Eiseley, Loren**. *Darwin's Century.* Garden City, N.Y.: Anchor Books/ Doubleday & Company, Inc., 1961.

**Eldredge, Niles**. *Fossils.* New York: Harry N. Abrams, Inc., 1991.

**Faure, Gunter**. *Principles of Isotope Geology.* Somerset, N.J.: John Wiley and Sons, Inc., 1986.

**Fedoroff, Nina,** and **David Botstein**. *The Dynamic Genome.* Plainview, N.Y.: Cold Spring Harbor Laboratory Press, 1992.

**Flannery, Timothy F.** "Debating Extinction." *Science* 283 (January 8, 1999).

**Friend, Tim**. "Gene Defect Is Linked to Parkinson's." *USA Today* (June 27, 1997).

**Gentry, Robert V.** *Creation's Tiny Mysteries.* Knoxville, Tenn.: Earth Science Associates, 1992.

**Galton, Francis**. *Inquiries Into Human Faculty.* London: Macmillan, 1883.

**Gibbons, Ann**. "Genes Put Mammals in Age of Dinosaurs." *Science* (May 1, 1998).

**Gibson, James**. Letter to Warren L. Johns. August 28, 1997.

**Gilpin, M.E.** and **M.E. Soule**. *The Science of Scarcity and Diversity.* Sunderland, Massachusetts: A. Sinauer, 1986.

**Gish, Duane T**. *Evolution: The Fossils Still Say No!* El Cajon, Calif.: Institute for Creation Research, 1995.

_____. *Creation Scientists Answer Their Critics.* El Cajon, Calif.: Institute for Creation Research, 1993.

**Glausiusz, Josie**. "Fast Forward Aging." *Discover* (November 1996).

_____. "The Genes of 1996." *Discover* (January 1997).

**Gore, Rick**. "Extinctions." *National Geographic* (June 1989).

**Hall, Wilbur**. *Partner of Nature.* CITY: Appleton-Century, 1939.

**Hayward, Alan**. *Creation and Evolution.* Minneapolis: Bethany House Publishers, 1995.

**Hitching, Francis**, *The Neck of the Giraffe.* London: Pan, 1982.

***Holy Bible New International Version.*** Grand Rapids: Zondervan Bible Publishers, 1979

**Howells, William**. "*Homo erectus* in Human Descent: Ideas and Problems." Homo erectus: *Papers in Honor of Davidson Black,* **Becky A. Sigmon** and **Jerome S. Cybulski**, editors. Toronto: University of Toronto Press, 1981.

**Hoyle, Sir Fred**, *New Scientist* (November 19, 1981).

_____, *The Universe: Past and Present Reflections.* Cardiff: University College, 1981.

_____, and **N. A. Wickramasinghe.** *Evolution From Space.* London: Dent, 1981.

**Hubbard, Ruth,** and **Elijah Wald**. *Exploding the Gene Myth.* Boston: Beacon Press, 1997.

**Johnson, Phillip E**. *Darwin on Trial.* Washington, D.C.: Regnery Gateway, 1995.

_____. *Objections Sustained.* Downers Grove, Illinois: InterVarsity Press, 1998.

**Kerkut, G.A**. *Implications of Evolution.* New York: Pergamon Press, 1965.

**Kidd, Benjamin**. *The Science of Power.* CITY: PUBLISHER, 1918.

**Kunzig, Robert**. "The Face of an Ancestral Child." *Discovery* (December 1997).

**Leakey, Mary D**. "Footprints in the Ashes of Time." *National Geographic* (April 1979).

**Leberg, P.L.** *J. Fish Biol,* 37, 1990.

**Lester, Lane P.**, and **Raymond G. Bohlin**. *The Natural Limits to Biological Change.* Dallas, Texas: Probe Books, 1989.

**Løvtrup, Søren**. *The Refutation of a Myth.* New York: Croom Helm, 1987.

**Lubenow, Marvin L**. *Bones of Contention.* Grand Rapids, Mich.: Baker Books, 1992.

**Macbeth, Norman**. *Darwin Retried: An Appeal to Reason.* Boston: The Harvard Common Press, 1978.

**Mayr, Ernst**. *Systematics and the Origin of Species.* New York: Columbia University Press, 1942; Dover Publications paperback, 1964.

**McGuire, R**. "Eerie: Human Arctic Fossils Yield Resistant Bacteria." *Medical Tribune* (December 29, 1988).

**McHenry, Henry M**. "Fossils and the Mosaic Nature of Human Evolution." *Science* 190 (October 31, 1975).

**Menon, Shanti**. "*Neanderthal* Noses." *Discover* (March 1997).

**Milton, Richard**. *Shattering the Myths of Darwinism.* Rochester, Vt.: Park Street Press, 1997.

**Monastersky, Richard.** *Science News* 140, 1991.

**Morell, Virginia**. "A Cold, Hard Look at Dinosaurs." *Discover* (December 1996).

**Morris, Henry M**. "What They Say." *Back to Genesis.* N. Santee, Calif.: Institute for Creation Research, 1999.

**Morowitz, Harold T**. *Energy Flow in Biology.* New York: Academic Press, 1968.

**Muncaster, Ralph O**. *Creation Versus Evolution.* Mission Viejo, Calif.: Strong Basis to Believe, 1997.

**Nelson, Byron C**. *After Its Kind.* Minneapolis: Augsburg Publishing House, 1927.

**Newman, D**. and **L. Saccheri, et. al.** *Nature,* 392, 1998.

**Pennisi, Elizabeth.** "Genome Data Shakes Tree of Life." *Science* 280 (May 1, 1998.

_____."New Gene Found for Inherited Macular Degeneration." *Science* 281 (July 3, 1998).

_____. "Genetic Study Shakes Up Out-of-Africa Theory." *Science* 283 (March 19, 1999).

**Perloff, James.** *Tornado in a Junkyard.* Arlington, Mass.: Refuge Press, 1999.

**Pringle, Heather**, "Traces of Ancient Mariners Found in Peru." *Science* 281 (September 18, 1998).

**Radetsky, Peter**. "Immune to a Plague." *Discover* (June 1997). *Radiocarbon Journal* (1966).

**Raup, David**. *Zoologic Record.* Published in *Paleobiology* 2 (1976).

**Richardson, Michael K.** "Haeckel's Embryos, Continued." *Science* 28 (August 28, 1998).

**Richardson, Sarah**. "Tarzan's Little Brain." *Discover* (November 1996).

**Root-Bernstein, Robert S**. "Darwin's Rib." *Discover* (September 1995).

**Ross, Hugh**. *The Creator and the Cosmos.* Colorado Springs, Colo.: NavPress, 1993.

**Roth, Ariel A.** *Origins.* Hagerstown, Md.: Review and Herald Publishing Association, 1998.

**Sawyer, Kathy**. "Ancient Footprints Discovered." *The Washington Post* (August 15, 1997).

_____. "New Light on a Mysterious Epoch," *The Washington Post* (February 5, 1998).

**Schlesinger, Gordon** and **Stanley L. Miller**. "Prebiotic Synthesis in Atmospheres Containing CH, CO, CO2," *Journal of Molecular Evolution* 19 (1983).

**Schueller, Gretel**. "Australia's Ups and Downs." *Earth* (August 1998).

_____. "Death in the Dunes." *Earth* (June 1998).

**Schwartz, John**. "Paleontology: Another Aspect in the Bird Debate." *The Washington Post* (November 17, 1997).

_____."An Endangered Primate's Paradigm: 'Make Love, Not War.'" *The Washington Post* (July 14, 1997). *Science* (November 14, 1997).

**Shapiro, Robert**. *Origins: A Skeptic's Guide to the Creation of Life on Earth.* New York: Summit Books, 1986.

**Shelton, J. S.** *Geology Illustrated.* San Francisco and London: W. H. Freeman and Co., 1966.

**Shipman, Pat.** *Taking Wing.* New York: Simon & Schuster, 1998.

**Shaw, George Bernard.** *Back to Methuselah.* CITY: Penguin, 1921.

**Shotyk, W., D. Weiss, P. G. Appleby, A. K. Cheburkin, R. Frei, M. Gloor, J. D. Kramers, S. Reese,** and **W. O. Van Der Knaap,** "History of Atmospheric Lead Deposition Since 12,370 $^{14}$C yr BP from a Peat Bog, Jura Mountains, Switzerland." *Science* 281 (September 11, 1998) pp. 1635-1640.

**Simpson, George Gaylord.** *The Meaning of Evolution.* CITY: Yale University Press, 1967.

**Smith, Dita,** and **Laura Stanton.** "Population Momentum." *The Washington Post* (1996).

**Smith, John Maynard,** "Life at the Edge of Chaos?" *New York Review* (March 2, 1995).

**Snelling, Andrew A.** "The Cause of Anomalous Potassium-Argon 'Ages' for Recent Andesite Flows at Mt. Ngauruhoe, New Zealand, and the Implications for Potassium-Argon Dating." *ICC Technical Symposium Sessions,* **Robert E. Walsh,** Editor. Pittsburgh: Creation Science Fellowship, Inc. 1998.

**Souder, William.** "New Reports of Deformed Frogs Trigger U.S. Ecological Alarms." *The Washington Post* (January 29, 1997).

**Soule, Michael E.,** and **L. Scott Mills.** "No Need to Isolate Genetics." *Science* (November 27, 1998).

**Spetner, Lee M.** *Not by Chance.* Brooklyn: The Judaica Press, Inc., 1997.

**Steel, Karen P.,** and **Steve D. M. Brown,** "More Deafness Genes." *Science* 280 (May 29, 1998).

**Stein, Rob.** "Sex May Rid Us of DNA Flaws." *The Washington Post* (February 1, 1999).

**Suplee, Curt.** "DNA Suggests *Neanderthal* Not a Direct Human Ancestor." *The Washington Post* (July 11, 1997).

_____. "Modern Humans May Have Coexisted With Ancestor." *The Washington Post* (December 13, 1996).

**Svitil, Kathy.** "Breakthroughs: Old Bird," *Discover* (March 1997).

_____. *Discover* (January 1994).

**Taylor, Ian T.** *In the Minds of Men.* Minneapolis, Minn.: TFE Publishing, 1991.

_____. "The Ultimate Hoax: Archaeoptyrx Lithographica." *The Proceedings of the Second International Conference on Creationism*, vol. II. Pittsburgh: ICC, 1990.

**Thaxton, Charles B., Walter L. Bradley** and **Roger L. Olsen.** *The Mystery of Life's Origin: Reassessing Current Theories.* Philosophical Library.

**Thompson, Ida.** *National Audubon Society Field Guide to North American Fossils.* New York: Alfred A. Knopf, Inc., 1982 (1994 printing).

**Thulborn, R. A.** "The Avian Relationships of *Archaeoptyrx* and the Origin of Birds." *Zoological Journal of the Linnean Society* 82 (1984).

**USA Today**. June 27, 1997.

**United Nations Food and Agriculture Organization.** "Dwindling Diversity." *The Washington Post* (February 15, 1997), p. A30.

**Ventura Star.** Ventura, California (June 24, 1997).

**Wallace, Alfred Russell.** *The Geographical Distribution of Animals, With a Study of the Relations of the Living and Extinct Faunas as Elucidating Past Changes of the Earth's Surface.* New York: Harper, 1876.

**Ward, Peter.** "Coils of Time," *Discover* (March 1998).

**Washington Post.** "Modern Humans May Have Coexisted With Ancestor." (December 13, 1996).

**Weaver, Daniel C**. "The River of Life." *Discover* (November 1997).

**Weiss, Rick**. "Defect Tied to Doubling of Risk for Colon Cancer." *The Washington Post* (August 26, 1997).

**Whitcomb, John C.,** and **Henry M. Morris.** *The Genesis Flood.* Phillipsburg, N.J.: Presbyterian and Reformed Publishing Company, 1995.

**Wieland, Carl**. "Superbugs Not Super After All." *Creation* 20:1 (December 1997/February 1998).

**Will, George F.** "The Gospel of Science." *Newsweek* (November 9, 1998).

**Wood, Bernard,** and **Mark Collard.** "The Human Genus." *Science* 284 (April 2, 1999).

**Wuethrich, Bernice.** "Stunning Fossil Shows Breath of a Dinosaur." *Science* 283 (January 22, 1999).

# ABOUT THE AUTHOR

Editor of the online Creation Digest, Warren LeRoi Johns practiced law as a career in California, Maryland, and the District of Columbia until partial retirement in the summer of 1992. His previous non-fiction *Dateline Sunday, U.S.A.*, drew national attention as a legal history documenting blue law confrontation with the U.S. constitution's first California Law Center. Holder of La Sierra University's 1994 "Alumnus of the Year" award, the author's professional resume appears in Who's Who in American Law; Who's Who in America; and Who's Who in the World.